THE PLUMBING APPRENTICE STUDENT WORKBOOK

Written By
Ruth H. Boutelle
Richard E. White
Charles R. White

The Thomson Learning Inc. logo is a registered trademark used herein under license.

Printed in the United States of America
1 2 3 4 5 XX 08 07 06

For more information contact
Thomson Delmar Learning
Executive Woods
5 Maxwell Drive, PO Box 8007,
Clifton Park, NY 12065-8007
Or find us on the World Wide Web at
www.delmarlearning.com

ALL RIGHTS RESERVED. No part of this work covered by the copyright hereon may be reproduced in any form or by any means—graphic, electronic, or mechanical, including photocopying, recording, taping, Web distribution, or information storage and retrieval systems—without the written permission of the publisher.

For permission to use material from the text or product, contact us by
Tel.   (800) 730-2214
Fax   (800) 730-2215
www.thomsonrights.com

ISBN: 1418065110

PLUMBING-HEATING-COOLING
CONTRACTORS ASSOCIATION

EDUCATIONAL FOUNDATION

Copyright © 2005 Plumbing-Heating-Cooling Contractors--National Association Educational Foundation
and
Plumbing-Heating-Cooling Contractors--National Association

NOTICE TO THE READER

Publisher does not warrant or guarantee any of the products described herein or perform any independent analysis in connection with any of the product information contained herein. Publisher does not assume, and expressly disclaims, any obligation to obtain and include information other than that provided to it by the manufacturer.

The reader is expressly warned to consider and adopt all safety precautions that might be indicated by the activities herein and to avoid all potential hazards. By following the instructions contained herein, the reader willingly assumes all risks in connection with such instructions.

The publisher makes no representation or warranties of any kind, including but not limited to, the warranties of fitness for particular purpose or merchantability, nor are any such representations implied with respect to the material set forth herein, and the publisher takes no responsibility with respect to such material. The publisher shall not be liable for any special, consequential, or exemplary damages resulting, in whole or part, from the readers' use of, or reliance upon, this material.

The PHCC Educational Foundation

The Plumbing-Heating-Cooling Contractors--National Association Educational Foundation was incorporated in 1986 as a 501(c)(3) tax-exempt, charitable organization. The purpose of the Foundation is to help shape the industry's future by developing and delivering educational programs that positively impact every aspect of the p-h-c contractor's business. Major programs include apprentice, journeyman, and business management training.

The Foundation relies on contributions from many sources. Gifts to the Foundation Endowment Fund are non-invadable, ensuring the Foundation's financial integrity for generations to come. Donors include contractors, manufacturers, suppliers, and other industry leaders who are committed to preparing plumbing-heating-cooling contractors and their employees to meet the challenges of a constantly changing marketplace.

The PHCC Educational Foundation mission statement sets forth this vision:

"The Plumbing-Heating-Cooling Contractors--National Association Educational Foundation is the world class organization that enables plumbing-heating-cooling contractors to create a new, exciting future by funding research and delivering powerful educational opportunities for the success of the industry professional."

ABOUT PHCC

The Plumbing-Heating-Cooling Contractors--National Association (PHCC) has been an advocate for plumbing, heating, and cooling contractors since 1883. As the oldest trade organization in the construction industry, approximately 4,000 member companies nationwide put their faith in the Association's efforts to lobby local, state, and federal government, provide forums for networking and educational programs, and to deliver the highest quality of products and services.

PHCC's mission statement is the guiding principle of the Association:

**"The Plumbing-Heating-Cooling Contractors–National Association
is dedicated to the promotion, advancement, education and training
of the industry for the protection of the environment,
and the health, safety and comfort of society."**

A complete account of PHCC's history is included in the PHCC Educational Foundation publication *A Heritage Unique*.

These manuals are dedicated to and in honor of

Patrick J. Higgins

without whom they would not exist.

Pat died suddenly of a pulmonary embolism on November 28, 2001, at the age of 50.

A licensed plumber, Pat founded and was President of P. J. Higgins & Associates in Frederick, MD. For 20 years prior to his death, Pat's company provided consulting services on plumbing codes, standards, product approval and regulatory approval to numerous industry companies and trade organizations.

Pat had been Chairman of the Main A112 Plumbing Materials and Equipment Committee for the American Society of Mechanical Engineers for 18 years, and a member of ASME's Council of Codes and Standards for 3 years. The summer before his death he assumed the Chair of ASME's Board of Standardization, of which he had been a member for 10 years.

Patrick J. Higgins & Associates participated extensively in the development of model and state plumbing codes. He was liaison to numerous industry committees, including ASPE, ASSE, and PHCC. He authored many publications and wrote for various trade magazines.

Pat was widely known throughout our industry. As a young man he was the Technical Director of the National Association of Plumbing-Heating-Cooling Contractors (now the PHCC--National Association). In his capacity there, Pat developed the original outline for the plumbing manuals that now appear in this form.

Pat was a dear friend to the current authors and is sadly missed by his family, the authors and many others in the industry.

Acknowledgement

The Plumbing-Heating-Cooling Contractors--National Association Educational Foundation gratefully acknowledges the many companies in our industry that have contributed to this book by supplying drawings, information, and advice.

Disclaimer

The material presented in this publication has been prepared for the general information of the reader. While the information is believed to be technically correct, neither the authors or the PHCC Educational Foundation warrant the publication's suitability for applications other than as an information guide. It shall be the responsibility of the readers or users of this publication to conform with the requirements of all local, state, and federal regulations, codes, and standards.

The ***National Standard Plumbing Code — Illustrated*** (published by the Plumbing-Heating-Cooling Contractors--National Association) is quoted extensively throughout this manual. This work is used as a reference, so that the desired information to be imparted can be more easily understood. It is the responsibility of the teacher to analyze the information in these lessons and alter, if necessary, any information which does not conform with the local code.

FOREWORD

This is the first in a series of four plumbing apprentice manuals. The lessons are grouped into various subject areas for variety and interest and also to try to match the work experience of the first-year apprentice.

Illustrations are used to help clarify the text. However, the use of a particular illustration does not constitute endorsement of a particular product or method, nor does the view show the only representation of a product or method.

Where the masculine gender is used, the feminine may be substituted.

FOREWORD TO THE FOURTH EDITION

The PHCC Educational Foundation Plumbing Apprentice Curriculum Committee commissioned this revised edition of the first year textbook to add new material, clarify many technical points, and generally improve the presentation of the subjects covered.

The number of lessons has been reduced by combining related subject matter into a single lesson.

We would appreciate any comments or corrections, as well as suggestions for any improvements, from any user of these works. We thank all those who have made such comments in the past, and express thanks to the many, many people who have reviewed and assisted in this book.

Please send comments and corrections to:

> PHCC Educational Foundation
> P. O. Box 6808
> Falls Church, VA   22046-1148

TABLE OF CONTENTS

Guide to Illustrations... III

Guide to Tables... XIII

References... XV

Plumbing Code Definitions.. Definitions I

Lessons 1-7 Plumbing Background, Skills Required of Mechanic, and Tool Information
Lesson 1: Introduction to the Plumbing Professional........................... 1
Lesson 2: Plumbing Laws, Tools, and Safety.................................. 11
Lesson 3: Hand Tools Used in Plumbing Work................................ 23
Lesson 4: Rough-In Tools – Copper, Plastic, and Soil Pipe................... 31
Lesson 5: Rough-In Tools – Steel Pipe....................................... 39
Lesson 6: Finish and Repair Tools... 45
Lesson 7: Welding and Power Tools.. 51

Lessons 8-11 Basic Arithmetic Needed for Plumbing Work
Lesson 8: Review – Numbers, Fractions, and Decimals....................... 71
Lesson 9: Mathematical Operations – Fractions............................... 77
Lesson 10: Mathematical Operations – Decimals and Fractions................ 87
Lesson 11: Measuring Tapes, Folding Rules, and Scale Rulers................ 95

Lessons 12-15 Related Science
Lesson 12: Goals of Plumbing, Water Sources, Waste Disposal............... 105
Lesson 13: Sewage Disposal.. 117
Lesson 14: Introduction to Gases... 122
Lesson 15: Mechanical Properties of Materials and Structures............... 131

Lessons 16-20 Installation Practices
Lesson 16: Cutting, Drilling, and Nail Protection for Building Structural Elements..... 137
Lesson 17: Piping Materials Used in Plumbing Work – Pressure............... 141
Lesson 18: Piping Materials Used in Plumbing Work – DWV................... 151
Lesson 19: Joining Methods and Materials for DWV Piping.................... 167
Lesson 20: Pressure Pipe and Fittings...................................... 175

Lessons 21-24   Mathematics
Lesson 21:   Review of Lessons 8-11 191
Lesson 22:   Percents and Decimals 199
Lesson 23:   Squares, Square Roots, and Circles 205
Lesson 24:   Angles, Ratios, and Triangles 227

Lesson 25   First Aid Education
To be taught by the American Red Cross 255

Lessons 26-28   Safe Work Practices
Lesson 26:   Safety on the Job – OSHA – PHCC Safety Manual 257
Lesson 27:   Safety Manual (continued) 303
Lesson 28:   Safety Manual (continued) 309

Lessons 29-31   Installation Practices
Lesson 29:   Typical Plumbing Fixtures 325
Lesson 30:   Fixture Fittings (Faucets) 363
Lesson 31:   Valves 371

Lessons 32-36   "Blueprint" Reading and Sketching
Lesson 32:   Building Plans and Drawings 387
Lesson 33:   Scale Rulers 397
Lesson 34:   Sketching – Freehand and With Drafting Tools 405
Lesson 35:   Symbols and Detail Sketching 415
Lesson 36:   Projections Used for Drawings – Advantages and Disadvantages 429

Appendix A:
Overview of Hand Tools . . . . Appendix A-1

Appendix B:
Overview of Hand Tools (continued) . . . . Appendix B-1

Appendix C:
Lead Safety . . . . Appendix C-1
Occupational Safety and Health Administration Publication
3142-09R 2003, Lead in Construction . . . . Appendix C-3

Guide to Illustrations By Manufacturer . . . . Illustrations I

Plumbing Apprentice Student Workbook Year One
Fourth Edition

GUIDE TO ILLUSTRATIONS

Figure	Title	Page Number
1-A	***2003 National Standard Plumbing Code — Illustrated***	6
1-B	Typical Contractor's License	8
2-A	Typical "Not Approved" Tag	13
	St. Joseph County-South Bend Building Department, South Bend, Indiana	
2-B	Typical Plumbing Tag	14
	St. Joseph County-South Bend Building Department, South Bend, Indiana	
2-C	Typical Electric Tag	14
	St. Joseph County-South Bend Building Department, South Bend, Indiana	
2-D	Typical Heating/Air Conditioning Tag	14
	St. Joseph County-South Bend Building Department, South Bend, Indiana	
2-E	Typical Structural Tag	14
	St. Joseph County-South Bend Building Department, South Bend, Indiana	
2-F	Typical Application for Building Permit	15
	St. Joseph County-South Bend Building Department, South Bend, Indiana	
3-A	RIDGID Torpedo Level	23
3-B	RIDGID Aluminum Level	24
3-C	RIDGID Family of Pliers	25
3-D	RIDGID Snips	26
3-E	RIDGID Manual Knockout Kit	26
3-F	Plumber's Pad	26
	Bennette Design Group, Inc.	
3-G	Knee N' Back Pad	26
	Bennette Design Group, Inc.	
4-A	RIDGID Screw Feed Cutter	31
4-B	RIDGID Torch Accessories	32
4-C	RIDGID ProPress System	33
4-D	NIBCO Press-to-Connect Copper Joinery System	33
4-E	RIDGID Flaring Tool	33
4-F	RIDGID Geared Ratchet Lever-Type Tube Bender	34
4-G	RIDGID Spring-Type Tube Bender	34
4-H	RIDGID Heavy-Wall Conduit Bender	34
4-I	RIDGID Thin-Wall Conduit Bender	34
4-J	RIDGID Plastic Pipe Cutters/Scissors Cutter	35
4-K	RIDGID Deburring Tools	35
4-L	Hammer and Cold Chisel Cutting of Soil Pipe	35
4-M	RIDGID Soil Pipe Cutter	36
4-N	Set Up for Pouring Lead Joint in Horizontal Line	36
4-O	Mephisto Tool – Plumbing Lead Joint Caulking Iron Set	37
4-P	RIDGID Soil Pipe Assembly Tool	37
4-Q	RIDGID Torque Wrench	37

Plumbing Apprentice Student Workbook Year One
Fourth Edition

5-A	RIDGID Vise Examples	39
5-B	RIDGID Heavy-Duty Pipe Cutter	39
5-C	RIDGID Pipe Reamers	40
5-D	RIDGID Manual Pipe Threader	40
5-E	RIDGID Model 700 Power Drive	40
6-A	RIDGID Basin Wrench	46
6-B	Stanley Putty Knife	46
6-C	RIDGID Midget Tubing Cutter	46
6-D	RIDGID Hack Saw	47
6-E	RIDGID Toilet Auger	47
6-F	RIDGID Hand Spinner	48
6-G	Handle Puller	48
7-A	Typical Oxy-Acetylene Welding Tools	51
7-B	More Gas Welding Tools	52
7-C	Miller CST-250 Stick/TIG Welding Power Source	53
7-D	MillerMatic DVI	54
7-E	Milwaukee 8" Metal Cutting Saw	56
7-F	Milwaukee Reciprocating Saw	56
7-G	Milwaukee Abrasive Cut-Off Machine	56
7-H	Milwaukee Magnum® Dual Torque Hammer Drill	56
7-I	RIDGID Model 300 Power Drive	57
7-J	RIDGID Threading Machine	57
7-K	Milwaukee Portable Band Saw	58
7-L	Milwaukee Operator's Manual	61
11-A	Engineers and Architects Scale	98
11-B	Ametek U. S. Gauge 5" Industrial Glass Tube Thermometer	102
11-C	Ametek U. S. Gauge Series P-500 Low Cost Utility Gauge	103
12-A	Backflow Caused by Back-Pressure	105
12-B	Backflow Caused by Back-Siphonage	106
12-C-a	Elements of A Fixture Trap	107
12-C-b	Trap Seal Reduction from 1" Negative Pressure	108
12-C-c	Trap Seal Reduction from 1" Positive Pressure	108
12-D	The Water Service Pipe in a Public Water Supply	113
12-E	A Typical Private Water Supply System	113
12-F	Minimum Distances for Location of Components of a Private Sewage Disposal System	114
12-G	Cross Connection Control by Individual Outlet Protection	116
12-H	The Potable Water Supply to an Aspirating Device Protected by An Atmospheric Vacuum Breaker	116
13-A	Separate Sanitary and Stormwater Building Drains and Sewers	117
13-B	Location and Spacing of Manholes	118
13-C	Typical Private Sewage Disposal System	120

13-D	A Typical Septic Tank	121
13-E	Typical Distribution Box	121
13-F	A Leaching or Seepage Well or Pit	122
13-G	A Pipe Penetration of a Floor Above a Food Handling Area	123
13-H	Plumbing Fixtures Above a Food Handling Area	124
15-A	Three Categories of Loads	131
15-B	Structural Member Under Stress	132
15-C	Stress Lines "Tighten Up"	132
15-D	Stress Concentration at Square Hole	133
15-E	Simple Beam (Deflection Exaggerated)	133
15-F	Wood Joist Failure	135
15-G	Steel I-Beam Local Failure	135
15-H	Loads Applied to Light Structures	136
16-A	Holes in Beams	137
16-B	Notching & Boring Limitations – Exterior & Bearing Walls	138
16-C	Notching & Boring Limitations – Interior Partitions	138
16-D	Treatment of Notch in Plate of Exterior & Bearing Walls	139
16-E	Floor Joist – Notch Cuts	
	Floor Joist – End Notches	
	Rafter/Ceiling Joist Notches	139
17-A	A Solvent Cement Joint in Socket (Bell) End Plastic Pressure Pipe	144
17-B	An Insert Fitting Joint in Plastic Tubing	144
17-C	A Heat Fused Joint in Plastic Water Piping	145
17-D	A Threaded Pipe Joint	148
18-A	Single Hub and Double Hub Cast-Iron Soil Pipe	152
18-B	Typical Bend	152
18-C	Y Branch with Cleanout Plug	153
18-D	Sanitary T Branches, Single and Double	153
18-E	Tee Wye and Double Tee Wye	153
18-F	Tapped Vent Tee	153
18-G	Tee Cleanout	154
18-H	Offset	154
18-I	Reducer	154
18-J	Increaser	154
18-K	P-Trap	154
18-L	Running Trap	155
18-M	Ferrule and Plug	155
18-N	Closet Bend With Tapped Side Outlet	155
18-N(2)	How to Determine Right or Left Hand Inlets	155
18-O	A Solvent Cement Joint in Plastic DWV or Water Piping	161
18-P	An Elastomeric Gasket Joint for Underground Plastic DWV Piping	162

18-Q	A Shielded Coupling on Plastic DWV Piping	162
19-A	A Lead Caulked Joint in Cast-Iron Soil Pipe	168
19-B	A Cast-Iron Hubbed Joint with a Compression Gasket	168
19-C	A Shielded Coupling on Hubless Cast-Iron Soil Pipe	169
19-D	A Rubber Ring Transition Joint to Vitrified Clay Pipe	171
19-E	An Externally Clamped Transition Coupling Joint to Vitrified Clay Pipe	171
19-F	A Joint in Bell and Spigot Concrete Pipe	172
20-A	Companion Flange	176
20-B	Reducing Flange	176
20-C	Victaulic Grooved Coupling	176
20-D	Victaulic Grooved End Fittings	176
20-E	Typical Victaulic Fittings	177
20-F	NIBCO Typical Fittings	179
20-G	NIBCO Typical Fittings – Continued	180
20-H	A Mechanical Expansion Joint in Pressure Piping	181
20-I	NIBCO Cast and Wrought Copper Fittings	181
20-I(2)	Continued	182
20-J	A Plastic DWV Threaded Male Adapter	183
20-K	A Solvent Cement Joint in Plastic DWV or Water Piping	183
20-L	A Solvent Cement Joint in Socket (Bell) End Plastic Pressure Pipe	184
20-M	An Elastomeric Gasket Joint for Underground Plastic DWV Piping	184
20-N	An Insert Fitting Joint in Plastic Tubing	185
20-O	A Heat Fused Joint in Plastic Water Piping	185
20-P	NIBCO Copper Tube Fittings	186
20-Q	NIBCO What Makes a Plumbing System Fail?	187
20-R	NIBCO The Fine Art of Soldering	188
20-S	NIBCO The Fine Art of Brazing	189
23-A	Parts of a Circle	218
24-A	Angle EOF – Straight Angle	229
24-B	Angle GOH – Right Angle	229
24-C	Angle JOL – Obtuse Angle	230
24-D	Angle MON – Acute Angle	230
24-E	Labeling an Angle	231
24-F	Pipe Line 90-Degree Bend	233
24-G	Angle	237
24-H	Right Triangle	238
24-I	Right Triangle "On Its Side"	239
24-J	Isosceles Right Triangle	239
24-K	Square Formed by Two Isosceles Triangles	242
24-L	Rectangle Formed by Two Scalene Triangles	243
24-M	Isosceles Right Triangle (a = b)	244
24-N	45° Pipe Offset	247
24-O	45° Pipe Offset With Fittings	248

24-P	Fitting Allowance	253
26-A	Fire Extinguisher	268
26-B	Hazard Signs	268
26-C	Welded Frame Scaffold	270
26-D	Rolling Scaffold	271
26-E	Needle Beam Scaffold	272
26-F	Float	273
26-G	Two-Point Suspended Scaffold	274
26-H	Trenches	276
26-I	Protruding Reinforcing Steel	278
26-J	Hard Hat	279
26-K	Respiratory Equipment	279
26-L	Safety Belts	279
26-M	Gloves	280
26-N	Protective Suit	282
26-O	Lifting	284
26-P	Barricading of the Crane Swing Radius	286
26-Q	Crane	287
26-R	Grounding	288
26-S	Flemish Eye Splice	289
26-T	Slings	289
26-U	Choker	291
26-V	Connections	293
26-W	Wire Rope Clips	293
26-X	Eyebolts	295
26-Y	Suspending Needle Beams	295
26-Z	Wire Rope 1	296
26-AA	Wire Rope 2	296
26-BB	Wire Rope 3	296
26-CC	Wire Rope 4	296
26-DD	Wire Rope 5	296
26-EE	Wire Rope 6	297
26-FF	Wire Rope 7	297
26-GG	Wire Rope 8	297
26-HH	Wire Rope 9	297
26-II	Wire Rope 10	297
26-JJ	Wire Rope 11	298
26-KK	Wire Rope 12	298
26-LL	Wire Rope 13	298
26-MM	Wire Rope 14	298
26-NN	Welder	298
26-OO	Cylinders	300
26-PP	Ladder Angles	301
26-QQ	Ladders	302

27-A	Grinder	304
27-B	Drill Press	305
27-C	Metric Ruler	307
28-A	Chock Size	312
28-B	Standard Hand Signals – Hoist	317
28-C	Standard Hand Signals – Lower	317
28-D	Standard Hand Signals – Use Main Hoist	317
28-E	Standard Hand Signals – Use Whipline	318
28-F	Standard Hand Signals – Raise Boom	318
28-G	Standard Hand Signals – Lower Boom	318
28-H	Standard Hand Signals – Move Slopwly	319
28-I	Standard Hand Signals – Raise the Boom and Lower the Load	319
28-J	Standard Hand Signals – Lower the Boom and Raise the Load	319
28-K	Standard Hand Signals – Swing	320
28-L	Standard Hand Signals – Stop	320
28-M	Standard Hand Signals – Emergency Stop	320
28-N	Standard Hand Signals – Travel	321
28-O	Standard Hand Signals – Dog Everything	321
28-P	Standard Hand Signals – Travel (Both Tracks)	321
28-Q	Standard Hand Signals – Travel (One Track)	322
28-R	Standard Hand Signals – Extend Boom	322
28-S	Standard Hand Signals – Retract Boom	322
28-T	Standard Hand Signals – Extend Boom	323
28-U	Standard Hand Signals – Retract Boom	323
29-A	Siphon Jet Water Closet	326
29-B	Reverse Trap Water Closet	326
29-C	Blowout Water Closet	326
29-D	Washdown Water Closet	326
29-E	Siphon Vortex Water Closet	327
29-F	Siphon Wash Closet	327
29-G	A Pressure-Assisted Water Closet With a Flushometer Tank	327
29-H	Pneumatic Assist Water Closet	327
29-I	ECOFLUSH	328
	WDI International Inc.	
29-J	Flushing Cycle at Rest	330
29-K	Flushing Cycle Beginning	330
29-L	Flushing Cycle Full Siphon	330
29-M	Flushing Cycle Siphon Broken	330
29-N	Typical Flush Tank – Lift Wire Model	331
29-N(2)	Typical Flush Tank – Flapper Model	331
29-N(3)	Water Tower-2 piece W C	332
29-N(4)	Water Tower-1 piece W C	332
29-O	Sloan Royal Flushometer	334

29-P	Sloan G2 Optima Plus Flushometer	334
29-Q	Sloan Royal Flushometer Installation Instructions	335
29-R	KOHLER Bardon™ Superior Urinal	343
29-S	FALCON F-1000 Waterfree Urinal	345
29-T	KOHLER Portrait® Vitreous China Bidet	347
29-U	KOHLER Bon Vivant® Self-Rimming Kitchen Sink	349
29-V	KOHLER Sudbury™ Service Sink	350
29-W	KOHLER River Falls™ Self-Rimming Laundry Sink	350
29-X	KOHLER Pennington™ Countertop Lavatory	353
29-Y	PLUMBEREX Trap Gear™ Undersink Protector	354
29-Z	Halsey Taylor Contour™ Barrier-Free Fountain With Back Panel	355
29-AA	Halsey Taylor Barrier-Free Cooler	356
29-BB	KOHLER Memoirs® Bath	356
29-CC	A Whirlpool Tub	357
29-DD	KOHLER Sonata® 60" Shower Module	360
30-A	Typical Single-Bibb Faucet	364
30-B	Powers HydroGuard® T/P Series e700	366
30-B(2)	Symmons Pressure-Balancing Shower Valve	366
30-C	Typical Shower Diverter Installation Wolverine Brass	368
30-C(2)	Delta Bath Mixing Valve Single Handle	368
30-D	Woodford Model 65 Freezeless Wall Hydrant Flow Diagram	369
30-E	Woodford Yard Hydrant	370
30-F	Symmons Laundry-Mate W-600	370
31-A	Hammond 118-FP Brass Gate	372
31-B	Hammond Gate Valve	373
31-C	Hammond 8201 Forged Brass Ball Valve	374
31-D	Watts Resilient Seated Butterfly Valve	375
31-E	Lubricated Plug Valve	376
31-F	Hammond 1560CB2 Flanged Ends Globe Class 150	377
31-G	Hammond Bronze Angle Globe Valve	377
31-H	Parker SN6 Series Needle Valve	378
31-I	Stockham Class 125 Bronze Lift Check Valve	379
31-J	Stockham Class 125 Bronze Swing Check Valve	379
31-K	Ball Check Valve, Flow Left to Right	380
31-L	Typical Spring Check Valve, Flow Right to Left	380
31-M	Watts Series 25AUB-Z3 Water Pressure Reducing Valve	381
31-N	Watts Series 530C Calibrated Pressure Relief Valve	381
31-O	Watts Series 0L Temperature and Pressure Relief Valve	382
31-P	Watts Series 70A Hot Water Extended Tempering Valve	383
31-Q	J. R. Smith 7150 Series In-Line Manual Shut-Off Gate Valve	385
31-R	Josam Series 67400 No-Hub Swing-Check Type Backwater Valve	386

32-A	Object Line	390
32-B	Two Center Lines	391
32-C	Two Hidden (or Dashed) Lines	391
32-D	Dimension and Extension Lines	391
32-E	Leader Line	392
32-F	Cutting Plane Line	392
32-G	Phantom Line	392
32-H	Long Break and Two Types of Short Break	392
32-I	Before and After Contour Lines	392
32-J	Typical Cross-Hatch Patterns	393
32-K	Alphabet of Lines	396
33-A	Scale Ruler on "Blueprint"	397
33-B	The Alphabet of Lines, Scale, and Meaning of Notes	399
34-A	Floor Plan and Riser Diagram Rough Sketch	406
34-B	Plan Views of a Typical Dwelling	407
34-C	Elevation of Dwelling Shown in Figure 34-B	408
34-D	Layout for Student Problem	410
34-E	Freehand Sketch to be Developed by Student of Elevation of Drainage on Graph Paper	411
34-F	Beginning Drain Sketch for Figure 34-E	412
34-G	Drain Sketch – Continue With Freehand Sketch of Vents	413
34-H	Completed Sketch	414
34-I	Paper Mate Pink Pearl® Eraser	416
34-J	Start for Student Piping Layout Sketch Using Drafting Tools	416
35-A	Piping Symbols and Definitions for Draftsmen (1-6)	418-423
35-B	Office Layout	423
35-C	Office Layou DWV Piping Rough Sketch Plan View	425
35-D	Office Layout DWV Piping Rough Sketch Elevation View	426
35-E	Office Layout (Full Basement Below)	427
35-F	Office Layout Water Piping Detailed Sketch Plan View	428
35-G	Office Layout Water Piping Detailed Sketch Elevation View	429
36-A	Orthographic View Arrangement	432
36-B	Orthographic Projection of a Tee	433
36-C	Isometric Axes	434
36-D	Isometric Drawing of a Brick Chisel	434
36-E	Isometric Sketch of Waste and Vent Piping of a Typical Bathroom	435
36-F	Orthographic Views of Typical Bath DWV Piping	436
36-G	Isometric Sketch of Layout in Figure 36-F	437
36-H	Example of Isometric Graph Paper	438
36-I	Isometric Sketch of Bathtub	439
36-J	Oblique Axes	440
36-K	End of Chisel	441
36-L	Projected Lines From End	441

Plumbing Apprentice Student Workbook Year One
Fourth Edition

36-M	Oblique View of Chisel.	441
36-N	Preferred Oblique View.	441
36-O	Oblique View of Layout in Figure 36-F.	442

Appendix A

A-A	RIDGID Wood Folding Rule.	Appendix A-2
A-B	RIDGID Fiberglass Folding Rule	Appendix A-2
A-C	RIDGID Long Steel Tape.	Appendix A-3
A-D	RIDGID Locking Steel Tape.	Appendix A-3
A-E	Stanley Brass Plumb Bob.	Appendix A-4
A-F	IRWIN Wooden Handle Carpenter Saw.	Appendix A-5
A-G	Stanley Clamping Mitre Box with Saw.	Appendix A-6
A-H	Stanley Wood Chisel – Short Blade.	Appendix A-7
A-I	Stanley 6" Cold Chisel.	Appendix A-7
A-J	Stanley 4-Piece File Set.	Appendix A-8
A-K	RIDGID Model 122 Copper Cutting and Prep Machine.	Appendix A-9
A-L	RIDGID 14.4V Impact Driver.	Appendix A-9
A-M	RIDGID Reciprocating Saw.	Appendix A-10

Appendix B

B-A	Stanley 3/16" 100 PLUS® Square Blade Standard Tip Screwdriver.	Appendix B-1
B-B	Stanley 2-Point 100 PLUS® Stubby Phillips Tip Screwdriver.	Appendix B-1
B-C	Stanley ProDriver T20x4" TORX.	Appendix B-2
B-D	Stanley 4 oz Ball Pein Hammer.	Appendix B-2
B-E	Stanley 10 lb. Hickory Handle Sledge Hammer.	Appendix B-2
B-F	Stanley 7 oz Curved Claw Wood Handle Nail Hammer.	Appendix B-3
B-G	Lead Mall.	Appendix B-3
B-H	Rawhide Mallet.	Appendix B-3
B-I	RIDGID Adjustable Wrench.	Appendix B-4
B-J	SK 14 Piece SuperKrome Metric Open End Wrench Set.	Appendix B-4
B-K	IRWIN Curved Jaw Locking Plier.	Appendix B-5
B-L	Stanley 4-7/8" Center Punch.	Appendix B-5
B-M	Telephone.	Appendix B-5
B-N	Fire Extinguisher.	Appendix B-5
B-O	Two-Way Radio.	Appendix B-5
B-P	Allen Wrench Set.	Appendix B-5
B-Q	RIDGID SeeSnake Plus.	Appendix B-6
B-R	Stanley IntelliLaser™ Pro Laser Line Level/Stud Finder.	Appendix B-6
B-S	Brenelle Co., LLC "Jet Swet™".	Appendix B-6
B-T	RIDGID Extractors/Twist Drills.	Appendix B-6
B-U	Striking Safety Tool. Vaughn Manufacturing	Appendix B-7-8

Appendix C

OSHA — Lead in Construction. . . . . Appendix C-3-39

GUIDE TO TABLES

Table	Title	Page Number
2-A	Tool Safety.	19
3-A	Wrenches.	29
5-A	National Pipe thread Taper	41
7-A	Arc Welding Lenses for Protective Shields	53
8-A	Position Names.	72
12-A	Fixture and Equipment Hazards.	110
12-B	Minimum Distances for Location of Components of a Private Water Supply System.	114
17-A	Characteristics of Tapered Pipe Threads.	149
17-B	USS Standard Pipe and Line Pipe — Continuous Butt-welded Steel Pipe.	150
19-A	Wrench Size.	170
19-B	Summary of Connections.	173
21-A	Position Names.	191
21-B	Position Names.	195
23-A	Square Root Table.	215
24-A	Fitting Allowance in Inches-Drainage Fittings.	250
24-B	Fitting Allowances in Inches – Pressure Piping.	251
26-A	Size and Spacing of Members.	276
26-B	Equipment in Addition to Basic Required Safety Equipment.	281
26-C	Lens Shade Numbers for Welding and Cutting Operations.	282
26-D	Load Capacities for Various Hitch types.	290
26-E	Webbed Sling Load Capacities.	291
26-F	Visual Indications of Damage to Webbing.	292
26-G	Qualities of Nylon Slings.	292
26-H	Efficiency of Wire Rope Connections.	293
26-I	Clips and Spacing for Safe Application.	294
26-J	Chain Slings	295
27-A	Common Metric Equivalents.	307-308
28-A	Pipe Wall Thickness and Weight Per Foot.	313
28-B	Weight Per Foot for Common Larger Pipe Sizes	315

REFERENCES

The information available about our industry grows and changes daily. Rather than recommend specific books or publications, it is suggested that the student access information through the library and the internet.

Most states provide information and code books through the state government. For state and local information contact the governing body in which you are located.

Many associations and societies have a catalog of publications which are updated on a periodic basis. Some, but by no means all, of these associations are listed below. Lesson 18 lists code and other bodies where information is also available.

In addition most manufacturers publish guides and technical information about their products. Please check our listing of contributors to this text at the end of this book.

American Gas Association (AGA)
400 N. Capitol St., NW
Suite 450
Washington, DC, 20001
202-824-7000
http://www.aga.org

American Welding Society (AWS)
550 N.W. LeJeune Road
Miami, Florida 33126
800-443-9353
http://www.aws.org

Cast Iron Soil Pipe Institute (CISPI)
5959 Shallowford Road, Suite 419
Chattanooga, Tennessee 37421
615-892-0137
http://www.cispi.org

Copper Development Association, Inc. (CDA)
260 Madison Avenue
New York, New York 10016
212-251-7200
http://ww.copper.org

National Clay Pipe Institute (NCPI)
P. O. Box 759
Lake Geneva, WI 53147
262-248-9094
http://www.ncpi.org

U.S. Department of Labor
Occupational Safety and Health Administration (OSHA)
200 Constitution Avenue, N.W.
Washington, D.C. 20210
http://www.osha.gov

Plastic Pipe and Fittings Association (PPFA)
Building C, Suite 20
800 Roosevelt Road
Glen Ellyn, Illinois 60137
630-858-6540
http://www.ppfahome.org

Plastic Pipe Institute
Suite 680
1825 Connecticut Avenue NW
Washington, DC 20009
202-462-9607
http://www.plasticpipe.org

Plumbing-Heating-Cooling Contractors—National Association
180 S. Washington Street
P. O. Box 6808
Falls Church, Virginia 22046-1148
800-533-7694
http://www.phccweb.org

Plumbing-Heating-Cooling Contractors—National Association Educational Foundation
180 S. Washington Street
P. O. Box 6808
Falls Church, Virginia 22046-1148
800-533-7694
http://www.phccweb.org/foundation

Two excellent videos on the history of plumbing available from The History Channel are:

Modern Marvels, Plumbing the Arteries of Civilization, 50 minutes
Modern Marvels, Bathroom Tech, 50 minutes

PLUMBING CODE DEFINITIONS

Throughout your plumbing career, you will encounter a number of new terms which are unique to the plumbing trade. The correct use of these terms should be a primary objective in your daily tasks within the trade. The correct use of terms will be particularly important later in the study of codes, later in our studies.

Most codes have one chapter specifically established to list definitions of terms pertinent to plumbing. These definitions are used to pinpoint the meaning of specific terms used in code requirements so that each section is properly interpreted. Words used in accordance with their established normal dictionary meanings are not listed in the Definitions Section.

NOTE

This set of definitions should, if necessary, be altered to conform to local codes. Check your local code for any differences with these definitions.

The following **DEFINITION OF TERMS** is taken from the *National Standard Plumbing Code 2003*. Note that the *National Standard Plumbing Code — Illustrated* is also available and includes pictures of many of the definitions listed below.

The definitions are listed for the purpose of the Code and the following terms shall have the meaning indicated in this list of definitions. No attempt is made to define ordinary words which are used in accordance with their established dictionary meaning, except where it is necessary to define their meaning as used in the Code to avoid misunderstanding.

1. **Accessible and Readily Accessible**

   Accessible means having access thereto without damaging building surfaces, but which first may require the removal of an access panel, door or similar obstructions with the use of tools.

   Readily accessible means direct access without requiring the use of tools for removing or moving any panel, door or similar obstruction.

2. **Acid Waste**

   See "Special Wastes"

3. **Adopting Agency**

   The agency, board or authority having the duty and power to establish the plumbing code which will govern the installation of all plumbing work to be performed in the jurisdictions.

4. **Air Break (Drainage System)**

   A piping arrangement in which a drain from a fixture, appliance, or device discharges indirectly into a fixture, receptor, or interceptor at a point below the flood level rim and above the trap seal of the receptor.

5. **Air Chamber**

   A pressure surge absorbing device operating through the compressibility of air.

6. Air Gap (Drainage System)

The unobstructed vertical distance through the free atmosphere between the outlet of the waste pipe and the flood level rim of the receptor into which it is discharging.

7. Air Gap (Water Distribution System)

The unobstructed vertical distance through the free atmosphere between the lowest opening from any pipe or faucet supplying water to a tank, plumbing fixture or other device and the flood level rim of the receptor.

8. Anchors

See "Supports"

9. Anti-Scald Valve

See "Water Temperature Control Valve"

10. Approved

Accepted or acceptable under an applicable standard stated or cited in this Code, or accepted as suitable for the proposed use under procedures and powers of the Authority Having Jurisdiction as defined in Section 3.12.

11. Area Drain

A receptor designed to collect surface or storm water from an open area.

12. Aspirator

A fitting or device supplied with water or other fluid under positive pressure which passes through an integral orifice or "constriction" causing a vacuum.

13. Authority Having Jurisdiction

The individual official, board, department, or agency established and authorized by a state, county, city or other political subdivision created by law to administer and enforce the provisions of the plumbing code as adopted or amended.

14. Autopsy Table

A fixture or table used for the postmortem examination of a body.

15. Backflow Connection

Any arrangement whereby backflow can occur.

16. **Backflow Drainage**

   A reversal of flow in the drainage system.

17. **Backflow Preventer**

   A device or means to prevent backflow.

18. **Backflow (Water Distribution)**

   The flow of water or other liquids, mixtures, or substances into the distributing pipes of a potable supply of water from any source or sources other than its intended source. Back-siphonage is one type of backflow.

19. **Backpressure Backflow**

   A condition which may occur in the potable water distribution system, whereby a higher pressure than the supply pressure is created causing a reversal of flow into the potable water piping.

20. **Back-Siphonage**

   The flowing back of used, contaminated, or polluted water from a plumbing fixture or vessel or other sources into a potable water supply pipe due to a negative pressure in such pipe.

21. **Backwater Valve**

   A device installed in a drain pipe to prevent backflow.

22. **Baptistery**

   A tank or pool for baptizing by total immersion.

23. **Bathroom Group**

   For the purposes of this Code, a bathroom group consists of one water closet, one or two lavatories, and either one bathtub, one combination bath/shower, or one shower stall, all within a bathing facility in a dwelling unit. Other fixtures within the bathing facility shall be counted separately when determining the water supply and drainage fixture unit loads.

24. **Battery of Fixtures**

   Any group of two or more similar adjacent fixtures which discharge into a common horizontal waste or soil branch.

25. **Bedpan Steamer**

   A fixture used for scalding bedpans or urinals by direct application of steam.

26. **Boiler Blow-Off**

   An outlet on a boiler to permit emptying or discharge of sediment.

27. **Boiler Blow-Off Tank**

   A vessel designed to receive the discharge from a boiler blow-off outlet and to cool the discharge to a temperature which permits its safe discharge to the drainage system.

28. **Branch**

   Any part of the piping system other than a riser, main or stack.

29. **Branch, Fixture**

   See "Fixture Branch"

30. **Branch, Horizontal**

   See "Horizontal Branch Drain"

31. **Branch Interval**

   A distance along a soil or waste stack corresponding, in general, to a story height, but in no case less than 8 feet within which the horizontal branches from one floor or story of a building are connected to the stack.

32. **Branch Vent**

   See "Vent, Branch"

33. **Building**

   A structure having walls and a roof designed and used for the housing, shelter, enclosure or support of persons, animals or property.

34. **Building Classification**

   The arrangement adopted by the Authority Having Jurisdiction for the designation of buildings in classes according to occupancy.

35. **Building Drain, Combined**

   A building drain which conveys both sewage and storm water or other drainage.

36. Building Drain

That part of the lowest piping of a drainage system which receives the discharge from soil, waste and other drainage pipes inside the walls of the building and conveys it to the building sewer beginning three (3) feet outside the building wall.

37. Building Drain, Sanitary

A building drain which conveys sewage only.

38. Building Drain, Storm

A building drain which conveys storm water or other drainage, but no sewage.

39. Building Sewer

That part of the drainage system which extends from the end of the building drain and conveys its discharge to a public sewer, private sewer, individual sewage-disposal system, or other point of disposal.

40. Building Sewer, Combined

A building sewer which conveys both sewage and storm water or other drainage.

41. Building Sewer, Sanitary

A building sewer which conveys sewage only.

42. Building Sewer, Storm

A building sewer which conveys storm water or other drainage but no sewage.

43. Building Subdrain

That portion of a drainage system which does not drain by gravity into the building sewer or building drain.

44. Building Trap

A device, fitting, or assembly of fittings, installed in the building drain to prevent circulation of air between the drainage system of the building and the building sewer.

45. Cesspool

A lined and covered excavation in the ground which receives the discharge of domestic sewage or other organic wastes from a drainage system, so designed as to retain the organic matter and solids, but permitting the liquids to seep through the bottom and sides.

46. Chemical Waste

See "Special Wastes"

47. Circuit Vent

See "Vent, Circuit"

48. Clear Water Waste

Effluent in which impurity levels are less than concentrations considered harmful by the Authority Having Jurisdiction, such as cooling water and condensate drainage from refrigeration and air conditioning equipment, cooled condensate from steam heating systems, and residual water from ice making processes.

49. Clinical Sink

A sink designed primarily to receive wastes from bedpans, having a flushing rim, integral trap with a visible trap seal, and having the same flushing and cleansing characteristics as a water closet.

50. Code

These regulations, or any emergency rule or regulation which the Authority Having Jurisdiction may lawfully adopt.

51. Combination Fixture

A fixture combining one sink and laundry tray, or a two- or three-compartment sink or laundry tray in one unit.

52. Combination Thermostatic/Pressure Balancing Valve

See "Thermostatic/Pressure Balancing Valve, Combined"

53. Combination Waste and Vent System

A designed system of waste piping embodying the horizontal wet venting of one or more sinks or floor drains by means of a common waste and vent pipe adequately sized to provide free movement of air above the flow line of the drain.

54. Combined Building Drain

See "Building Drain, Combined"

55. Combined Building Sewer

See "Building Sewer, Combined"

56. Common Vent

See "Vent, Common"

57. Conductor

A pipe within a building that conveys stormwater from a roof to its connection to a building storm drain or other point of disposal.

58. Continuous Vent

See "Vent, Continuous"

59. Continuous Waste

A drain from two or more fixtures connected to a single trap.

60. Critical Level

The critical level marking on a backflow prevention device or vacuum breaker is a point established by the manufacturer, and usually stamped on the device by the manufacturer, which determines the minimum elevation above the flood level rim of the fixture or receptor served at which the device may be installed. When a backflow prevention device does not bear a critical level marking, the bottom of the vacuum breaker, combination valve, or the bottom of any approved device shall constitute the critical level.

61. Cross Connection

Any connection or arrangement between two otherwise separate piping systems, one of which contains potable water and the other either water of questionable safety, steam, gas, or chemical, whereby there may be a flow from one system to the other, the direction of flow depending on the pressure differential between the two systems. (See "Backflow" and "Back-Siphonage.")

62. Day Care Center

A facility for the care and/or education of children ranging from 2½ years of age to 5 years of age.

63. Day Nursery

A facility for the care of children less than 2½ years of age.

64. Dead End

A branch leading from a soil, waste, or vent pipe, building drain, or building sewer, and terminating at a developed length of 2 feet or more by means of a plug, cap, or other closed fitting.

65. Developed Length

The length of a pipe line measured along the center line of the pipe and fittings.

66. Diameter

See "Size of Pipe and Tubing"

67. Domestic Sewage

The water-borne wastes derived from ordinary living processes.

68. Double Check Valve Assembly

A backflow prevention device consisting of two independently acting check valves, internally force loaded to a normally closed position between two tightly closing shut-off valves, and with means of testing for tightness.

69. Double Offset

See "Offset, Double"

70. Downspout

See "Leader"

71. Drain

Any pipe which carries waste or water-borne wastes in a building drainage system.

72. Drainage Pipe

See "Drainage System"

73. Drainage, Sump

A liquid and air-tight tank which receives sewage and/or liquid waste, located below the elevation of the gravity system, which shall be emptied by pumping.

74. Drainage System

Includes all the piping, within public or private premises, which conveys sewage, rain water, or other liquid wastes to a point of disposal. It does not include the mains of a public sewer system or private or public sewage-treatment.

75. Drainage System, Building Gravity

A drainage system which drains by gravity into the building sewer.

76. Drainage System, Sub-Building

See "Building Subdrain"

77. Dry Vent

See "Vent, Dry"

78. **Dry Well**

See "Leaching Well"

79. **Dual Vent**

See "Vent, Common"

80. **Dwelling Unit, Multiple**

A room, or group of rooms, forming a single habitable unit with facilities which are used, or intended to be used, for living, sleeping, cooking and eating; and whose sewer connections and water supply, within its own premise, are shared with one or more other dwelling units. Multiple dwelling units include guest rooms in hotels and motels.

81. **Dwelling Unit, Single**

A room, or group of rooms, forming a single habitable unit with facilities which are used, or intended to be used, for living, sleeping, cooking and eating; and whose sewer connections and water supply, are, within its own premise, separate from and completely independent of any other dwelling.

82. **DWV**

An acronym for "drain-waste-vent" referring to the combined sanitary drainage and venting systems. This term is technically equivalent to "soil-waste-vent" (SWV).

83. **Effective Opening**

The minimum cross-sectional area at the point of water supply discharge, measured or expressed in terms of (1) diameter of a circle, or (2) if the opening is not circular, the diameter of a circle of equivalent cross-sectional area.

84. **Equivalent Length**

The length of straight pipe of a specific diameter that would produce the same frictional resistance as a particular fitting or line comprised of pipe and fittings.

85. **Existing Work**

A plumbing system, or any part thereof, installed prior to the effective date of this Code.

86. **Family**

One or more individuals living together and sharing the same facilities.

87. **Fixture**

See "Plumbing Fixture"

88. **Fixture Branch, Drainage**

A drain serving one or more fixtures which discharges into another drain.

89. Fixture Branch, Supply

A branch of the water distribution system supplying one fixture.

90. Fixture Drain

The drain from the trap of a fixture to the junction of that drain with any other drain pipe.

91. Fixture Supply Tube

A flexible or soft temper water supply tube (or riser), typically 3/8" or ½" nominal size O.D., connecting a water closet ballcock, faucet, appliance or similar fixture to its stop valve and/or fixture supply branch pipe.

92. Fixture Unit (Drainage — d.f.u.)

An index number that represents the load of a fixture on the drainage system so that the load of various fixtures in various applications can be combined. The value is based on the volume or volume rate of drainage discharge from the fixture, the time duration of that discharge, and the average time between successive uses of the fixture.

One d.f.u. was originally equated to a drainage flow rate of one cubic foot per minute or 7.5 gallons per minute through the fixture outlet.

93. Fixture Unit (Water Supply — w.s.f.u.)

An index number that represents the load of a fixture on the water supply system so that the load of various fixtures in various applications can be combined. The value is based on the volume rate of supply for the fixture, the time duration of a single supply operation, and the average time between successive uses of the fixture.

Water supply fixture units were originally based on a comparison to a flushometer valve water closet, which was arbitrarily assigned a value of 10 w.s.f.u.

94. Flood Level

See "Flood Level Rim"

95. Flood Level Rim

The edge of the receptor or fixture which water overflows.

96. Flooded

The condition which results when the liquid in a receptor or fixture rises to the flood level rim.

97. Flow Pressure

The pressure in the water supply pipe near the faucet or water outlet while the faucet or water outlet is fully open and flowing.

98. Flushing Type Floor Drain

A floor drain which is equipped with an integral water supply connection, enabling flushing of the drain receptor and trap.

99. Flush Valve

A device located at the bottom of a tank for flushing water closets and similar fixtures.

100. Flushometer Tank

A device integrated within an air accumulator vessel which is designed to discharge a predetermined quantity of water to the fixture for flushing purposes.

101. Flushometer Valve

A device which discharges a predetermined quantity of water to fixtures for flushing purposes and is closed by direct water pressure or other mechanical means.

102. Force Mains

Force mains deliver waste water discharged from a sewage ejector or pump to its destination which may be a public or private disposal system, or a higher point in the drainage/sewage system.

103. Grade

The fall (slope) of a line of pipe in reference to a horizontal plane. In drainage, it is usually expressed as the fall in a fraction of an inch per foot length of pipe.

104. Grease Interceptor

See "Interceptor"

105. Grease Trap

See "Interceptor"

106. Ground Water

Subsurface water occupying the zone of saturation:

(a) Confined Ground Water — a body of ground water overlaid by material sufficiently impervious to sever free hydraulic connection with overlying ground water.

(b) Free Ground Water — ground water in the zone of saturation extending down to the first impervious barrier.

107. Half-Bath

For the purposes of this Code, a half-bath or powder room consists of one water closet and one lavatory within a dwelling unit.

108. Hangers

See "Supports"

109. Hazard, High

In backflow prevention, an actual or potential threat of contamination to the potable water supply of a physical or toxic nature that would be a danger to health.

110. Hazard, Low

In backflow prevention, an actual or potential threat to the physical properties or potability of the water supply, but which would not constitute a health or system hazard.

111. High Hazard

See "Hazard, High"

112. Horizontal Branch Drain

A drain branch pipe extending laterally from a soil or waste stack or building drain, with or without vertical sections or branches, which receives the discharge from one or more fixture drains and conducts it to the soil or waste stack or to the building drain.

113. Horizontal Pipe

Any pipe or fitting which makes an angle of less than 45 with the horizontal.

114. Hot Water

Potable water at a temperature of not less than 120 F and not more than 140 F.

115. House Drain

See "Building Drain"

116. House Sewer

See "Building Sewer"

117. House Trap

See "Building Trap"

118. Indirect Connection (Waste)

The introduction of waste into drainage system by means of an air gap or air break without having a direct connection to the drainage system.

119. Indirect Waste Pipe

A waste pipe which does not connect directly with the drainage system, but which discharges into the drainage system through an air break or air gap into a trap, fixture, receptor or interceptor.

120. Individual Vent

See "Vent, Individual"

121. **Industrial Wastes**

Liquid or liquid borne wastes resulting from the processes employed in industrial and commercial establishments.

122. **Insanitary**

Contrary to sanitary principles — injurious to health.

123. **Installed**

Altered, changed or a new installation.

124. **Interceptor**

A device designed and installed so as to separate and retain deleterious, hazardous, or undesirable matter from normal wastes while permitting normal sewage or liquid wastes to discharge into the drainage system by gravity.

125. **Invert**

The lowest portion of the inside of a horizontal pipe.

126. **Leaching Well or Pit**

A pit or receptor having porous walls which permit the contents to seep into the ground.

127. **Leader**

An exterior vertical drainage pipe for conveying storm water from roof or gutter drains.

128. **Load Factor**

The percentage of the total connected fixture unit flow which is likely to occur at any point in the drainage system.

129. **Local Ventilating Pipe**

A pipe on the fixture side of the trap through which vapor or foul air is removed from a fixture.

130. **Loop Vent**

See "Vent, Loop"

131. **Low Hazard**

See "Hazard, Low"

132. **Main**

The principal pipe artery to which branches may be connected.

133. **Main Sewer**

See "Public Sewer"

134. May

The word "may" is a permissive term.

135. Medical Gas System

The complete system used to convey medical gases for direct application from central supply systems (bulk tanks, manifolds and medical air compressors) through piping networks with pressure and operating controls, alarm warning systems, etc., and extending to station outlet valves at use points.

136. Medical Vacuum Systems

A system consisting of central-vacuum-producing equipment with pressure and operating controls, shut-off valves, alarm warning systems, gauges and a network of piping extending to and terminating with suitable station inlets to locations where suction may be required.

137. Multiple Dwelling

Building containing two or more dwelling units.

138. Non-Potable Water

Water not safe for drinking or for personal or culinary use.

139. Nuisance

Public nuisance at common law or in equity jurisprudence; whatever is dangerous to human life or detrimental to health; whatever building, structure, or premises is not sufficiently ventilated, sewered, drained, cleaned, or lighted, in reference to its intended or actual use; and whatever renders the air or human food or drink or water supply unwholesome.

140. Offset

A combination of elbows or bends which brings one section of the pipe out of line but into a line parallel with the other section.

141. Offset, Double

Two changes of direction installed in succession or series in a continuous pipe.

142. Offset, Return

A double offset installed so as to return the pipe to its original alignment.

143. Oil Interceptor

See "Interceptor"

144. Person

A natural person, his heirs, executors, administrators or assigns; including a firm, partnership or corporation, its or their successors or assigns. Singular includes plural; male includes female.

145. **Pitch**

See "Grade"

146. **Plumbing**

The practice, materials and fixtures within or adjacent to any building structure or Conveyance, used in the installation, maintenance, extension, alteration and removal of all piping, plumbing fixtures, plumbing appliances, and plumbing appurtenances in connection with any of the following:

a. Sanitary drainage system and its related vent system,
b. Storm water drainage facilities, venting systems,
c. Public or private potable water supply systems,
d. The initial connection to a potable water supply upstream of any required backflow prevention devices and the final connection that discharges indirectly into a public or private disposal system,
e. Medical gas and medical vacuum systems,
f. Indirect waste piping including refrigeration and air conditioning drainage,
g. Liquid waste or sewage, and water supply, of any premises to their connection with the approved water supply system or to an acceptable disposal facility.

NOTE: The following are excluded from the definition:

1. All piping, equipment or material used exclusively for environmental control.
2. Piping used for the incorporation of liquids or gases into any product or process for use in the manufacturing or storage of any product, including product development.
3. Piping used for the installation, alteration, repair or removal of automatic sprinkler systems installed for fire protection only.
4. The related appurtenances or standpipes connected to automatic sprinkler systems or overhead or underground fire lines beginning at a point where water is used exclusively for fire protection.
5. Piping used for lawn sprinkler systems downstream from backflow prevention devices.

147. **Plumbing Appliance**

Any one of a special class of plumbing fixture which is intended to perform a special plumbing function. Its operation and/or control may be dependent upon one or more energized components, such as motors, controls, heating elements, or pressure or temperature-sensing elements. Such fixtures may operate automatically through one or more of the following actions: a time cycle, a temperature range, a pressure range, a measured volume or weight; or the fixture may be manually adjusted or controlled by the user or operator.

148. **Plumbing Appurtenance**

A manufactured device, or a prefabricated assembly, or an on-the-job assembly of component parts, which is an adjunct to the basic piping system and plumbing fixtures. An appurtenance demands no additional water supply, nor does it add any discharge load to a fixture or to the drainage system. It is presumed that an appurtenance performs some useful function in the operation, maintenance, servicing, economy, or safety of the plumbing system.

149. Plumbing Fixture

A receptacle or device which is either permanently or temporarily connected to the water distribution system of the premises, and demands a supply of water therefrom, or discharges used water, liquid-borne waste materials, or sewage either directly or indirectly to the drainage system of the premises, or which requires both a water supply connection and a discharge to the drainage system of the premises. Plumbing appliances as a special class of fixture are further defined.

150. Plumbing Inspector

See "Authority Having Jurisdiction"

151. Plumbing System

Includes the water supply and distribution pipes, plumbing fixtures and traps; soil, waste and vent pipes; sanitary and storm drains and building sewers, including their respective connections, devices and appurtenances to an approved point of disposal.

152. Pollution

The addition of sewage, industrial wastes, or other harmful or objectionable material to water. Sources of sewage pollution may be privies, septic tanks, subsurface irrigation fields, seepage pits, sink drains, barnyard wastes, etc.

153. Pool

See "Swimming Pool"

154. Potable Water

Water free from impurities present in amounts sufficient to cause disease or harmful physiological effects and conforming in its bacteriological and chemical quality to the requirements of the Public Health Service Drinking Water Standards or the regulations of the public health authority having jurisdiction.

155. Powder Room

See "Half-Bath"

156. Pressure Balancing Valve

A mixing valve which senses incoming hot and cold water pressures and compensates for fluctuations in either to stabilize its outlet temperature.

157. Private Sewage Disposal System

A system for disposal of domestic sewage by means of a septic tank or mechanical treatment, designed for use apart from a public sewer to serve a single establishment or building.

158. Private Sewer

A sewer not directly controlled by public authority.

159. Private Water Supply

A supply, other than an approved public water supply, which serves one or more buildings.

160. Public Sewer

A common sewer directly controlled by public authority.

161. Public Toilet Room

A toilet room intended to serve the transient public, such as in, but not limited to, the following examples: service stations, train stations, airports, restaurants, and convention halls.

162. Public Water Main

A water supply pipe for public use controlled by public authority.

163. Receptor

A fixture or device which receives the discharge from indirect waste pipes.

164. Reduced Pressure Principle Backpressure Backflow Preventer

A backflow prevention device consisting of two independently acting check valves, internally force loaded to a normally closed position and separated by an intermediate chamber (or zone), in which there is an automatic relief means of venting to atmosphere internally loaded to a normally open position between two tightly closing shut-off valves and with means for testing for tightness of the checks and opening of relief means.

165. Relief Vent

See "Vent, Relief"

166. Return Offset

See "Offset, Return"

167. Revent Pipe

See "Vent, Individual"

168. Rim

An unobstructed open edge of a fixture.

169. Riser

A water supply pipe which extends vertically one full story or more to convey water to branches or to a group of fixtures.

170. Roof Drain

A drain installed to receive water collecting on the surface of a roof and to discharge it into a leader or a conductor.

171. Roughing-In

The installation of all parts of the plumbing system which can be completed prior to the installation of fixtures. This includes drainage, water supply, and vent piping; the necessary fixture supports; and/or any fixtures that are built into the structure.

172. Safe Waste

See "Indirect Waste Pipe"

173. Sand Filter

A treatment device or structure, constructed above or below the surface of the ground, for removing solid or colloidal material of a type that cannot be removed by sedimentation from septic tank effluent.

174. Sand Interceptor

See "Interceptor"

175. Sand Trap

See "Interceptor"

176. Sanitary Sewer

A sewer which carries sewage and excludes storm, surface and ground water.

177. SDR

An abbreviation for "standard dimensional ratio" which relates to a specific ratio of the average outside diameter to the minimum wall thickness for outside controlled diameter plastic pipe.

178. Seepage Well or Pit

See "Leaching Well"

179. Septic Tank

A watertight receptacle which receives the discharge of a building sanitary drainage system or part thereof, and is designed and constructed so as to separate solids from the liquid, digest organic matter through a period of detention, and allow the liquids to discharge into the soil outside of the tank through a system of open joint or perforated piping, or a seepage pit.

180. Service Sink

A sink or receptor intended for custodial use that is capable of being used to fill and empty a janitor's bucket. Included are mop basins, laundry sinks, utility sinks, and similar fixtures.

181. Sewage

Liquid containing human waste (including fecal matter) and/or animal, vegetable, or chemical waste matter in suspension or solution.

182. Sewage Ejectors, Pneumatic

A device for lifting sewage by air pressure.

183. Sewage Pump

A permanently installed mechanical device, other than an ejector, for removing sewage or liquid waste from a sump.

184. Shall

"Shall" is a mandatory term.

185. Shock Arrestor (mechanical device)

A device used to absorb the pressure surge (water hammer) which occurs when water flow is suddenly stopped.

186. Short Term

A period of time not more than 30 minutes.

187. Side Vent

See "Vent, Side"

188. Sink, Commercial

A sink other than for a domestic application. Commercial sinks include, but are not limited to:

1. Pot sinks;
2. scullery sinks;
3. and side sinks used in photographic or other processes.

189. Size of Pipe and Tubing

The nominal inside diameter in inches as indicated in the material standards in Table 3.1.3. If outside diameter is used, the size will be followed by "o.d."

190. Size of Pipe and Tubing, Incremental

Where relative size requirements are mentioned, the following schedule of sizes is recognized, even if all sizes may not be available commercially: ¼, , ½, ¾, 1, 1¼, 1½, 2, 2½, 3, 3½, 4, 4½, 5, 6, 7, 8, 10, 12, 15, 18, 21, 24.

191. Slip Joint

A connection in drainage piping consisting of a compression nut and compression washer that permits drainage tubing to be inserted into the joint and secured by tightening the compression nut. Slip joints are typically used in trap connections for lavatories, sinks, and bathtubs. They permit the trap to be removed for cleaning or replacement, and to provide access to the drainage piping.

192. Slope

See "Grade"

Definitions

193. Soil Pipe or Soil Stack

Pipe which conveys sewage containing fecal matter to the building drain or building sewer.

194. Special Wastes

Wastes which require special treatment before entry into the normal plumbing system.

195. Special Waste Pipe

Pipes which convey special wastes.

196. Stack

A general term for any vertical line including offsets of soil, waste, vent or inside conductor piping. This does not include vertical fixture and vent branches that do not extend through the roof or that pass through not more than two stories before being reconnected to the vent stack or stack vent.

197. Stack Group

A group of fixtures located adjacent to the stack so that by means of proper fittings, vents may be reduced to a minimum.

198. Stack Vent

The extension of a soil or waste stack above the highest horizontal drain connected to the stack.

199. Stack Venting

A method of venting a fixture or fixtures through the soil or waste stack.

200. Storm Drain

See "Drain, Storm"

201. Storm Sewer

A sewer used for conveying rain water, surface water, condensate, cooling water, or similar liquid wastes.

202. Subsoil Drain

A drain which collects subsurface or seepage water and conveys it to a place of disposal.

203. Suction Line

The inlet pipe to a pump on which a negative pressure may exist under design conditions.

204. Sump

A tank or pit which receives liquid wastes only, located below the elevation of the gravity system and which shall be emptied by pumping.

205. Sump, Drainage

A liquid and air-tight tank which receives sewage and/or liquid waste, located below the elevation of the gravity system, and which shall be emptied by pumping.

206. Sump Pump

A permanently installed mechanical device for removing clear water or liquid waste from a sump.

207. Supports

Devices for supporting and securing pipe, fixtures and equipment.

208. Swimming Pool

Any structure, basin, chamber or tank containing an artificial body of water for swimming, diving, or recreational bathing.

209. Tempered Water

Water at a temperature of not less than 90 F and not more than 105 F.

210. Thermostatic/Pressure Balancing Valve, Combination

A mixing valve which senses outlet temperature and incoming hot and cold water pressure and compensates for fluctuations in incoming hot and cold water temperatures and/or pressures to stabilize its outlet temperatures.

211. Thermostatic (Temperature Control) Valve

A mixing valve which senses outlet temperature and compensates for fluctuation in incoming hot or cold water temperatures.

212. Toilet Facility

A room or combination of interconnected spaces in other than a dwelling that contains one or more water closets and associated lavatories, with signage to identify its intended use.

213. Trap

A fitting or device which provides a liquid seal to prevent the emission of sewer gases without materially affecting the flow of sewage or waste water through it.

214. Trap Arm

A trap arm is that portion of a fixture drain between a trap and its vent.

215. Trap Primer

A trap primer is a device or system of piping to maintain a water seal in a trap.

216. Trap Seal

The maximum vertical depth of liquid that a trap will retain, measured between the crown weir and the top of the dip of the trap.

217. Vacuum

Any pressure less than that exerted by the atmosphere.

218. Vacuum Breaker

See "Backflow Preventer"

219. Vacuum Breaker, Non-Pressure Type (Atmospheric)

A vacuum breaker which is not designed to be subject to static line pressure.

220. Vacuum Breaker, Pressure Type

A vacuum breaker designed to operate under conditions of static line pressure.

221. Vacuum Breaker, Spill-proof (SVB)

A pressure-type vacuum breaker specifically designed to avoid spillage during operation, consisting of one check valve force-loaded closed and an air inlet vent valve force-loaded open to atmosphere, positioned downstream of the check valve, and located between and including two tightly closing shut-off valves and a means for testing.

222. Vacuum Relief Valve

A device to prevent vacuum in a pressure vessel.

223. Vent, Branch

A vent connecting one or more individual vents with a vent stack or stack vent.

224. Vent, Circuit

A vent that connects to a horizontal drainage branch and vents from two to eight traps or trapped fixtures connected in a battery.

225. Vent, Common

A vent connected at a common connection of two fixture drains and serving as a vent for both fixtures.

226. Vent, Continuous

A vertical vent that is a continuation of the drain to which it connects.

227. Vent, Dry

A vent that does not receive the discharge of any sewage or waste.

228. Vent, Individual

A pipe installed to vent a fixture drain. It connects with the vent system above the fixture served or terminates outside the building into the open air.

229. Vent, Loop

A circuit which loops back to connect with a stack vent instead of a vent stack.

230. Vent, Relief

An auxiliary vent which permits additional circulation of air in or between drainage and vent systems.

231. Vent, Side

A vent connecting to the drain pipe through a fitting at an angle not greater than 45° to the vertical.

232. Vent, Sterilizer

A separate pipe or stack, indirectly connected to the building drainage system at the lower terminal, which receives the vapors from non-pressure sterilizers, or the exhaust vapors from pressure sterilizers, and conducts the vapors directly to the outer air. Sometimes called vapor, steam, atmosphere or exhaust vent.

233. Vent, Wet

A vent which receives the discharge of wastes other than from water closets and kitchen sinks.

234. Vent, Yoke

A pipe connecting upward from a soil or waste stack to a vent stack for the purpose of relieving pressures in the stack.

235. Vent Pipe

Part of the vent system.

236. Vent Stack

A vertical vent pipe installed to provide circulation of air to and from the drainage system and which extends through one or more stories.

237. Vent System

A pipe, or pipes, installed to provide a flow of air to or from a drainage system or to provide a circulation of air within such system to protect trap seals from siphonage and back pressure.

238. Vertical Pipe

Any pipe or fitting which makes an angle of 45° or more with the horizontal.

239. Wall Hung Water Closet

A water closet installed in such a way that no part of the water closet touches the floor.

240. Waste

Any liquid or liquid-borne material or residue intended to be discarded that remains after any activity or process but not including any such materials that contain animal or human fecal matter.

241. Waste Pipe

A pipe which conveys only waste.

242. Waste Stack, Pipe or Piping

Pipes which convey the discharge from fixtures (other than water closets), appliances, areas, or appurtenances, which do not contain fecal matter.

243. Water Distribution Pipe

A pipe within the building or on the premises which conveys water from the water-service pipe to the point of usage.

244. Water Lifts

See "Sewage Ejector"

245. Water Main

A water supply pipe for public use.

246. Water Outlet

A discharge opening through which water is supplied to a fixture, into the atmosphere (except into an open tank which is part of the water supply system), to a boiler or heating system, to any devices or equipment requiring water to operate, but which are not part of the plumbing system.

247. Water Riser Pipe

See "Riser"

248. Water Service Pipe

The pipe from the water main, or other source of potable water supply, to the water distributing system of the building served.

249. Water Supply System

The water service pipe, the water-distributing pipes, and the necessary connecting pipes, fittings, control valves, and appurtenances in or adjacent to the building or premises.

250. Water Temperature Control Valve

A valve of the pressure balance, thermostatic mixing, or combination pressure balance/thermostatic mixing type, which is designed to control water temperature to reduce the risk of scalding.

251. Wet Vent

See "Vent, Wet"

252. Whirlpool Bathtub

A plumbing appliance consisting of a bathtub fixture which is equipped and fitted with a circulation piping system, pump, and other appurtenances and is so designed to accept, circulate, and discharge bathtub water upon each use.

253. Weir (trap or crown)

Discharge overflow of the trap outlet.

254. Yoke Vent

See "Vent, Yoke"

LESSON 1

INTRODUCTION TO THE PLUMBING PROFESSION

PLUMBING WORKERS

The following definition of **plumbing** is reprinted from the *Plumbing Dictionary, Fifth Edition*, with permission from the American Society of Sanitary Engineering.

1. includes the work and/or practice, materials and fixtures used in the installation, removal, maintenance, extension, and alterations of a plumbing system of all piping, fixtures, fixed appliances and appurtenances in connection with any of the following: sanitary drainage, storm drainage facilities, special wastes, the venting system and the public or private water supply systems, within or adjacent to any building, structure, or conveyance to their connection with any point of public disposal or other acceptable terminal within the property line.

2. the pipes, fixtures and all other apparatus concerned in the introduction, distribution and disposal of water in a building.

3. the pipes, fixtures and other apparatus of a water, gas or sewage system.

4. the work of a plumber.

A **plumber** is defined as follows:

1. one who installs and repairs piping fixtures, appliances, appurtenances in conjunction with water supply, and drainage systems, etc., both inside and outside of buildings.

2. a worker of lead or similar materials.

3. a person trained and experienced in the art of plumbing, design, fabrication, engineering and installation.

The **National Standard Plumbing Code** defines Plumbing as:

The practice, materials and fixtures within or adjacent to any building structure or conveyance, used in the installation, maintenance, extension, alteration and removal of all piping, plumbing fixtures, plumbing appliances, and plumbing appurtenances in connection with any of the following:

a. Sanitary drainage system and its related vent system,
b. Storm water drainage facilities, venting systems,
c. Public or private potable water supply systems,
d. The initial connection to a potable water supply upstream of any required backflow prevention devices and the final connection that discharges indirectly into a public or private disposal system,
e. Medical gas and medical vacuum systems,
f. Indirect waste piping including refrigeration and air conditioning drainage,
g. Liquid waste or sewage, and water supply, of any premises to their connection with the approved water supply system or to an acceptable disposal facility.

NOTE: The following are excluded from the definition:

1. All piping, equipment or material used exclusively for environmental control.

2. Piping used for the incorporation of liquids or gases into any product or process for use in the manufacturing or storage of any product, including product development.

3. *Piping used for the installation, alteration, repair or removal of automatic sprinkler systems installed for fire protection only.*

4. *The related appurtenances or standpipes connected to automatic sprinkler systems or overhead or underground fire lines beginning at a point where water is used exclusively for fire protection.*

5. *Piping used for lawn sprinkler systems downstream from backflow prevention devices.*

As indicated in the definitions of plumbing above, a plumber may work on a wide variety of systems with many components. Thus, the range of skills required for a competent plumber is quite broad, ranging from the ability to visualize the details of a new project before it starts, recognizing the required performance of an existing system, to being able to diagnose the reason(s) for improper operation (when that occurs) of a fixture or device that is part of a system. The plumber must also have the mechanical understanding to carry out the necessary changes. These requirements vary for new work, remodeling, and repair; and from very small projects to very large ones. To describe this great range is not to say that such broad talent is found in most individual plumbers — it is to suggest a complete standard that each of us can aspire to approach as close as possible!

You can see that the plumber has many opportunities for challenging, satisfying work in a large range of types and sizes of organizations.

Workers trained as plumbers are also needed in related fields such as pipe-fitting, boiler installation, solar energy, and an extensive range of process piping projects.

In common with all construction trades, plumbing is first of all concerned with safety — safety for the public, the customer, fellow workers on the project, **and you**. Safety is both short- and long-term. You must use safe work practices at the time of execution as well as materials and methods that will perform as intended for years into the future.

BECOMING A PLUMBER

To become a plumber requires a considerable amount of training. The apprentice training method has proven to be highly successful in producing skilled mechanics. You are embarking on a learning experience that will equip you to be an effective and competitive plumber in the construction industry. The experiences you gain will be an excellent foundation for many possible paths of development in your future.

In this first year of apprenticeship your classmates may include individuals who have been working with tools for many years or individuals who are just learning what tools to use and how to use them. This is a four-year program which guides you from basic information to more complex information. Much of what you learn in future years has a base in what you will study this year.

You are here to take part in the school portion of your apprenticeship. This is only a part of your total apprenticeship experience. Your on-the-job training through your employer is essential for a well-rounded plumber. Your school will require certain things from you such as consistent attendance at school (no matter how tired you are after a full day), paying attention in class, work records, and the beginning of an understanding of our industry.

You will be expected to contribute in class as well as at your job. The more information you are exposed to, the further you will be able to go in our industry.

POTABLE WATER

The most important responsibility of a plumber is to ensure the safety of our potable water supply. ***Potable*** water is defined in the ***National Standard Plumbing Code*** as:

Plumbing Apprentice Student Workbook Year One
Fourth Edition

Water free from impurities present in amounts sufficient to cause disease or harmful physiological effects and conforming in its bacteriological and chemical quality to the requirements of the Public Health Service Drinking Water Standards or the regulations of the public health authority having jurisdiction.

The following definition of **potable water** is reprinted from the *Plumbing Dictionary, Fifth Edition*, with permission from the American Society of Sanitary Engineering.

Water which is suitable for drinking, culinary, and personal purposes. 2. Water free from impurities present in amounts sufficient to cause disease or harmful physiological effects. Quality is normally controlled by public health regulations.

Proper plumbing correlates to public health more closely than the entire health care profession. This statement puts a heavy burden of responsibility on those of us who work as plumbers.

HISTORY OF PLUMBING

Some natural features of the earth act as plumbing systems — water is purified by passing over rocks and stones in stream beds and is aerated by various means; it is transported by clouds and delivered as rain; and materials are carried away by water flowing in streams and rivers. As plumbers, however, we are concerned mainly with man-made systems.

No one knows the name of the first plumber, but we know the craft was performed in ancient times. Water cisterns have been discovered that were used by the Babylonians 5000 years ago. Made of masonry, these cisterns were most likely used to collect rainwater for distribution to the people. Even before that, man used simple things like cups and trenches to move water from one place to another.

About 4000 years ago, vented toilets were used in Crete. The Romans had the most extensive plumbing system of the ancients, and they also gave us the word plumber which comes from the Latin "plumbum", meaning lead. Originally a plumber was a leadworker, so the Latin name was applied to such workers. In modern times very little lead work remains in American plumbing work, partly for reasons of cost, but mostly for reasons of safety and health.

After the era of the domination of the Roman Empire, plumbing declined in use; this decline is frequently cited as the cause of the dreaded epidemics that occurred during the Middle Ages. Plumbing enjoyed a comeback during the late Middle Ages — effective plumbing systems are now known to be essential to the survival of large population centers.

Plumbing fixtures similar to those we know today were developed in the nineteenth century — the bathtub by Lord Russell of England in the 1830's and the siphonic water closet by Sir R. Harrington and improved by Sir Thomas Crapper (although there is much controversy over the part actually played by him) in the 1880's. In the United States, plumbing systems were developed with the assistance of plumbers beginning in the 1850's.

In the twentieth century, three fields of work in the United States combined to produce the safe water systems that we generally expect, and try to encourage all the populations of the world to develop:

1. The production of pure water for distribution to the public

2. The total field of plumbing

3. The sewage and waste treatment industry

As plumbers, we are in the middle of this chain: our responsibility is to keep the pure water safe and easy to use, and to keep waste products securely separated until those products are delivered to the waste treatment system. For the large majority of the population, the waste treatment system is a public sewage treatment plant, but many people are served by (usually smaller) private systems.

If you would like to learn more about the history of plumbing, there are two videos available through The History Channel: Modern Marvels, *Plumbing the Arteries of Civilization*, 50 minutes, and Modern Marvels, *Bathroom Tech*, 50 minutes.

BASIC PRINCIPLES OF PLUMBING

Modern plumbing systems are designed and installed to conform to plumbing codes. Most of the modern codes in use in the United States are based upon certain Basic Principles. We will explore the Principles fully during your studies in this apprenticeship course. Listed below are the Principles as they appear in the *National Standard Plumbing Code,* 2003 Edition.

Principle No. 1 — All occupied premises shall have potable water

All premises intended for human habitation, occupancy, or use shall be provided with a supply of potable water. Such a water supply shall not be connected with unsafe water sources, nor shall it be subject to the hazards of backflow.

Principle No. 2 — Adequate water required

Plumbing fixtures, devices, and appurtenances shall be supplied with water in sufficient volume and at pressures adequate to enable them to function properly and without undue noise under normal conditions of use.

Principle No. 3 — Hot water required

Hot water shall be supplied to all plumbing fixtures which normally need or require hot water for their proper use and function.

Principle No. 4 — Water conservation

Plumbing shall be designed and adjusted to use the minimum quantity of water consistent with proper performance and cleaning.

Principle No. 5 — Safety devices

Devices for heating and storing water shall be so designed and installed as to guard against dangers from explosion or overheating.

Principle No. 6 — Use public sewer where available

Every building with installed plumbing fixtures and intended for human habitation, occupancy, or use, and located on premises where a public sewer is on or passes said premises within a reasonable distance, shall be connected to the public sewer.

Principle No. 7 — Required plumbing fixtures

Each family dwelling unit shall have at least one water closet, one lavatory, one kitchen-type sink, and one bathtub or shower to meet the basic requirements of sanitation and personal hygiene.

All other structures for human habitation shall be equipped with sufficient sanitary facilities. Plumbing fixtures shall be made of durable, smooth, non-absorbent and corrosion resistant material and shall be free from concealed fouling surfaces.

Principle No. 8 — Drainage system

The drainage system shall be designed, constructed, and maintained to guard against fouling, deposit of solids and clogging, and with adequate cleanouts so arranged that the pipes may be readily cleaned.

Principle No. 9 — Durable materials and good workmanship

The piping of the plumbing system shall be of durable material, free from defective workmanship, and so designed and constructed as to give satisfactory service for its reasonable expected life.

Principle No. 10 — Fixture traps

Each fixture directly connected to the drainage system shall be equipped with a liquid seal trap.

Principle No. 11 — Trap seals shall be protected

The drainage system shall be designed to provide an adequate circulation of air in all pipes with no danger of siphonage, aspiration, or forcing of trap seals under conditions of ordinary use.

Principle No. 12 — Exhaust foul air to outside

Each vent terminal shall extend to the outer air and be so installed as to minimize the possibilities of clogging and the return of foul air to the building.

Principle No. 13 — Test the plumbing system

The plumbing system shall be subjected to such tests as will effectively disclose all leaks and defects in the work or the material.

Principle No. 14 — Exclude certain substances from the plumbing system

No substance which will clog or accentuate clogging of pipes, produce explosive mixtures, destroy the pipes or their joints, or interfere unduly with the sewage-disposal process shall be allowed to enter the building drainage system.

Principle No. 15 — Prevent contamination

Proper protection shall be provided to prevent contamination of food, water, sterile goods, and similar materials by backflow of sewage. When necessary, the fixture, device, or appliance shall be connected indirectly with the building drainage system.

Principle No. 16 — Light and ventilation

No water closet or similar fixture shall be located in a room or compartment which is not properly lighted and ventilated.

Principle No. 17 — Individual sewage disposal systems

If water closets or other plumbing fixtures are installed in buildings where there is no sewer within a reasonable distance, suitable provision shall be made for disposing of the sewage by some accepted method of sewage treatment and disposal.

Principle No. 18 — Prevent sewer flooding

Where a plumbing drainage system is subject to backflow of sewage from the public sewer or private disposal system, suitable provision shall be made to prevent its overflow in the building.

Principle No. 19 — Proper maintenance

Plumbing systems shall be maintained in a safe and serviceable condition from the standpoint of both mechanics and health.

Principle No. 20 — Fixtures shall be accessible

All plumbing fixtures shall be so installed with regard to spacing as to be accessible for their intended use and for cleansing.

Principle No. 21 — Structural safety

Plumbing shall be installed with due regard to preservation of the strength of structural members and prevention of damage to walls and other surfaces through fixture usage.

Principle No. 22 — Protect ground and surface water

Sewage or other waste shall not be discharged into surface or sub-surface water unless it has first been subjected to some acceptable form of treatment.

MODERN-DAY PLUMBING

Plumbing is essential for public health. Our bodily functions require potable water for consumption and cleaning — plumbing systems provide the means of bringing potable water for personal use as well as carrying away the waste after use. Large population centers would be prone to epidemics of disease without satisfactory means to protect potable water and to carry away waste. It is equally true that the most remote family home must have proper plumbing if the family's health is to be maintained.

To assure responsible plumbing installations, many areas have adopted a plumbing code (Figure 1-A) to set forth minimum plumbing designs; a system of licensing persons to work as plumbers; permits to give notice to building departments that plumbing work is to be done and permission to do it; and a way to inspect the work after it has been done.

In many areas, apprentices must be licensed or otherwise recognized by the licensing body. Apprentices gain experience working with journeymen and contractors, as well as by attending classes and studying textbooks. Of course, journeymen and contractors are also subject to licensing provisions. In this way, the public can be assured that plumbing facilities are installed, serviced, and maintained by skilled workers. Apprentice training programs produce the skilled workers of the future.

THE CHARACTERISTICS OF A GOOD MECHANIC

You are beginning the process which will help you to become a skilled mechanic in an important part of the construction industry — plumbing. You will be involved with work that is essential to the health and safety of everyone in your community. You will need physical ability, mental alertness, experience, and ambition to measure up to high industry standards.

The ultimate mark of a true mechanic, however, is pride — pride in oneself, and pride in your craft. This pride makes you do proper work, not only because an inspector is going to look at it, but because your own standards require it.

This pride is quiet, born of competence and confidence in yourself, and causes you to maintain high standards in your performance (both in class and at work), in your appearance, and in your consideration of other people and their property.

PLUMBING IS A PROUD CRAFT

Plumbing is a proud craft and you should be proud of it. You will help your own development if you will

Figure 1-A – 2003 National Standard Plumbing Code Illustrated

honor these guidelines regarding yourself as a skilled worker:

1. Treat the owner's property *better* than you would your own

2. Be considerate of others

3. Be neat in appearance — start each day in clean clothes

4. Do not bring contamination from your job site home

5. Keep your work area clean and safe

6. Avoid needless conversation

7. Do not use profanity — it diminishes you in the eyes of others

Further examples of pride concerning yourself as a valuable employee are these:

8. Report to work on time and be prepared to work hard

9. Realize that usually you must be one member of a group of workers — be an agreeable, effective member of the group

10. Use proper methods

11. Keep your truck and your tools neat and clean

12. Respect your employer

13. Keep growing in your knowledge and skills

FURTHER OPPORTUNITIES

You can progress in this industry as far as your skill, talent, and ambition will permit:

The *journeyman* is a skilled mechanic who is able to work alone and make decisions about how to proceed with the work.

The *foreman* plans the work to be done and supervises the actions of several journeymen.

The *superintendent* supervises several foremen on very large projects.

Estimators, project managers, and *salesmen* design, estimate, and negotiate projects.

Contractors own the businesses that employ some or all of the above types of employees.

The apprenticeship training program is the beginning of these careers.

What is involved in becoming a plumbing contractor? First, become an apprentice. As you know, an apprentice works for a licensed contractor during the day and attends related instruction, usually in the evening. Length of apprenticeship varies in different localities, usually influenced by local custom and practice, but it may be dictated by local, state, or federal law.

Such legal conditions will usually require that the apprentice is registered with the appropriate licensing division of your government. See **STEPS TO BECOMING A REGISTERED PLUMBING APPRENTICE** at the end of this lesson for an example of what is necessary in the State of Indiana.

In many localities, after serving your apprenticeship, you will be expected to take a test, both written and **hands-on** to acquire your journeyman's license. A licensed journeyman works for a plumbing contractor.

At this point in your training you may also choose to study further and obtain your contractor's license. You will then not only be a plumber, you may choose to become a businessperson as well.

See Figure 1-B for an example of the license

issued by one state to recognize a professional plumber.

PLUMBING BUSINESS

The plumber, as a businessperson, is faced with dealing with the costs of performing plumbing work and producing value for the clientele of that business. Your success is a reflection of your business; your business is a reflection of your workmanship; your workmanship is a reflection of your training. Experience, study, and training all combine to assure your success. Every person employed by that business has contact with the public, *and must be responsive to the requirements to deal fairly and professionally with the public.*

The plumbing trade requires a combination of mechanical aptitude and the ability to solve problems logically. Many operations require manual ability and dexterity. Planning and design require an understanding of hydraulics and many other specialties.

Operating a business requires knowing and dealing with fiscal controls such as buying, selling, scheduling, invoicing, and estimating.

All facets require study to keep up with new methods and to move forward. Improved performance is a continuing demand on each of us in this interesting, vital field of plumbing!

As a businessperson you may choose to become involved in an association such as the Plumbing-Heating-Cooling Contractors--National Association. Associations are important sources of information and education as well as a conduit to our legislative processes.

STATE PROFESSIONAL LICENSING AGENCY
123 Main St., Anytown, USA 00000
Receipt No.:
License No.:

SAMPLE

is duly licensed, as prescribed by law, as a

EXPIRES:
Unless suspended or revoked   EXECUTIVE DIRECTOR

Figure 1-B – Typical Contractor's License

EXAMPLE OF THE REQUIREMENTS TO BECOME A REGISTERED PLUMBING APPRENTICE

The individual must be employed by a licensed plumber.

The employer must have an agreement, called *Apprenticeship Standards*, with the U. S. Department of Labor, Office of Apprenticeship Training, Employer and Labor Services (OATELS). Each employer must have an individual agreement negotiated by the employer and the OATELS.

After the *Apprenticeship Standards* are completed, an Apprenticeship Agreement must be completed for each apprentice. This completed form is sent to the local Office of Apprenticeship Training, Employer and Labor Services. It must be signed by the employer, the apprentice, and an authorized person from the OATELS.

When the *Apprenticeship Agreement* is completed, a copy of it along with the completed *Application for Registration as an Apprentice Plumber* and a check for the fee (if any) must be sent to the appropriate Agency of the State. The State Application must be filled out entirely, signed by both the employer and employee, and notarized.

After processing the papers, and if everything is in order, the Licensing Agency will assign the apprentice a Plumbing Apprentice License number and send the apprentice (to his/her home address) a plumbing apprentice card to be carried. This Plumbing Apprentice License must be renewed every year.

The Plumbing Apprentice License means that the apprentice can *only* work under the direction and immediate personal supervision of a licensed plumbing contractor or a licensed journeyman plumber and such supervisor is physically present on the project while the apprentice is performing plumbing work.

The employer should keep a record of the Plumbing Apprentice License number for company records.

Plumbing-Heating-Cooling Contractors – North Central Indiana Association

LESSON 2

PLUMBING LAWS, TOOLS, AND SAFETY

WATER AND LIFE

It took many centuries of observation and experience for the following facts to be generally recognized:

Water is essential to all life: water is the major part of the makeup of nearly all living things.

Whenever water is present, some sort of plant or animal life will appear unless considerable precautions are taken to prevent it.

Human beings cannot tolerate many of the possible water-supported life forms within or even on our bodies.

The methods that minimize the presence of such organisms in our water supplies also help keep out physical and chemical impurities.

We are still learning how to recognize and, therefore, how to prevent those circumstances that permit impurities to enter our water or food supplies.

Note that we use definitions from both the *National Standard Plumbing Code* and the *ASSE Plumbing Dictionary*. You should be aware that different codes may have slightly different definitions of terms. You should be aware of the definitions used in your code jurisdiction.

WATER AS A SOLVENT

A characteristic of water that must be recognized is that water likes to dissolve things! This means that water will become involved with many, many products if the opportunity is present — and many of these products are lethal to human beings.

Therefore, for us to have and to use pure water in as routine a way as is expected in the United States, we must guard it jealously from all the many things that water would like to become a part of. Thus, it is essential that we keep the pure water that we receive from the water purveyor (either a municipal utility, or a private source) isolated from the "outside world." This further means that we contain our pure water in clean containment (piping) and safe from mixing with any impure or questionable material until the water is released at the point of use.

POSSIBLE PROBLEMS

Some of the problems the plumber must guard against are these:

CROSS CONNECTIONS

The following definition of **cross connection** is reprinted from the *Plumbing Dictionary, Fifth Edition*, with permission from the American Society of Sanitary Engineering.

1. *a connection between two pipes in the same water supply system or between two water supply systems containing potable water.*
2. *a fitting used to allow two different pipes to cross at a right angle in the same plane.*

The National Standard Plumbing Code – Illustrated 2003 defines a cross connection as:

Any connection or arrangement between two otherwise separate piping systems, one of which contains potable water and the other either water of questionable safety, steam, gas, or chemical, whereby there may be a flow from one system to the other, the direction of flow depending on the pressure differential between the two systems.

Many forms of cross connection are possible. Two arrangements often seen are faucet spouts below the rim of a fixture, or a hose connected to a lawn faucet on one end with the other end lying open on the ground, or — even worse — submerged in a container of stagnant water or chemicals that a human being cannot tolerate. Another common situation is one where water is needed in a process, so the potable supply is piped (for convenience of the operation) to the process system. Unless this connection includes necessary protective devices, a pressure loss in the water system could lead to backflow contamination.

LEAKAGE

Leakage can occur due to material failure or poor workmanship. Piping locations should be planned to minimize the consequences of a leak, should one occur. Leakage from a soil or waste line results in, at the very least, a nuisance; but it could endanger not only the water supply but food, food-handling equipment, and any material which could be damaged by being wetted.

Waste pipes installed over food bins or food serving utensils can leak and cause contamination. A leak from a waste pipe which was located over a food bin caused twenty cases of typhoid fever in New York City in July of 1946. The potential for this type of problem must be guarded against at all times.

INADEQUATE SYSTEMS

Sewage and potable water may mix due to inadequate space or size of components, or failure due to overload. Private systems are frequently complicated by space limitations, high costs, poor maintenance, and insufficient monitoring.

Overloaded systems may so frustrate the user that unsafe *temporary* arrangements may be used to *solve* a problem until a permanent, safe solution can be put in place — but the *temporary fix* is never removed.

A septic system located too close to a well may jeopardize the well contents by allowing contaminated water to enter the well system. In extreme cases, the contamination could proceed to the aquifer and spoil the water source for many people.

PHYSICAL FAILURES

Physical failures are the cause of great loss every year in terms of injuries, deaths, and property destruction. Examples of such failures include the following:

1. Water heaters unprotected by relief valves may explode

2. Inadequate supports of fixtures, piping, or appliances may fail

3. Buried pipes may be crushed or broken by imposed loads

4. Extremes of temperatures can produce pipe and/or joint failures

CODES

A community usually responds to all these problems and considerations by adopting a *Plumbing Code*. The code is a manual of minimum requirements for installations in that community (which could be a city, county, or state). The code process is not much more than one hundred years old, so it should not be surprising that this process is still developing, and that some people are impatient with some aspects of that process.

The code protects the public by requiring safe installation of plumbing fixtures. As a result, the building owner has his investment protected and the entire community is safeguarded from the hazards that could arise from faulty work.

Codes are written by experienced people with one primary objective: to merge the present capabilities of plumbing materials and installation skills with the needs of the public. Most codes are under more or less constant public review; consequently, only those provisions based on sound engineering principles and common sense will survive.

LICENSES, PERMITS, INSPECTIONS

Licenses, permits, and inspections are devices that the community uses to assure that proper designs are approved (permit) before experienced persons (licensed) perform work that **proves** to be acceptable when completed (inspection).

For reasons of public health and safety, governing bodies prohibit most persons from performing certain types of work. Exceptions to this prohibition are granted in the form of a **license**. Thus, a license is a waiver that exempts its holder from a **general prohibition** to perform certain types of work.

Plumbers must study the field for a prescribed time, possess the required experience, and make evident their knowledge and skills in an examination. If a licensed plumber performs bad work, the license can be withdrawn and that person is prohibited from working as a plumber.

A **permit** is a written document allowing a licensed plumber to perform plumbing work **according to a plan** at a **specified location**. Permits must be taken before work is started; they serve as notice to inspection agencies and their inspectors that the work is going to be performed soon. For a project that will last for more than a few days, most permit regulations require that the permit is prominently displayed at the project site.

Figure 2-A shows a typical **Not Approved** tag (colored red), and Figure 2-B shows a typical **Plumbing** tag (colored green), hence the term **red tagged** meaning a course of action is stopped, or an installation is not approved.

Inspection is performed after work is completed to assure that the work fulfills code requirements. The inspector issues an **acceptance** or **approved** certificate when the project is completed satisfactorily. Inspections may be made on a portion of the work if such partial inspection serves the interests of all concerned. Concealed piping is inspected during the "rough-in" inspection. Completed work is approved during the "final" inspection prior to building occupancy.

As your training progresses, you will study the physical properties and materials that are the foundations of codes. You will also learn methods and techniques required to perform work to code standards, how to take out a permit, and why the inspection process is important to your development, as well as vital to the community.

The **approved** certificate for work you have done should reinforce your pride in yourself and your work.

Figure 2-F shows a typical application for permit to perform work. This example shows that the same form is used for all types of construction applications. Other cities may have separate application forms for different classes of work.

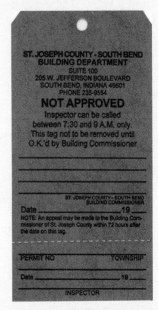

Figure 2-A – Typical "Not Approved" Tag
St. Joseph County-South Bend Building Department

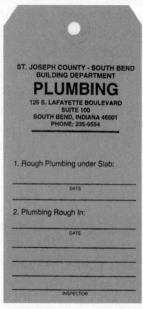

Figure 2-B – Typical Plumbing Tag
St. Joseph County-South Bend Building Department

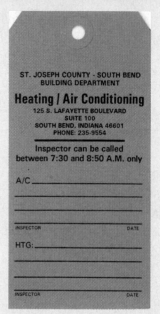

Figure 2-D – Typical Heating/Air Conditioning Tag
St. Joseph County-South Bend Building Department

Figure 2-C – Typical Electric Tag
St. Joseph County-South Bend Building Department

Figure 2-E – Typical Structural Tag
St. Joseph County-South Bend Building Department

APPLICATION FOR BUILDING PERMIT
ST. JOSEPH COUNTY/SOUTH BEND BUILDING DEPARTMENT

DATE_____ TOWNSHIP:_____

SEWAGE/WATER PERMIT:_____ SECTION:_____ TOWNSHIP:_____ RANGE:_____

VALUE:_____ ZONING:_____

STATE PROJECT NO:_____ PREVIOUS ZONING:_____

OWNER'S
NAME:_____ PHONE:_____

PRESENT MAILING ADDRESS:_____

Plans to erect:_____ Demolish:_____

Dimensions: First Floor:_____ S.F.:_____

    Second Floor:_____ S.F.:_____

    Basement:_____ S. F.:_____

    Garage:_____ Att:_____ Unatt:_____ S.F.:_____

LOCATED AT:_____

LOT NO.:_____ SUBDIVISION:_____ ACREAGE:_____

DEED RECORDED: Date:_____ Book & Page or Microfilm:_____

BUILDING LINES: Front:_____ Sides:_____ Rear:_____

BUILDING CONTRACTOR:_____ Phone No._____ FEES:_____

    Electric Service:_____ amp. Temp. Ser.:_____

    Number of Circuits:_____ Electric Company:_____

ELECTRICAL CONTRACTOR:_____ FEES:_____

    No. Traps/Floor Drains:_____ Water Heater:_____ Gas Outlets:_____

PLUMBING CONTRACTOR:_____ FEES:_____

    Kind of Heating:_____ Air Conditioning:_____

HEATING CONTRACTOR:_____ FEES:_____

                                                           SUB-TOTAL:_____

                                                             PENALTY:_____

                                                           GRAND TOTAL:_____

MANUFACTURED HOUSE INFORMATION:

    Make:_____ Date of Manufacture:_____

    HUD No.:_____ Serial No.:_____

    Roofing Materials:_____ Pitch:_____

    Siding Materials:_____

COMMENTS:_____
_____
_____

Figure 2-F – Typical Application for Building Permit
St. Joseph County-South Bend Building Department

INTRODUCTION TO TOOLS AND SAFETY

TOOL OVERVIEW

A brief, general introduction to the tools used in plumbing is presented here. Lessons three through seven pursue these topics in considerable detail.

A tool is an instrument or device that makes it possible for us to do something we otherwise could not do with our bare hands or other body parts. Most tools increase the force we can exert — a hammer hits a nail harder than we can with our hand, a wrench turns a pipe with greater force than we can by hand — but some tools help us see (magnifying glass), reach (ladder), or perform tasks dangerous to us (soldering torch).

Tools have been traced back to the beginnings of human beings. The progression of mankind, in fact, is often told in the development and complexity of tools which have been used in certain historical periods. In our own time, we are simply at a point in this long history of tool development. Lasers, backhoes, computers, laptop computers, personal electronic devices, and cell phones reflect the tools of this generation.

At this point, we see very few new tools: nearly everything is modification or specialization of relatively few tool concepts. But what a difference many of these modifications make! Using exactly the right tool for a task can increase productivity many times over.

SAFETY

GENERAL

Since many tools are made to increase our muscle force, we must use them carefully or we risk being badly hurt. Improper use of tools can also hurt other persons around us, or damage property.

Using tools safely is only part of our obligation as professional plumbers. On construction projects there are many more risks than we encounter doing repair work in our homes or offices. Regardless of the setting, we must be alert to hazards of all sorts.

Common sense is the key to safe working conditions. Think before you act, and anticipate your needs. Above all, include safety in your planning. Consider others around you. Do not endanger them or their property. Dollars spent for safety are more productive than those spent for hospitalization or worse.

Many tools are designed to meet industry safety standards. Check for seals or other information attesting to the safety status of the tool. Carefully read and observe operating instructions. Check each tool for condition; make certain that all the parts are intact. Do not use a jury-rigged tool or any tool which appears unsafe.

OSHA

To reduce the number and severity of accidents in the workplace, the Congress of the United States established the Occupational Safety and Health Administration (OSHA). Most states have established similar organizations. OSHA develops and enforces rules of safe operation in many types of workplaces; their rules heavily influence tool manufacturers, and strongly affect the design and use of tools.

OSHA is an excellent source of information on workplace safety. OSHA's mission is to assure the safety and health of America's workers by setting and enforcing standards; providing training, outreach, and education; establishing partnerships; and encouraging continual improvement in workplace safety and health. You can look up many safety issues at www.OSHA.gov.

CLOTHING

It is important to wear proper clothing, but be aware

that the rules regarding clothing and safety may conflict depending on the task to be accomplished. Note the following rules regarding proper clothing.

1. Working with rotating devices, wear short sleeve shirts or carefully roll up sleeves.

2. When welding or exposed to extreme heat or cold, wear long sleeved shirts.

3. With chemicals which will burn or with tools that will develop blisters, wear appropriate gloves.

4. With revolving machinery, do not wear gloves unless suited for such use (e.g., snaking mitts).

5. When working with hot objects, wear gloves.

6. Gloves should be worn to protect hands when holding chisels or where sharp metal is encountered.

7. When working on hot water or steam piping, do not wear gloves — a sudden leak on gloves may prevent rapid removal of the glove, thereby making the burn worse.

8. On the typical job site where the heaviest item likely to fall on your feet is a piece of pipe, safety shoes are preferred.

9. Protect your eyes, face and head — on most construction projects these parts are required to be protected at all times.

10. Doing repair work, or if you are the only person using tools at the site, you may not need this protection.

11. Leave rings and jewelry home and keep your watch in your pocket.

12. Wear proper clothing commensurate with the weather. Hypothermia and heat stroke — the two extremes — can be very dangerous.

13. Control long hair so it doesn't get pulled into machinery.

14. Use common sense — do not take chances with your body. About 90% of our input comes through our eyes — don't lose even one!

15. Wear proper clothing including clean clothing, not only for looks but for comfort and the ability to move more easily. Washing your hands often helps control the build-up of chemicals and/or materials such as lead which are potentially harmful.

Leave contaminated clothing at the jobsite; change into street clothes before you go home. Remember that whatever you get into on the jobsite goes home to your family with you unless you plan to leave it at work.

Remember, relatively minor head injuries can be incapacitating or even fatal.

Table 2-D discusses hazards and the precautions you should take with many types of tools.

JOBSITE SAFETY RULES

1. In general, use safety equipment yourself and encourage others to so.

2. Avoid horseplay or practical jokes — these can be deadly.

3. Do not run on the job.

4. Take care of injuries, report them at once, no matter how "minor."

5. Correct any unsafe condition at once.

6. Remove oil, scraps, and debris from the work area.

7. Keep hands away from moving parts.

8. Do not use fingers to remove chips or foreign items from your work.

9. Wear special clothing when working with hot or corrosive materials.

10. Work only in well-ventilated and lighted areas.

11. Warn others when you are doing hazardous work, and if necessary, rope off the work area.

12. Use care while working near exposed wiring — it could be hot.

13. Preheat any material to assure that it is dry before it is added to melting pots.

14. Keep your face away while pouring molten lead.

15. Clamp any work you are drilling or sawing.

16. Know your tools — use them properly.

TABLE 2-A
TOOL SAFETY

TOOLS	HAZARDS	PRECAUTION
Irons, chisels, hammers, screwdrivers	Case hardened steel can break off chips that may become flying projectiles	Use safety glasses Never use tool for purposes other than for which it was made Chip with chisel pointing away from you and others Grind away all *mushroomed* heads on chisels Keep chisels sharp
Files	Case hardened metal can chip Tang protrudes dangerously	Do not pry with file Place a secure handle over tang
Wrenches	Improper direction of use Wrong wrench for the job	Make sure it fits the bolt (pipe) securely Use the proper size and type Never use a *cheater* — get a larger wrench Never use a wrench as a hammer
Lead work — ladles, pots	Explosions occur when wet lead is placed into molten lead, when a wet ladle is placed in a molten pot of lead, or if molten lead is poured into a wet joint	Only dry lead should be added to the pot or ladle; pre-heat or warm lead before adding to the pot Warm up ladle before pouring Keep face away from directly over the pouring area
Lead melting	Fire or explosion from fuel gas or gas at high pressure Burns	Check all connections to make sure that they are tight Use only in ventilated area Keep tanks in cool area Turn off furnaces when they are not being used for a prolonged time Store furnace in a secure area to protect others

TABLE 2-A (Con't)
TOOL SAFETY

TOOLS	HAZARDS	PRECAUTION
Torches and welding tanks	Destruction of surroundings	Avoid exposure of flame to wood or anything flammable. Do not direct flames at a concrete surface: concrete can explode
	Fires	Use a metal shield to protect flammable surfaces. Have an adequate fire extinguisher available when working with flames
	Explosions	Avoid placing cylinders near furnace heat, radiators, open fire, or sparks from a torch. Never use oxygen as a compressed air source — pure oxygen can produce spontaneous combustion with many materials
	Broken tank valves	Keep welding cylinders in a vertical position and chained to a cart, work table, or wall to prevent knocking cylinder over. Place protective caps on oxy-acetylene or other tanks while the tanks are being moved, whether across the room or across town. Do not lift the tank by the protective caps
	Regulator leakage	Send regulator valves to manufacturer or authorized repair facility for maintenance
	Regulator blowout or gauge failure propelling fragments	Stand to the side when opening cylinder valves. Release spring tension on regulators before turning on.
	Hot metal burn	Avoid placing your body in line with falling metal or the flame
	Oil and oil container explosion	Store all oil away from any flame source

TABLE 2-A (Con't)
TOOL SAFETY

TOOLS	HAZARDS	PRECAUTION
Threaders — hand die stocks	Metal chips	Use brush or rag to clean up chips Oil carefully — gears are *fingereaters* — keep oil can spout away from geared mechanisms Keep hands away from gear mechanisms
Oil drippings	Slips and falls	Balance weight carefully, avoid hasty movements Clean, contain, or absorb drippings
Vises	Tipping	Secure vise to floor or ceiling
	Tripping from cut pipe pieces	Keep area clear of debris and cut ends of pipe
Ladders	Tipping	Secure the footing of the ladder prior to ascent Tie in place if necessary Move the ladder only from the ground position DO NOT *walk* the ladder If possible, have a *spotter* at the base of the ladder to secure its footing while you are working on it Use the proper ladder *feet* to go with the working surfaces Place at 4-1 pitch, and extend the ladder top a minimum of 36" above landing
	Slipping	Never reach sideways on a ladder
	Electrical shock	Be aware of electrical hazards while using any metal ladder Look for overhead electrical lines before moving ladder
Air tools	Blow-outs	Check all connections before each use Make certain that safety clips are used to prevent hose and fittings from forceful discharge Do not exceed manufacturer's recommended pressure

TABLE 2-A (Con't)
TOOL SAFETY

Powder actuated	Misfires	Check all tools daily
		Train all users properly
		Do not load tools until ready
		Do not leave loaded tools unattended
	Accidental injury or death	Do not fire blankly into a wall without knowing its structural composition; nails can penetrate through hollow walls and kill or maim
Electrical tools	Electrical shock	Use ground fault interrupters (GFI) properly
		Check all tools for proper grounding and insulation
		Use 3-wire extension cords
		Do not fasten by wires, staples, or hang by nails
		Do not use a spliced extension cord
		Repair or replace all frayed cords
		Work in a dry area with dry footing
Drain cleaning	Shock	Use properly grounded equipment, and ground fault interrupters
	Clothing attacks	Keep loose clothing away from spinning snake or drums
		Wear recommended snaking mitts only

NOTE

No safety list can include every possible precaution. Consult tool and manufacturers' recommendations as well as safety bulletins for additional safety precautions.

LESSON 3

HAND TOOLS USED IN PLUMBING WORK

INTRODUCTION

Many of the tools described in this lesson are *sized*, that is, based on a system of measurement. For many years, the system of measurement used in the United States has been the English system, using inches and feet, ounces and pounds, etc. As this is written (2005), the construction industry is trying to come to grips with using the ISO system: meters, kilograms, liters, etc. We will make some metric references in this lesson, but despite some enthusiasts who have tried to change our ways, very little has happened in the last several years, and it seems unlikely that we will see many differences for several years to come.

The competent mechanic knows that tools are necessary to perform plumbing tasks, and he/she therefore treats tools properly and respectfully. In this way, the plumber obtains maximum life and utility from the tools.

This section presents a view of the range of tools that are especially used in plumbing work. General purpose tools that are typically known to most people are described in Appendices A and B at the end of this book.

We divide manual hand tools into seven categories:

- Measuring tools (linear)
- Testing tools (horizontal and vertical)
- Sawing and cutting tools
- Assembly tools
- Boring tools
- Cutting (metal) tools
- Ergonomic tools

MEASURING (LINEAR)

A linear measuring tool is used to establish length. It is usually marked in feet and inches, subdivided to 1/16 inches, and they are available in lengths up to 300'. Typical examples are shown in Appendix A.

METRIC NOTE

The basic configurations of measuring devices are available in metric calibration. In this system, measurements in building applications are in millimeters (**1/1000** meter). You should realize that metric calculations, where everything is a decimal, are generally *much easier* than English units of subdivided fractions: $1/2$, $1/4$, $1/8$, $1/16$, etc., and 12"=1'. Even with this "sales point," however, the metric system has made essentially zero headway in the construction field in the United States since the early 1980's, when many leaders felt that the move to metric was inevitable. It seems that the supposed advantages of the metric system are not as significant as some people thought they were.

TESTING (Horizontal and Vertical)

Testing tools are used to check measurements and workmanship. While levels in one form or another are used throughout construction, the following discussion centers on versions that are especially useful to plumbers:

SPIRIT LEVEL

Nine-inch level (also called torpedo) used to check horizontal and vertical lines for slope or plumb condition. A small bubble in a vial of alcohol will center between two lines on the tube when the level is absolutely horizontal. See Figure 3-A.

Figure 3-A – RIDGID Torpedo Level

Most levels also can show vertical (plumb) alignment, and many levels are equipped with a vial at 45° to measure that vertical angle.

Longer levels are available for determining that a wall is plumb, or that a floor is level, or for other uses.

Some of these levels are also equipped with a feature which immediately indicates 1/8" per foot and 1/4" per foot slope. See Figure 3-B.

Figure 3-B – RIDGID Aluminum Level

LASERS

Lasers are a concentrated energy source that produce rays that do not "spread-out" as they travel away from the source. In general, they are dangerous because the beams contain enough energy to damage any living tissue that they strike. Laser tools suitable for construction work are relatively safe, however, because the energy level is very low. Even so, the laser source should not be looked at directly!

The earliest laser applications in plumbing work involved placing a laser source in a manhole in a new sewer, and setting the source so that it projected a line along the centerline of the desired path of the pipe. The installer then positioned the next section of sewer pipe so that the laser beam struck the center of a target temporarily placed on the end of the piece to be installed. After placing same, the target would then be put on the end of the next piece, and the process repeated. When finished, the pipe line would be straight in both the left-right and up-down directions. One advantage of this method is that a sewer can be laid with a very small slope because the line will not have any sags or dips—conditions that are impossible to avoid when laying a sewer using only a level on each joint.

Lasers for the above-described sewer work were bulky and expensive. They have since evolved to smaller and less expensive models for a variety of applications, such as establishing level and plumb lines to provide references for the tops of foundations, right-angle relationships for laying out partitions, the route of plumbing stacks, and many others.

SAWING AND CUTTING

SAWS

Sawing and cutting tools are used to cut wood, plastic, or metal. Saws are available with different number of teeth per inch and different hardness of the saw material itself. Check the manufacturer's recommendations for best performance with each material.

Manual cutting is performed with gentle, smooth down strokes using nearly the full length of the blade for the cut. For best results and personal safety, keep a firm stance when cutting, keep blade sharp and lubricated, watch for metal embedded in wood, clamp or hold all work firmly to prevent injury, and clean up all debris when finished. Examples of saws with applications for plumbers include:

KEYHOLE (COMPASS)

Tapered blade is used to cut any shape hole in wood or sheet rock. The cutting starts in a drilled hole or other opening in the work and can proceed in any direction because of the narrow width of the blade.

BORING

Boring tools are used to form relatively small round holes. Examples include the following:

MANUAL DRILLS

There are many forms of devices that hold drill bits and that are turned manually. Appendix A describes various types.

POWER DRILLS

Power drills will be described in later lessons that deal with power tools.

ASSEMBLY TOOLS

Assembly tools include screwdrivers, hammers, and wrenches in many forms. Many of these are listed in Appendix A.

Many special purpose screw or bolt heads are available, including the square shank, Torx, or star with matching drive tools. Many of you will have encountered these in motor vehicles, etc. Tamper resistant screws are frequently used in fixtures and trim in public restrooms and similar locations where they are used to minimize vandalism.

WRENCHES

While wrenches are used in many trades, they are very important in plumbing work. Wrenches come in many sizes and shapes. The proper wrench should be used for each job. Size is indicated by measuring the distance from end of the handle to the jaw.

Pipe wrenches are made to grip circular objects. Pipe wrenches are described in more detail in Lesson 5.

Table 3-A shows the characteristics of many wrench types.

Examples of other types include the following: Adjustable, Fixed types (box end and open), and Vise-Grip™. These are all shown in Appendix A.

PLIERS

Pliers and cutters should be used only where a wrench cannot be used or where the use of a wrench may be damaging. Examples include:

Tongue and groove, shown in Figure 3-C, are adjustable, for general small or light work. The closing force with these is developed by the strength of the user's hand, unlike a pipe wrench where the closing force is a function of the force applied to the handle.

Diagonal and long nose (needle nose) are used for appliance and control wiring and other fine work. Lineman's pliers are used for appliance wiring and for small bolt cutting. These are powerful tools that are capable of cutting copper conductors up to at least size 6AWG.

Figure 3-C – RIDGID Family of Pliers

Wire cutters — this function is included on many plier types — are also available in single-purpose tools, commonly called "diagonal cutters." These tools are the most efficient for appliance and control wiring.

METRIC NOTE

Most of these assembly tools will not be affected in the event of a change to metric units, except for wrenches, especially box- and open-end types. Screwdrivers and hammers may be identified differently, but they will still be the same basic devices.

CUTTING SHEET METAL

Tools are used on sheet metal to shape, bend, and put holes in various forms of the work. Special care must be taken when using these tools as the metal edges left on the material being worked are extremely sharp. Gloves should be worn as a precaution against cuts. Examples include:

Tinner's snips, which are available in straight cutting, left hand, or right hand shapes. Also available are center strip models, which cut a strip out of the material being worked. Center strip snips are vital for ease in cutting round or wide shapes. These tools are used to cut sheet metal up to about 22 gauge. They should be sharp to shear rather than bend the metal. See Figure 3-D.

Figure 3-D – RIDGID Snips

Chassis punch set, which are used to cut circular holes in stainless steel sinks, metallic studs, etc. These are

Figure 3-E – RIDGID Manual Knockout Kit

more commonly thought of as electrician's tools, but they are very useful for the plumber. Figure 3-E shows an example.

ERGONOMIC TOOLS

This general class of tools, which could also be considered safety devices, are intended to make the worker comfortable when performing work in difficult locations or awkward positions. The intention is to keep the worker free from injury and to make the work proceed more efficiently and productively.

One manufacturer produces pads which support the body and head when working inside sink or lavatory cabinets. This pad eliminates the height differential between the floor and the raised surface inside the cabinet. It also covers the sharp edge of the cabinet opening which is uncomfortable and adds stress to the back.

Another offering is a knee and/or back support pad that aids in comfort and therefore productivity when the worker must be in a kneeling position.

KEEPING TRACK OF TOOLS

It should be noted that every company has its own method of dealing with tools, both hand tools and larger tools which remain in the shop unless in use. A method of **checking out** tools is used quite often. This involves the plumber actually signing for the tool which

Figure 3-F – Plumber's Pad Bennette Design Group, Inc.

Figure 3-G – Knee N' Back Pad Bennette Design Group, Inc.

will be used. This individual then becomes responsible for seeing that the tool is returned when it is no longer being used. Lost, misplaced, or stolen tools are very expensive to the individual plumber or the company. It is advisable to mark (etch, paint, etc.) your hand tools so you know which are yours on a job.

TOOLS — BUT USUALLY NOT RECOGNIZED!

TIME CARDS AND TIME SHEETS

Time cards and time sheets are vitally important to the office so that you can be paid correctly, and so that work performed on the job can be properly allocated. Pens and pencils are also essential to this function. See an example of a time card at the end of this lesson. Each company may prefer its own style of time card with its own choice of information to be printed.

MISCELLANEOUS ITEMS

Telephones, two-way radios, and safety devices (like fire extinguishers) are included in this category. A remarkable device that has become almost universally used is the cell phone. There are areas in the United States where coverage has gaps, particularly with one service provider as opposed to another. It is highly likely that cell phone companies will continue to enlarge their service areas. Thus, cell phone usage will continue to grow.

An additional factor is the extreme competitive nature of that business — so it is also likely that the number of different companies in the business will be drastically reduced.

Besides cell phones, other "tools" such as Personal Data Assistants (PDA) exist and future developments are to be expected. One of the current crop is the Blackberry, which is a device capable of receiving and sending phone calls, e-mail, keeping your calendar, and being your personal organizer.

Other items in this category are digital cameras, buried-line tracing equipment, and buried-line locating equipment.

SAFETY CONSIDERATIONS WITH TELEPHONE USAGE

Telephone usage at the jobsite will often lead to inattention to what is going on around you. This is especially true when using cell phones or wireless remote handsets because this usage is often done outside the job trailer. Inside the job trailer, with the conventional phone tethered to a line cord, the hazard is lessened. As always, conditions on the job at construction projects demand your full attention, so if you must use the phone for an important conversation, place yourself in a protected area where you can give your attention to the phone conversation.

WHILE OPERATING A VEHICLE

Cell phone usage should be avoided whenever you are operating a vehicle, especially if you are on a public right-of-way. Only a few jurisdictions make such usage illegal at this time, but in most cases it is hazardous. If you must make calls while driving, pull over to a safe stopping place so that you can devote your full attention to the phone conversation.

REPAIRS, inc.   TIME SHEET

March 29, 2005   Fill out one sheet for each job each day

EMPLOYEE NAME _____

DATE _____ _____ _____   TIME ENDED _____
         Month   Day   Year   TIME STARTED _____

                                                TOTAL TIME _____

JOB NAME _____   **CUSTOMER P. O. #** _____

COMPANY NAME _____   **JOB #** _____

CONTACT NAME _____   **PHONE #** _____

ADDRESS _____   **FAX #** _____

_____

DESCRIPTION OF WORK PERFORMED

_____
_____
_____
_____
_____
_____

MATERIALS USED

Quantity	Item	Supplier	Unit Cost	Total Cost

Plumbing Apprentice Student Workbook Year One
Fourth Edition

TABLE 3-A
WRENCHES

TYPES	SIZES	USES
Straight pipe wrench	6, 8, 10, 12, 14, 18, 24, 36, 48, 60-inch handle lengths	General piping assembly Jaw perpendicular to handle
End pipe wrench	6, 8, 10, 12, 14, 18, 24, 36-inch handle lengths	Jaw parallel to handle
Offset wrench	6, 8, 10, 12, 14, 18, 24, 36-inch handle lengths	Offset jaw offers more angle accessibility
Chain wrench	12, 14, 18, 24, 36, 48, 60-inch handle lengths	Chain allows gripping pipe set in close quarters
Strap wrench	1/8 - 5-inch pipe size	Nylon strap excellent for gripping plastic pipe or pipe with polished finish surface
Socket set	1/4, 3/8, 1/2, 3/4 inch shank sizes sockets assorted	Fast bolt and nut handling
Basin wrenches	Variable or standard 10" length variable or standard size jaws	Hard-to-reach positions such as faucet basin nuts
Internal wrench	1 size — 1" through 2"	Holds inside of pipe (basket strainers, closet spuds)
Adjustable end wrench	4, 6, 8, 10, 12, 15, 18, 24-inch handle lengths	All hex or square bolts Jaw size adjusts with thumb screw
Spud wrench (Monkey wrench)	1 or 2 sizes	Utility work
Hex wrenches	Handle length varies with hex nut sizes	Jaws fit all hex nuts, unions, etc.
End wrench	3/16 through 2-inch jaw size	Fixed open jaws — slide onto work
Box-end wrench	3/16 through 2-inch jaw size	Fixed closed jaws — goes over work Hex shape only or square box can be 12 point

NOTE: All these tools are made of cast or forged steel. Except for the jaws, pipe wrenches are also available in cast aluminum in many sizes.

LESSON 4

ROUGH-IN TOOLS — COPPER, PLASTIC, AND SOIL PIPE

The safe, proper use of tools is essential for your own protection and the safety of others near you.

The tools described in this lesson will also be studied in later lessons.

COPPER TOOLS

CUTTING

Copper tubing is cut either by using a wheel cutter or by sawing. The cutter consists of a slide frame which holds two rollers opposite a cutter wheel, and a knurled tightening knob. The tool is rotated around the tube, and the roller frame is advanced forward, pressing the cutting edge of the wheel into the tube wall. The copper tube material is displaced sideways and inward as the cutter wheel is pressed into the wall of the tube. After a few rotations around the tube (advancing the roller frame on each rotation), the tubing separates, leaving cut ends with a restricted inner diameter because of the material displaced inward (see reaming and deburring). The cutter wheel, rollers, and adjusting screw should be lubricated for proper operation.

Cutters are available in various tube or pipe diameter ranges and, if necessary, to cut from the inside of the tube. Tubing can also be sawed with a hacksaw, a powered rotary swing saw, or abrasive cut-off wheel. Figure 4-A shows a manual copper tubing cutter with reaming attachment.

REAMING AND DEBURRING

Cutting the pipe with a cutter or saw leaves a sharp, rough edge. Reaming the inside diameter and deburring the outside diameter of the tubing end are needed for best flow performance and for easier fitting assembly. Many wheel cutter tools are fitted with a reamer blade that can be positioned for use after the cutting is completed as shown in Figure 4-A, or a separate tool may be used. Deburring can be done with a file or a sharp blade on the outside of the tubing end.

ASSEMBLY

Assembly of copper tubing may be accomplished by any of the following methods:

1. Soldering or brazing with socket fittings
2. Compression tools and fittings
3. Flaring
4. Swedging and sizing

DETAILS OF SEVERAL METHODS

SOLDERING AND BRAZING (Size range: capillary tube to largest diameters available)

The tubing ends are cleaned with sand cloth or a special tool; and immediately covered with a light coating of soldering paste (flux). The sockets of the fittings are also cleaned with special brushes or (for larger sizes) the same abrasive material used for the tubing. The cleaned sockets are also lightly coated with flux.

Note: Some references state that the tubing (and/or fitting) may be cleaned with steel wool. **The Copper**

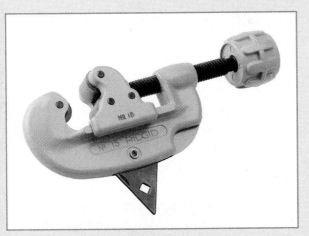

Figure 4-A – RIDGID Screw Feed Cutter

Development Association advises against this because small steel fibers are likely to be left after the "cleaning." Such steel fibers contaminate the joint and thus increase the chance of an inferior solder joint.

The tubing and fittings are then assembled and supported in proper position and the joints are soldered. The usual heat source is an acetylene-air or propane-air torch which is used to heat the tubing-fitting assembly above the solder melting point. Various tank sizes are available from plumbing wholesalers or welding suppliers. See Figure 4-B for typical equipment.

Figure 4-B – Torch Accessories

The solder is applied in wire form to the hot joint. The solder wets the clean copper surfaces and fills the space between the tubing and fitting socket, even if the opening between the two is facing downward.

The most commonly used solder in the past is known as 50-50, which is 50% lead and 50% tin. This solder has a large "pasty" range from 450°F to 550°F. This large range makes working with this solder relatively easy. However, this solder has been banned (in the USA for several years) for any tubing that will be used for potable water. The reason for the ban is that some testing performed in new houses showed that lead from the solder leached out into the water in the first draw off after several hours of non-use. Since lead is a serious thread to the health of children and other persons in fragile health, it was decided that 50-50 solder could not be used for potable water tubing.

Using lead solder on a potable water line may result in a stop work order for the job and an expensive replacement of work already completed. Because of the serious consequences of using the wrong solder on a potable piping installation, many contractors will not purchase nor use lead-alloy solders (even though perfectly legal) on non-potable projects because of the temptation to use the wrong material on "that last joint, because I ran out of the correct material" at the last minute! If you don't have it on the truck, you can't be tempted!

If extremely strong joints are required, brazing may be used. The joint material has a melting point above 1000°F. Therefore, use an oxy-acetylene torch for all but the smallest sizes of tubing. Some brazing alloys do not require that you clean the copper tubing or fittings to produce satisfactory joints (check with your supervisor) – if the material is good and clean to start with. Brazing techniques are similar to the methods required for soldering, but there is greater demand for care and precision because the tolerances for tubing and fitting temperatures and braze material addition to the work are much tighter than for lower temperature soldering.

Alternate heat sources for soldering joints include electric heaters and natural gas-air torches. Electric heating tools are used where the open flame of a torch could be hazardous.

Safe practices are vital when using any torch or other heating device.

1. Avoid burning yourself or others around you.

2. Do not damage property. Shield flammable surfaces from the flame.

3. Absolutely no horseplay can be tolerated.

4. If you are using a torch in an area where sparks may go down a pipe chase or otherwise be out of your sight, it is important that

a spotter be located on the next floor. Fire is possible from one spark landing in a pile of debris.

5. Have fire extinguishing equipment by your side.

Always remember that the tools you are using in this process get very hot. They must be cooled down before being stored to prevent injury or a fire.

Figure 4-C RIDGID ProPress System

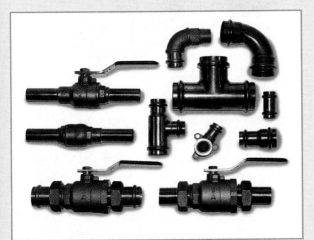

Figure 4-D – NIBCO Press-to-Connect Copper Joinery System

PRESS-CONNECTION JOINING METHOD

Compression fittings are patented by the manufacturer, which results in relatively limited sources for the products. The method uses a special hydraulic tool to squeeze matching fitting ends (of special fittings) onto copper tubing. The attractions of this method include extremely fast joint completion, no flame or temperature hazard to the surrounding space, and less possibility of a leak due to poor soldering technique. Size range: 1/4" through 4".

FLARING (Size Range: 1/8" to 2")

Flaring is accomplished with soft copper tubing as follows: An enlarged end (the flare) is developed in the soft copper tubing and a nut is used to hold this end against a matching male fitting for a leak-proof joint. Smaller soft copper tubing is flared by holding the tubing in a clamping block and forcing a tapered flaring plug into the tubing end, which pushes the tubing against the desired flare shape which is formed in the clamping block. The tapered plug is mounted on a threaded shaft. Advancing the threaded shaft pushes the plug into the tubing end. See Figure 4-E.

Figure 4-E – RIDGID Flaring Tool

Large tubing (1" to 2") is flared by hammering a plug (which has the desired flare shape) into the end of the tubing.

Always remember to place the flare nut on the tubing before making the flare.

With either method, it is possible that the tube end will split during the flaring operation unless the tubing is very soft. Copper tubing will work-harden from normal vibrations encountered in a moving vehicle. Consequently, copper tubing will not remain *soft* if carried for an extended period in your truck or van. Thus, it is best practice to flare only copper that is recently taken from your shop or a vendor's warehouse.

SWEDGING AND SIZING

Swedging and sizing both involve forming the pipe around a plug. The swedge is a two-diameter tool with a taper between the two sizes used to enlarge a pipe end to a female socket. Swedging produces a socket end on copper tube, permitting the connection of two pieces of tube without using a coupling. The sizing tool is used to restore a round shape to an end so that the tube end can be used for satisfactory joints.

Thus, this is a special case of the soldering or brazing method described above.

BENDING

For some types of installation, it is necessary to minimize the number of joints in the line. This goal is met by bending the tubing, rather than using ells or 45's.

Bending tools are available in various forms to aid in making uniform bends in soft copper tubing. They all

Figure 4-G – RIDGID Spring-Type Tube Bender

Figure 4-H – RIDGID Heavy-Wall Conduit Bender

Figure 4-I – RIDGID Thin-Wall Conduit Bender

involve supporting the tube so that it is not flattened when making short radius bends. Spring benders slip over the pipe and maintain the round shape when the pipe is bent by hand. Other benders use semi-circular forms over which the tubing is bent by a system of levers. See Figure 4-F and Figure 4-G for typical bending tools. Some conduit benders are suitable for bending tubing or pipe. See Figure 4-H and Figure 4-I.

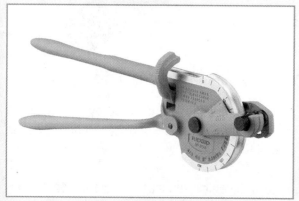

Figure 4-F – RIDGID Geared Ratchet Lever-Type Tube Bender

PLASTIC PIPE AND TUBING

CUTTING

Plastic pipe is cut with tools similar to those used for copper tube, although the cutter wheels for plastic are thinner. See Figure 4-J for a scissors-like tool that can cut plastic tubing quickly. Sawing methods may also be used. It is important when cutting plastic pipe that the end be cut squarely so that the pipe may be fitted firmly into the fitting socket. Failure to seat the pipe properly may result in a failed joint later.

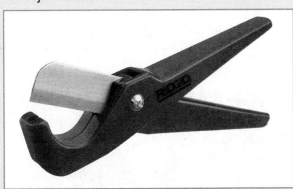

Figure 4-J – RIDGID Plastic Pipe Cutter/Scissors Cutter

Figure 4-K – RIDGID Deburring Tools

REAMING AND DEBURRING

Reaming and deburring plastic pipe is required for best flow performance and to ease fitting assembly. Special combination reaming-deburring tools are made for plastic pipe, see Figure 4-K; however, you can perform these operations separately by using files or a knife blade to remove unwanted material.

ASSEMBLY

Assembly of plastic pipe is most often done by using solvent cement. Each type plastic requires the proper joint cement. Some plastic materials require special pre-treatment, others do not. These variations will be discussed in later lessons.

Other methods of assembly for plastic pipe include flaring, insert fittings with and without clamping devices, heat fusion, compression fittings, and threading. The type of plastic and wall thickness dictate the type of joint to use. All these variations will be studied later.

CAST-IRON SOIL PIPE

Cast-iron soil pipe is a very heavy, strong material used for plumbing drainage and vents. Therefore, the implements you will be working with are the heaviest and most rugged of the rough-in tools. Be especially careful with these tools as there is considerable risk of injury.

CUTTING CAST-IRON SOIL PIPE

Cutting soil pipe can be accomplished in two ways — controlled cracking of the circumference, or by sawing the pipe.

There are several ways to crack soil pipe; the most

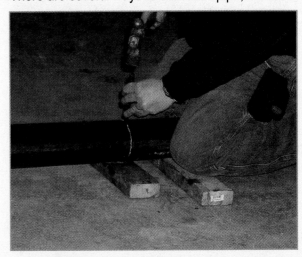

Figure 4-L – Hammer and Cold Chisel Cutting of Soil Pipe

elementary is with a cold chisel and ball peen hammer. A chalk line is drawn around the pipe. Indentations are made along this chalk line with the chisel and hammer until the pipe cracks apart. This method is very slow and can only be used for cutting pipe on the floor or on a bench. Therefore, it is unsuitable for cutting into an existing pipe line. See Figure 4-L. Some spun soil pipe cannot be cut with this method. This method is very rarely used any more and is not recommended.

A more versatile and productive cutter is the snap chain cutter. This tool uses a series of piercing wheels chained together and a scissors action lever to snap the pipe. The chain is tightened around the pipe so that the piercing wheels are simultaneously forced into the pipe wall. The pipe soon breaks.

The variations in this method are in the arrangements used for developing the tightening force — simple leverage, ratchet and screw, or hydraulic piston. There is also a close-quarters compression cutter which incorporates a ratchet and chain to progressively tighten the chain until the pipe breaks. See Figure 4-M.

Sawing soil pipe can be done with a hacksaw, portable band saw, portable abrasive cut-off saw, or large swing saw with abrasive cut-off blade.

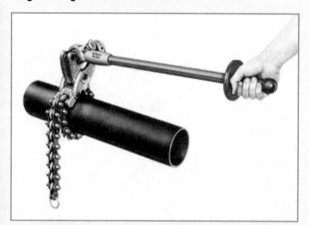

Figure 4-M – RIDGID Soil Pipe Cutter

DEBURRING

Deburring should be done with either a file or a series of light hammer blows on the cut end. Deburring is necessary to facilitate entry into joints assembled with gaskets.

CAST-IRON PIPE ASSEMBLY- HUB AND SPIGOT

Assembly of cast-iron soil pipe is made with hub-and-spigot joints or hubless joints.

Hub-and-spigot joints are made either with lead and oakum as the filler material or with a preformed, lubricated elastomeric sealing gasket.

LEAD AND OAKUM JOINTS

The tools required for lead and oakum joints include yarning and packing irons to place the oakum; furnace, lead pots, and ladles to melt and handle the

Figure 4-N – Set Up for Pouring Lead Joint in Horizontal Line

lead; joint runner rope to hold the molten lead in place until chilled (needed for joints in horizontal piping); and caulking irons to set the lead in the joint. These joints are not used very often in the field any more but it may be necessary that you learn how to **pour a joint** in order to make repairs on older lines. Many licensing authorities require that a joint be poured to pass the plumbing licens-

Figure 4-O – Mephisto Tool – Plumbing Lead Joint Caulking Iron Set

ing exam. *Do not attempt to pour a joint until you have been thoroughly instructed by an experienced plumber.* There are many pitfalls and hazards when working with molten lead. It is not practical to use lead and oakum any longer in most circumstances. The OSHA lead rules (see Appendix C) should be followed at all times.

GASKET HUB AND SPIGOT JOINTS

The lubricated, elastomeric gasket joint requires tools to drive or pull the pipe into the lubricated gasket in the hub so that the pipe is properly seated. With small-sized fitting assemblies, this task can be done manually with a lead maul or by hammering on a scrap piece of lumber placed against the pipe. Larger sizes require a special tool that clamps on the spigot and hub, and then, using leverage, draws them together. See Figure 4-P.

CAST-IRON PIPE ASSEMBLY-HUBLESS JOINTS

The hubless joint uses an elastomeric sealing sleeve with a corrosion-resistant supporting sleeve with drawbands. Unshielded couplings may also be encountered.

The only assembly tool needed is a torque wrench commonly fitted with a $1/4$" or $5/16$" socket. The clutch on this wrench is set to release at 60, 80, or 120 inch-pounds depending upon coupling design, thus assuring a tight joint and minimizing the risk of over-tightening the drawbands. Always use the proper wrench with the coupling which is specified. Over-torquing can break the bands; under-torquing may result in a leak. See Figure 4-Q for an example of a torque wrench for hubless soil-pipe assembly.

Figure 4-Q – RIDGID Torque Wrench

SAFETY

Wear eye protection when cutting soil pipe; chips often fly. Use safety shoes or steel toe protection when working this material, as a dropped fitting or piece of pipe can cause serious injury. Wear your hard hat (unless everyone around you is working on floor-level jobs — job restrictions may require you to wear your hard hat in any case). Use proper gloves for the type of work you are doing.

Figure 4-P – RIDGID Soil Pipe Assembly Tool

Use a face shield when pouring lead, but always avoid being directly over the joint.

A fire extinguisher should be available at any workplace, but be sure it is *immediately* available when working with any torch or furnace.

A first aid kit is also required equipment — remember to have it handy and keep it fully stocked with fresh supplies.

LESSON 5

ROUGH-IN TOOLS — STEEL PIPE

Steel pipe is the strongest and toughest piping material commonly used. Consequently, the tools for steel pipe are heavy, strong, and produce high local forces — characteristics that call for careful and safe procedures.

Steel pipe tools are designed to hold, cut, ream, thread, and assemble steel pipe components.

HOLDING

Holding devices are required for steel pipe because of the high torque forces involved in working on the pipe. The principal holding tool is the pipe vise. Vises are available in ranges up to 4" in quick-opening yoke types, and up to 8" in configurations with a roller chain upper-tightening member. Different mounting arrangements are available: bench mounts, three legged and four legged stands, post mounts, and truck mounts. See Figure 5-A for examples.

Figure 5-A – RIDGID Vise Examples

CUTTING (see figure 5-B)

Cutting is done by displacement cutters (similar to those discussed with copper tools), sawing, or by using powered abrasive cutoff tools. The common cutter has two rollers and one cutting wheel. This arrangement requires that the cutter travel 360° around the pipe. When that clearance is not available, tools are available with three or four cutter wheels. The multiple wheel cutter is more difficult to use than the common cutter; since any wear on a wheel or shaft makes it unlikely that the multiple wheels will cut in the same track. In addition, the multiple cutter wheels cause a considerable increase in the outside diameter (OD) of the pipe — the rollers in the conventional cutter keep the OD from increasing — thus complicating subsequent operations. Heavy filing may be necessary to reduce the expanded OD prior to moving to the threading task.

Support pipe when you are cutting pieces of any significant length. Do not use worn out or dull cutter wheels. Replace them with new, sharp wheels; and keep moving parts lubricated.

REAMING

Reaming is required, as for other materials, to obtain best flow performance and to remove a shoulder that can start a blockage or build-up in a pipe line. Reamers are wedge- or cone-shaped with cutting edges to remove the undesired metal. They are available in 1/8" to 2" size, and 2½" to 4" size. If the pipe was cut with a multiple wheel cutter, the outside burr will also have to be removed. A rat tail file may also be used to remove burrs. Figure 5-C shows one type of reamer.

Figure 5-B – RIDGID Heavy-Duty Pipe Cutter

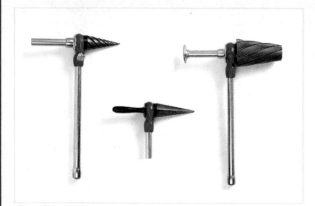

Figure 5-C – RIDGID Pipe Reamers

The cut end must be reamed just enough to remove the erupted metal from the cutting operation, but not more. In other words, the pipe end should have the same wall thickness as the run of the pipe.

THREADING

The National Pipe thread-Taper (NPT) dimensions are shown in Table 5-A. Both the fitting and the pipe must have the same taper. Consequently, the thread engagement is loose at the beginning (for easy starting), but proceeds to an interference fit as the joint is made up. This tapered fit produces a leak-tight joint that won't loosen.

Threading is done by turning a die clockwise on the pipe end until the necessary length of the thread is produced. Dies are usually fixed in square blocks for pipe sizes 1/8"-1", cylinder mounts for pipe sizes 1/8"-2", or in adjustable segments for 1"-2" in one tool, 2½"-4" in one tool, and 4"-6" in one tool. Figure 5-D shows manual threaders. Figure 5-E shows a hand-held power drive.

Figure 5-D – RIDGID Manual Pipe Threader

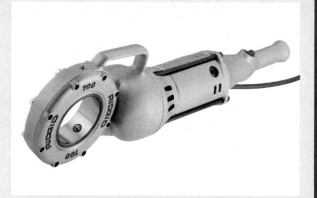

Figure 5-E – RIDGID Model 700 Power Drive

Threading steel pipe requires a continuous supply of oil on the dies to develop clean, full threads and to extend the life of the die. The oil both lubricates and cools the die and the threads as they are formed. This cutting oil is a special product which contains additives which cause the oil to adhere to the pipe during threading.

The oil and chips from the thread cutting should be collected to avoid staining the floor, to minimize slipping hazards, and to eliminate having to sweep up scattered chips. Threading chips have sharp edges — avoid clearing them with your bare hands. Oil pans which have a hand oil pump and which collect the chips and excess oil are available. The use of such pans is recommended.

Power vises combine the vise holding function with a turning operation. When you ream and thread the steel pipe with a power vise, you will use the same tools as you did with the manual methods. The only difference is that with power vises, you need only hold the tool against the moving pipe – usually a bar holds the die handle. The resulting labor saving greatly increases productivity.

So-called pipe machines combine cutting, reaming, threading, oil delivery, and oil collection in one total unit for even greater productivity. More about these power tools in Lesson 7.

TABLE 5-A
NATIONAL PIPE thread TAPER
Taper: ¾"/ft on the diameter

PIPE SIZE (INCHES)*	THREADS PER INCH	TOTAL LENGTH OF THREAD (inches)
⅛	27	0.3924
¼	18	0.4018
⅜	18	0.4078
½	14	0.7815
¾	14	0.7935
1	11.5	0.9845
1¼	11.5	1.0085
1½	11.5	1.0252
2	11.5	1.0582
2½	8	1.5712
3	8	1.6337
4	8	1.7337
5	8	1.8400
6	8	1.9462

*NOTE: 3½ and 4½ pipe may be found in older work, but are generally unavailable today. Current sizes 1¼ and 2½ are approaching that status, but all will change *if and when* metric sizes become commonplace.

ASSEMBLY

Assembly tools are wrenches (which were discussed briefly in Lesson 3) and vises.

Pipe wrenches are made to grip circular objects. They differ from the usual adjustable wrench in two ways: the hardened-steel jaw faces are directionally grooved with sharp-edged ridges, and the adjustable jaw is arranged so that it can swivel through a small angle (that is parallel to the wrench handle). For best results, keep jaws clean and sharp. When the jaws are placed on a pipe and rotated in the direction of the fixed jaw, the other jaw closes on the pipe wall with a force that is proportional to the force being applied to the handle. Note that the turning effort of the wrench cannot be reversed by simply pushing the handle in the opposite direction. The wrench must be removed and placed on the other side of the pipe.

Pipe wrenches are made with handle lengths from

6" to 5'. The standard tools are made of special cast-iron, and the larger sizes are also available made of an aluminum/titanium alloy. These lighter wrenches are much less tiring to use. The aluminum wrenches have steel jaw faces, so the gripping characteristics are the same for either material version.

An oversized pipe wrench will tend to flatten an undersized pipe (called **egging**) rather than turn it. This will occur because the large wrench will exert a very large force on the diameter of the pipe if the small pipe (that is to be removed) is very tightly seated in the fitting, or if the pipe (that is to be installed) is over-tightened into the fitting. This last happens when the joint drips when the assembly is tested, and tightening the joint is attempted to stop the drip. Such a procedure is seldom successful — the joint almost always has to be disassembled and reassembled with new thread sealant. In addition, the larger wrench is heavier and more tiring to use.

Do not use wrenches as hammers and do not extend the wrench handles with a **cheater** (pipe slipped over handle) — you can break the jaw and severely injure someone, and/or cause much property damage.

VISES include the following types:

YOKE VISE (See inset, Figure 5-A)

1. Used for pipe up to a diameter of 4"

2. Vise should have a quick-release lever.

CHAIN VISE (See inset, Figure 5-A)

Used for pipe with a larger diameter — available up to 8"

To use a chain vise, follow these directions:

1. Release the chain and let it hang freely away from the vise jaws.

2. Back off the tightening handle to allow the most chain slack.

3. Place pipe in the lower jaws and wrap the chain around the top half of the pipe.

4. Set the tightest link inside the link holder.

5. Tighten the tightening handle until the pipe is securely seated in the lower jaws.

6. Pipe is now held tightly in the lower jaws and by the chain.

VISE OPTION

1. Bench clamping model

2. Post clamping model

3. Extra support stands

4. Tripod stand assembly (usually equipped with two or three pipe bender slots and a ceiling screw support). See Figure 5-A. This is the most commonly used in the field because it can be set up quickly. When used with a piece of pipe of appropriate length, the ceiling screw support allows you to lock the tripod firmly at four support points. This may be necessary, as other wise long lengths of pipe may cause the tripod to tip. Such long lengths should be separately supported for best stability when doing the work.

SAFETY

Steel pipe is heavy — you should store it in or on substantial racks. You need personal protection when working with it. Always carry the long lengths carefully.

Use steel-toe shoes and hard hats.

Protect your eyes and skin from chips and cutting oils.

Remember that steel pipe is a rolling and tripping hazard — from full lengths down to the short stubs that accumulate around the pipe vise. You can seriously injure yourself should you fall because you tripped over a piece of pipe. Therefore, pick up and safely store all pieces — including threading chips — as the workday proceeds. Don't wait for day's end to do it!

LESSON 6

FINISH AND REPAIR TOOLS

Finish and repair tools are usually smaller and lighter than rough-in tools and require less force. These tools are also less expensive, item-by-item, than rough-in tools. However, more finish tools are needed: furthermore, these tools are more precise instruments. As a result, a reasonable assortment of them will require a considerable investment. You can protect this investment by caring for these tools properly.

FINISH TOOLS

Finish work, also called **setting fixtures and trim**, is the final work on a project, performed after most of the other project work is completed, including wall and floor covering and any cabinetry for plumbing fixtures. No fixture should be set if it has to be pulled up shortly. Unnecessary fixture setting wastes time and money and increases the probability of leaks and fixture damage.

The tools generally required are described below:

BOX-END WRENCH (See also Lesson 3)

Closed end wrenches in 1/16" increments are available. They are ideal for working hexagon nuts or bolt heads in various situations. Each end is offset at a 15° angle. These wrenches can be used on square nuts or heads if they are 12 point type. These tools are plated to minimize corrosion.

Metric Note: This wrench type is also made to metric sizes. There are many products sold in the U.S. that are made to metric measurements.

FLARE NUT WRENCH

Similar to a box end wrench with a section removed, these wrenches are used to slip over tubing to tighten flare nuts.

OPEN-END WRENCH (See also Lesson 3)

This wrench is similar to the box-end wrench except that one side is open. This open side allows you to slide the wrench directly onto the nut or head. These can be applied to square or hex- heads bolts or nuts. These wrenches are also plated. See **Metric Note** above.

ADJUSTABLE-END WRENCH (See also Lesson 3)

One side of the jaw is movable so that you can slide the wrench onto the hexagon (or square) nut or head and then bring the movable side up for proper fit. Because the movable side is not as well held as a fixed wrench would be, these wrenches are not as strong as open-end, and, especially, box-end wrenches.

When using the adjustable wrench, always place the adjustable side facing you, and pull toward yourself — the wrench is less likely to distort and slip. Adjustable wrenches are available in many sizes to 24" handle length. These wrenches are also plated or finished to minimize corrosion.

Metric Note: Since these are adjustable, the metric conversion is not so important. The only metric impact will be in terminology — e.g., a 12" wrench will be 300 mm.

BASIN WRENCH (See also Lesson 3)

These are used to tighten (or loosen) the nuts on the underside of fixtures where conventional wrenches cannot be used. See Figure 6-A.

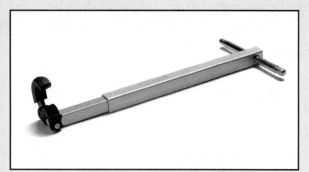

Figure 6-A – RIDGID Basin Wrench

TUBING BENDERS (See also Lesson 4)

As discussed in Lesson 4, tubing benders are used to form bends in small diameter copper tubing. These are handy for closet and lavatory supplies made from copper tubing.

PUTTY KNIFE

The putty knife is used to clean up fixture or mounting areas to prepare for fixture installation. A stiff, wide blade is best for these purposes. See Figure 6-B.

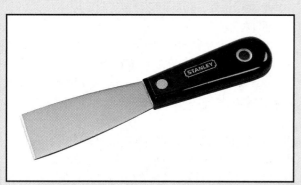

Figure 6-B – Stanley Putty Knife

SCREWDRIVERS (See also Lesson 3 and Appendix A)

Screwdrivers come in various sizes and types. Screwdriver bits are available for brace mounting for high-torque screw applications. Screwdriver reduction gear attachments are also available for power drills. These attachments can result in high production in the right situation.

The greatest productivity is provided by battery-powered drills with screw-driver bit attachments. See Lesson 7 (Power Tools) for a detailed description.

TONGUE AND GROOVE PLIER

(See also Lesson 3) Tongue and groove pliers are available in several sizes, each one with a range of adjustable gripping spacing. The jaws should be adjusted to be parallel when gripping the work. These pliers cannot develop the torque of pipe wrenches, but they can get into smaller places. When the tools wear, the jaws will slip or jump out of adjustment — either eventuality will result in injury. Therefore, do not use old or worn pliers.

FLASHLIGHT

Needed for installation, emergency, and repair work. Keep spare battery cells and lamp on hand. Flashlights are available in sizes from penlights to heavy duty multi-cell fluorescent. Some flashlights are available with continuity test leads, which may be helpful in service and repair work. Continuity leads are not to be used on live circuits!

SMALL COPPER CUTTER (POCKET)

Suitable to cut supply tubes to most fixtures and appliances. See figure 6-C.

Figure 6-C – RIDGID Midget Tubing Cutter

TIN SNIPS (See also Lesson 3)

Used to cut sheet metal for water heater vents and similar applications. Specific snips cut in specific directions. They are available in left, right, or straight cutting capacities.

Pipe and duct snips are equipped with two cutting blades to cut a strip out of the metal. This type cut does not deform the cut edge as regular snips tend to do.

CLOSET SEAT WRENCH

Used exclusively to mount closet seats to the bowl, it is available in two or three nut sizes. With top mounted seat posts becoming most common, the use of this tool is generally limited.

CHASSIS PUNCH SET (See also Lesson 3)

This tool is used to cut large diameter holes in thin metals. It is available in clearance pipe sizes from ½" through 4", manual or hydraulic.

HACKSAW

A hacksaw is used for a variety of finishing tasks. A small frame model is ideal for cutting in confined spaces. See Figure 6-D for standard frame model.

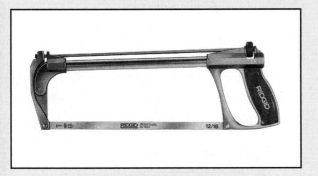

Figure 6-D – RIDGID Hack Saw

UTILITY KNIFE

Used to open shipping boxes and for other non-metallic cutting.

RIM SCREW TOOL

Used to mount sink frame screws to the clips that hold the sink to the counter top. This tool has a screw holding device to aid in the assembly.

STRAP WRENCH

A strap wrench uses a resin-coated strap and tightening arrangement to turn polished pipe nipples into fittings. The strap will not leave wrench jaw marks on the pipe as conventional pipe wrenches do.

Various specialty tools will be added to your tool roster as you find the need. The above list will give you a good start.

SERVICE TOOLS

Additional tools for service and repair work include some or all of the following:

PLUNGER

A plunger consists of a rubber cup and handle used to clear fixture trap stoppages. This tool is nearly useless for stoppages beyond the trap arm because the pressure impulse provided simply diverts up the vent if the stoppage is down the line.

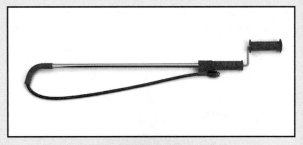

Figure 6-E – RIDGID Toilet Auger

CLOSET AUGER

A closet auger is a flexible cable, with a crank handle on one end, and a short corkscrew on the other. The cable is held in a short tubular sleeve used to direct the end into a water closet trap. It can then

Figure 6-F – RIDGID Hand Spinner

be advanced into the trap to clear the bowl of obstructions.

HAND SNAKE

A 10' to 25' flexible cable with a corkscrew head. The cable is coiled inside a drum for convenient handling. Longer, larger diameter snakes are installed in larger drums and mounted on wheeled stands for larger drain or sewer cleaning.

FISH TAPE

The small fish tape is used in a similar fashion to the hand snake above. It is a thin, flat steel strap rather than a round cable.

Large tapes are used to clean larger drain lines. They are 1" to 2" wide and about an 1/8" thick. They are usually not more than 50' long — longer lengths dissipate the pushing force along their length, so they are not effective against stubborn stoppages.

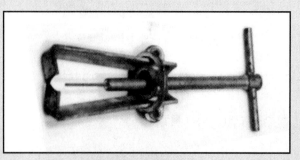

Figure 6-G – Handle Puller

All the metal tape or cable drain snakes should be cleaned and oiled after every use.

NEEDLE NOSE PLIER (See also Lesson 3)

Used to reach into small spaces to remove foreign objects, retrieve dropped objects, or to serve as pliers where a long straight jaw is needed.

HANDLE PULLER

Used to remove stubborn faucet handles.

SEAT TOOL

Either hexagon or square in shape, a seat tool comes in various sizes to remove and replace renewable faucet seats.

FILE (See also Lesson 3)

Files are needed in various sizes to smooth out rough or deteriorated areas in many types of equipment.

SOCKET SET WRENCH (See also Lesson 3)

Needed in standard and special designs for disassembly and reassembly of many plumbing fixtures, appliances, and appurtenances.

SCRATCH AWL

Useful for marking parts, retrieving small parts, and removing faucet washers. Be careful! Awls can cause painful, dangerous puncture wounds.

TAP AND DIE SET

Used to clean or repair damaged threads.

EASY-OUTS AND NIPPLE EXTRACTORS

Used to remove broken-off screws or nipples to prepare for replacement.

VOLTAGE TESTER

Used to ascertain if power circuits to plumbing

appliances are energized. Be familiar with its safe use before actually using it to test a circuit.

SUMMARY AND SAFETY CONSIDERATIONS

The skilled mechanic is as proud of his tools as he is of himself. Keep your tools in good repair and stay alert for new tool developments. Many minor tool improvements result in substantial increases in production.

Remember that clean, sharp tools are safer.

Use your tools properly. Do not overload them and have regard for your own and others' safety.

LESSON 7

WELDING AND POWER TOOLS

WELDING EQUIPMENT

This introduction to welding covers manual methods applied to mild steels used in typical piping applications.

The welding equipment must develop temperatures above the melting point of the metals to be joined (usually steel) in two adjacent edges, causing the molten edges to join together. Simultaneously, filler metal (a similar alloy to the pieces to be joined) is added to the molten area so that it fuses with the melted edges. This filler metal addition permits the joint area to contain more material than either of the base materials, if that result is desired by the mechanic.

In manual welding, the elevated temperatures required are obtained by high temperature flame or electric arc. We can subdivide these methods further after we discuss one more consideration: steel and most metals are extremely active chemically at the melting temperatures involved in welding. Therefore, the steel would combine quickly with the oxygen in the air unless some method is employed to prevent it. Generally, an inert atmosphere immediately above the molten area is used. With some welding systems, a slag cover (a hard deposit material) is laid over the just-completed weld.

HIGH-TEMPERATURE FLAME

High-temperature flames are usually developed with acetylene-oxygen torches. (In some circumstances, natural-gas/oxygen can be used.) Heavy steel cylinders are the containers most often used for portable welding using acetylene and oxygen. Because the gas is dissolved in acetone, the full-tank pressure of acetylene is moderate (about 250 psi). The full-tank pressure of oxygen is over 2000 psi. The welding tanks should be secured to a supporting cart or building column to prevent them from tipping over.

Acetylene should be drawn from the tank only when it is upright; otherwise acetone may be lost. Tanks should not be moved without the protective transport cap in place to guard the tank valve. Welding tanks should be stored in a place away from heat and traffic, and be sure to keep full tanks separated from empty ones.

The flow of gas from the tank is controlled by a regulator mounted on the tank and the gas is sent by hose to the torch, where the two gases are combined in the flame. The welding gas regulators have two pressure gauges: one to observe tank pressure (an indication of gas remaining in the tank), and one to show pressure to the hose. When not in use for an extended time, remove the gas regulators and hoses and store them in a safe, secure place. Bleed the pressure off the hoses and regulators during periods of non-use.

When welding, you must use the proper flame size and type to produce satisfactory welds. The flame

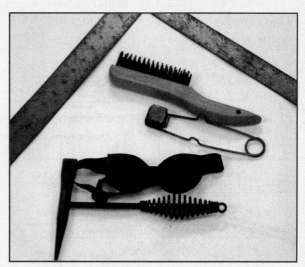

Figure 7-A – Typical Oxy-Acetylene Welding Tools

size is controlled primarily by torch tip size and (somewhat) by gas pressure. A neutral flame is used to produce the inert atmosphere to protect the steel from oxidizing.

The flame temperature is sufficient to melt the steel. You must move the torch tip in a small circular (or oscillating) motion to develop a molten area, called a *puddle*. You must see the puddle to control its size and rate of movement. The amount of added filler metal is controlled manually. Filler metal is usually provided in 1/8 " diameter by 36" long steel rods.

Figure 7-B – More Gas Welding Tools (Victor)

Because the puddle is extremely hot, it is very bright. Consequently, you should wear heavily tinted goggles while watching the puddle to protect your eyes from the bright light and from the spatter of metal particles. Use a lens made for welding; do not use sun glasses.

Wear heat and fire-resistant clothing and gloves.

Do not use any welding equipment until you have been thoroughly instructed in the methods required for satisfactory performance of that equipment. See Figures 7-A and 7-B for typical equipment.

Do not allow any oil to contact oxygen. Oil and 100% oxygen will explode instantly when they are exposed to each other.

Do not use oil or pipe dope on the oxygen regulator or the threads on hoses. Pipe dope contains oil.

COMMENTS ON OXY-ACETYLENE WELDING

The oxy-acetylene gas welding procedure is slow, and produces considerable heat distortion in the work pieces. On the other hand, it involves the least amount of equipment and time for set-up to begin welding. Thus, for one or two welds, it is frequently the most productive method.

MANUAL ELECTRIC ARC

Manual electric arc welding can be done by shielded-metal arc, metal arc inert gas (MIG), or tungsten inert - gas (TIG). Common characteristics and safety procedures for all electric arc welding methods are as described below:

PROCEDURES

An electric arc is made which produces the *puddle*. The filler material is added and the arc must advance to avoid melting through the base materials.

All electric arcs produce harmful radiation in addition to extremely high heat. Therefore, you must shield your entire face with a covering hood. Use a lens made for arc welding. Anything less can result in serious eye injury. (See Table 7-A)

Any skin exposed to the welding arc will be *sunburned* — the longer the exposure, the more severe the "sunburn" will be. An hour of welding will produce a noticeable burn; four to eight hours can produce painful burns. Shield the work area, if possible, as eye burns can also be produced in fellow workers from arc rays reflected from a wall!

To maintain proper welding currents, keep the welding cable and connections in good repair. Use welding cable no longer than job conditions require. The *ground* connection is made with a heavy clamp connection to the work piece. Like all welding current components, this clamp must be kept in good order.

Large magnetic fields exist with electric welding. To avoid magnetizing your watch, lock it safely away from the welding area.

TABLE 7-A ARC WELDING LENSES FOR PROTECTIVE SHIELDS	
Shade 5	Spot Welding
Shade 6	30 amps or less
Shade 8	30-75 amps
Shade 10	75-200 amps
Shade 12	200-400 amps
Shade 14	400 amps or more

SHEILDED METAL ARC

In shielded-metal arc welding a coated electrode is used to develop the arc. The alloy in the electrode is the filler material, and the coating chemicals form the inert atmosphere over the puddle. Usually these chemicals provide a *slag* covering over the completed weld as it cools. The power supply for this welding method can be an alternating current (A.C.) transformer, or, for superior welds, a direct current (D.C.) generator. See Figure 7-C for a typical D.C. welding supply.

Figure 7-C – Miller CST-250 Stick/TIG Welding Power Source

These machines are adjustable current, adjustable voltage type. The magnitude of the current affects the electrode melting rate:

1. 70-100 amps for light work
2. 125-150 amps for general shop welding
3. 175-250 amps for structural work

D.C. welding involves **polarity**. For straight polarity, the electrode is negative and the work piece is positive. This produces average quality welds with higher production. For reverse polarity, the electrode is positive and the work piece is negative. This produces the highest quality welds, but reduces production rates.

The electrode (welding rod) holder is a simple clamp device, insulated to minimize flashover; it is light in weight for ease in controlling weld-rod position. Never lay the holder on a metal bench as it may complete the welding circuit and produce an intense, hazardous arc flash.

The diameter and alloy of the weld rod determine the welding current range. For high-quality pipe work in sizes up to 6" or 8", use $1/8$ " rod, electrode positive, work pieces negative (called reverse polarity), and 100 amp welding current. For substantial weld deposit thickness, you must use multiple passes. You must remove all slag and foreign material from each welding pass before adding another layer of filler material.

For heavy cross-sections of structural steel, use 5/32" rod with 135 amp, or 3/16" rod with 180 amp.

For general welding (brackets, etc.), A.C. welding or straight polarity D.C. welding is satisfactory. Somewhat higher welding currents are used for greater productivity.

METAL - ARC INERT - GAS (MIG)

MIG uses a special characteristic D.C. power supply, an inert gas flowing over the work, and a spool of filler material in wire form to develop the arc and provide the supplementary material. The wire (usually 0.035" diameter) is continuously fed into the weld area through the hand-held *gun*. Simultaneously, the inert shielding gas is delivered to the weld area. The usual shielding gas is a mixture of argon and carbon dioxide. See Figure 7-D for typical equipment.

Figure 7-D – MillerMatic DVI

This is a high-production welding procedure, offset by higher investment costs for equipment needed to achieve the greater production. This method is also suitable for some *problem* welds, such as stainless steels. Although this procedure produces less heat distortion in the work than the shielded metal arc, it is less suitable for repair work where dirt, rust scale, or paint present more problems than they do for shielded-metal arc. This method is also more difficult to use outdoors because even low-velocity winds will remove the shielding gas from protecting the weld.

TUNGSTEN INERT- GAS (TIG)

TIG welding uses a tungsten electrode to maintain the arc; however, the tungsten does not contribute to the weld material. Instead, a separate filler material is added to the puddle. The inert shielding gas (argon or helium) is made to flow over the arc area to protect the base metal. This procedure is similar to flame welding in that the heat buildup and filler material addition are independently controlled by the operator. Thus, this method is suitable for many materials and alloys that would otherwise be impossible to weld.

Flame welding is best for light work or minor welding where electric power is not available. Shielded-metal arc is best for repairs, or for a limited amount of work where saving set-up time is more important than welding production saving. MIG is best for new work and *usual* steels where welding production is most important. TIG is best for problem alloys or materials (various stainless steel or aluminum alloys), very thin sections, or where thick and thin materials are to be joined.

POWER-ASSISTED TOOLS

Power-operated tools are available to perform many of the operations described in earlier lessons. Power sources are electricity, explosive powder, compressed air, or pressurized liquids (hydraulic). A very rapidly growing segment of this type is in battery-powered tools. New, more powerful models are introduced almost continuously by the various manufacturers, and the competition between the manufacturers is brisk, indeed.

Power-operated tools usually result in increased production, improved quality, and/or enhanced safety in the sense that otherwise hazardous tasks can be accomplished with the operator somewhat removed from the work. However, since these tools amplify forces, they can be more hazardous if used carelessly.

Do not use any powered tool unless you have been

instructed in its proper use. For some tools, OSHA regulations require that you are certified to use the tool after receiving training from a certified instructor.

Even if you have been instructed in proper use of equipment, accidents can happen. Be aware at all times of possible injury. A construction worker in Littleton, Colorado, in January of 2005 went to the dentist after complaining about a toothache. The dentist immediately sent him to the hospital when x-rays showed a 4" nail lodged in the roof of his mouth. He underwent four hours of surgery to remove the nail which barely missed his right eye and had plunged 1-1/2" into his brain. A nail gun had backfired and sent a nail into a piece of wood nearby. He did not realize that a second nail had shot through his mouth. He is doing well now – but not sure he is going to continue in the construction industry.

ELECTRIC

Electric tools are the least expensive and most widely available of the general group of power tools. Sources of electricity may be the public utility, portable generators, or batteries (more about battery-powered tools below). The power supplied to a tool must be sufficient for the tool to operate at normal capacity or the tool will overheat and not perform properly.

Most electric power tools are equipped with a thermal switch to prevent tool damage. Some tools have a manual-reset over-temperature switch, while others reset automatically after cooling down.

In using electric tools that are powered from usual building receptacles, you are exposed to mechanical and shock hazards. Mechanical hazards arise from high-speed operation and large torque reaction of some rotating parts. Follow these safety tips:

1. Keep your hands, hair, and clothing from rotating parts.

2. Do not wear rings and other jewelry when using power tools.

3. Do not use the tool until you are on solid footing and are holding the tool properly.

Torque (twisting) power is dependent upon the gear-ratios of the tool. Torque **kickback** can knock you off a ladder or scaffold. Furthermore, if you lose your hold on the tool, it can flail around wildly, possibly causing severe injury.

Electric shock hazards can be directly and immediately lethal. But even if these shocks are electrically minor, you could fall off a ladder or lose your hold on the tool exposing yourself to the mechanical risks described above. Therefore, be aware of these considerations when using these tools:

1. Be sure to use a three-wire cord in grounded receptacles.

2. Do not cut off the grounding prong.

3. Do not use an adaptor for two-prong power.

4. Do not stand in water or hold onto a grounded surface (such as a pipe).

5. Use ground-fault interrupters (GFI) to further assure your safety.

6. If you are not certain the tool is properly grounded, do not use it!

Note that some tools are double insulated and only have two (2) prongs.

Nearly all electric tools have trigger switches to control the operation. Many of these switches have a means to hold the switch on after you have turned the motor on. ***Do not use this feature unless you have solid footing and can turn off the tool if any problem develops!***

Be sure extension cords are of sufficient capacity for the tool being used. Cord sized 18-2 with ground is okay for a trouble light, but use at least 16-2 with ground for light duty tools. Use not less than 14-2 with ground for general purpose electric tools and pipe machines. Use 12-2 with ground if you must go over 100' to the power outlet.

COMMON ELECTRIC TOOLS INCLUDE THE FOLLOWING:

CIRCULAR SAW

Blades are available for cutting wood, metal, or concrete. Saw may be bench-mounted, table-mounted, or portable. Blade turns at high speed for clean, square cuts.

Figure 7-E – Milwaukee 8" Metal Cutting Saw

Portable saws are useful for cutting wood, but any uses except cutting framing lumber and similar items is seldom satisfactory. Saws solidly mounted in tables, however, are excellent for all types of precise finish work.

RECIPROCATING SAW

Used for rough-in work; the blade moves in a back-ward-forward slicing motion. Blades are available to cut fiberglass sheets, sheet metal, pipe, wood, etc. Blades are often fragile and will break if forced.

ABRASIVE CUT-OFF SAW

Looks like a bench-mounted circular saw with an arbor feeder arm to control the feed speed of the cutting blade. Pipe material is anchored to the table, and the blade is advanced manually into the work material at a rate to match the cutting rate. The blades are brittle and may shatter if they are abused. This cutting method develops the highest cutoff productivity.

Never work this tool with the guard removed. Never stand in the throw path of the blade.

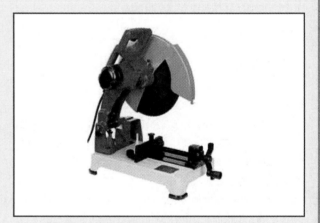

Figure 7-G – Milwaukee Abrasive Cut-Off Machine

HAMMER DRILL

The tool rams at the material with high speed blows while drilling. Some models can be adjusted to hammer only or drill only.

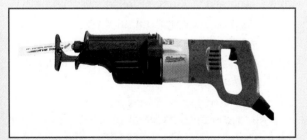

Figure 7-F – Milwaukee Reciprocating Saw

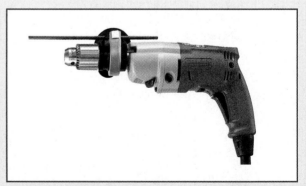

Figure 7-H – Milwaukee Magnum® Dual Torque Hammer Drill

They are used primarily in medium and heavy duty construction for drilling holes in concrete or masonry.

Bits are available in a multitude of sizes and shapes such as bullpoints, spades, and drills to drill or hammer.

POWER VISE

Used to speed threading time for commercial piping.

The vise jaws grip the pipe which has been inserted into the head. Most models can be operated in forward or reverse directions to cut normal threads, left-hand threads, or to back the threaded section out of the die.

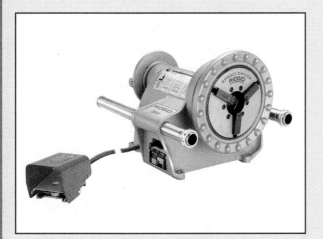

Figure 7-I – RIDGID Model 300 Power Drive

Figure 7-J – RIDGID Threading Machine

Using this machine requires the use of conventional cutting tool, reamer, threader, and oiler

THREADING MACHINE

These machines provide sophisticated high speed threading operation. The cutting tool, reamer, oiling system, threader, and power vise are incorporated into one tool.

SOLDERING GUN

Used for safe, flameless soldering of pipe. Two mating clamps wrap around the pipe; sufficient heat to melt solder is generated by current passing from clamp to clamp.

TEE DRILL

This tool is capable of drawing out a solder cup from a drilled hole in a copper tube. It saves the cost of a reducing tee (and its installation) and provides flexibility in where side branches are located when installing tubing mains in either plumbing projects or heating, cooling, process-piping jobs.

PLASMA CUTTERS

This tool uses an electric arc to cut almost any metal, including cast iron. It is very productive when any cut is needed in a metal object.

JIG SAW

Used mostly for finish work such as cutting sink openings in counter tops. Blades are available to cut metal, wood, Formica®, and plastic.

BAND SAW

Used to cut steel, plastic or copper materials. They are excellent for high speed production.

HUBLESS WRENCH SCREW

This tool appears similar to an electric drill. It is

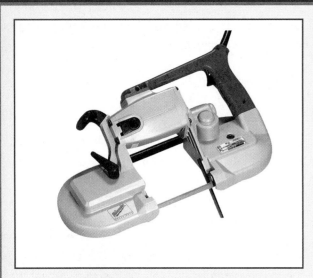

Figure 7-K – Milwaukee Portable Band Saw

used to screw hubless clamps to 60 inch-pounds torque.

IMPACT WRENCH

Used for work on systems requiring multiple bolting, reaming, tapping, screwing, etc. These are more commonly found as compressed air tools, but can be electric and cordless also.

ANCHORING DRILL SYSTEM

Used to develop holes for anchors into concrete. They are equipped with a depth adjustment accessory to assure that the holes will be deep enough to accept the full length of the anchor, and minimize over-drilling. Drilling a deeper hole than is needed wastes time and shortens tool life.

DRAIN ROD OR CABLE

Used to breakup and move along stoppages in drain pipes — they are available in many sizes. Cable should correspond to drain size. Check manufacturer's recommendations for correct sizes.

SPECIAL PURPOSE COPPER ASSEMBLY TOOL

At least one manufacturer makes a tool for compressing the special end of the "Ridgid Pro Press" fitting for use with copper tubing. The advertised advantages of this joining system is nearly zero time to make a joint. After the tubing is cut, chamfered, and reamed, no other preparation is needed, and the fitting is placed on the tube end. The power tool completes the joint in [literally] a few seconds. There is no need to have the tubing dry as is required with soldering, and no elevated temperature (flame or electric contact jaw), so fire hazard or personal injury from burns are non-existent. There is also a reduced exposure to leaks using this method.

ELECTRIC — BATTERY SOURCE

Battery-powered tools are a special class of electric tools. In appropriate circumstances, these tools greatly increase productivity. Tool types include rotating output (drills and screw-setting), hammer/rotating (hammer drills), hammers, circular saws, reciprocating saws, and even ProPress copper tubing assembly tools. Offsetting the need to recharge batteries, there are these advantages: immediate availability, no shock hazard, no need to find a power source, and no extension cord to string out, to trip over, or to be retrieved after the job is finished.

Battery-powered tools in the past few years have seen many improvements and a large variety of types. Where in the early days they were light powered, with limited charge capacity, short battery life, and long recharge times, now with battery voltages as high as 24 volts these tools are very powerful, with high charge capacity, and quick recharge times. Most models include a two-speed switch for high-production, lesser torque jobs; or slower, high torque applications.

The usual tool case includes the tool itself, two batteries, a battery charger, instructions, and — often — accessories items like drills, chisel bits, or similar things.

The batteries are the weak-link with these tools. To many craftsmen, they seem to have a relatively short life, and are generally expensive. The workers, and their contractor employers, must evaluate these tools to determine whether they are advantageous for the work to be performed.

EXPLOSIVE POWDER

It is imperative that you are fully instructed in the use of these special tools before you operate them. OSHA rules require that you have an operator's certificate to use them.

Powder tools are used to set anchors in concrete or steel construction. The tools are used with various energy level charges, depending on the nature and weight of the construction and the type of anchor used. The tools use either .22 or .38 caliber cartridges.

Never shoot any powder activated nail gun into a wall or floor without knowing the structure of the assembly. The nail could penetrate the assembly and injure or kill.

STAIR-CLIMBING HAND TRUCKS

This battery-powered appliance makes it possible for one person to move loads as heavy as 1,500 pounds up or down a flight of stairs safely.

COMPRESSED AIR

The first tool that comes to mind of this type is the heavy pavement breaker and trailer-mounted compressor use to make street cuts. These are extremely noisy and rugged, but they do a great deal of work!

Shop — or assembly — compressed-air-operated tools include grinders, impact wrenches, impact hammers, demolition hammers, circular saws, tamping hammers, and pumping equipment. These tools are more rugged and more expensive than the corresponding electric tools. They are inherently very long-lived, offsetting the higher initial cost. Although these tools are less energy efficient than electric tools, they present no electric shock or spark hazards in explosive atmospheres. They are less likely to be stolen, as they are not usable in most private or small shops.

For typical construction jobsite conditions, these tools require the presence of an air compressor of sufficient capacity for the type and number of tools to be used at any one time. This compressed air source represents a large added investment, depending on overall requirements.

For a fabrication shop where a large air system is probably already installed, air tools could be the lowest cost choice when tool additions are needed.

One of the most common uses for an air compressor (or compressed air system) is to keep the tires on pneumatic-tired vehicles properly inflated. Even the smallest contractor organization can use a compressed air source for this purpose alone. Other uses include air testing tanks or pre-fabricated piping assemblies.

COMPRESSED AIR SAFETY

NEVER indulge in horseplay with compressed air. Do not aim a stream of compressed air at a person — there have been horrible tragedies resulting from such fooling around!

Remember that compressed air contains stored energy, so a ruptured hose — or worse yet, a ruptured tank — can produce serious problems. Compressed air offers many advantages, but it requires serious attention to use it properly and safely.

HYDRAULIC

Hydraulic tools include large-diameter cut off devices, benders, vises, jacks, hole punchers, and boiler assembly equipment. Many electric and some pneumatic tools incorporate hydraulic operating systems as the final output mechanism of the tool.

The essential characteristic of hydraulic systems is that they contain virtually no stored energy. This means that although they can develop very large forces (in the 1000's of psi), a leak or other system failure is not a significant hazard. This is not the case with compressed air systems, steam systems, or hot water systems. Such a hydraulic leak can be annoying, can result in the loss or contamination of hydraulic fluid, or it could result in lost time for the operation that is underway — but it is usually not hazardous.

GENERAL

All of the above tool types require careful maintenance. Repair or replace any deteriorated component. Remember, most of these tools magnify forces to produce their effect, which can result in serious injury or property damage if things go wrong.

Do not use any power tool without instruction for proper operation. These tools are designed to be labor savers; be sure all details of their application are in good order to operate them safely and to realize their full usefulness.

Figure 7-L – Milwaukee Operator's Manual

GENERAL SAFETY RULES — FOR ALL POWER TOOLS

 WARNING!

READ ALL INSTRUCTIONS

Failure to follow all instructions listed below may result in electric shock, fire and/or serious injury. The term "power tool" in all of the warnings listed below refers to your mains-operated (corded) power tool or battery-opearted (cordless) power tool.

SAVE THESE INSTRUCTIONS

WORK AREA SAFETY

1. **Keep work area clean and well lit.** Cluttered or dark areas invite accidents.

2. **Do not operate power tools in explosive atmospheres, such as in the presence of flammable liquids, gases, or dust.** Power tools create sparks which may ignite the dust or fumes.

3. **Keep children and bystanders away while operating a power tool.** Distractions can cause you to lose control.

ELECTRICAL SAFETY

4. **Power tool plugs must match the outlet. Never modify the plug in any way. Do not use any adapter plugs with earthed (grounded) power tools.** Unmodified plugs and matching outlets will reduce risk of electric shock.

5. **Avoid body contact with earthed or grounded surfaces such as pipes, radiators, ranges and refrigerators.** There is an increased risk of electric shock if your body is earthed or grounded.

6. **Do not expose power tools to rain or wet conditions.** Water entering a power tool will increase the risk of electric shock.

7. **Do not abuse the cord. Never use the cord for carrying, pulling, or unplugging the power tool. Keep cord away from heat, oil, sharp edges, or moving parts.** Damaged or entangled cords increase the risk of electric shock.

8. **When operating a power tool outdoors, use an extension cord suitable for outdoor use.** Use of a cord suitable for outdoor use reduces the risk of electric shock.

PERSONAL SAFETY

9. **Stay alert, watch what you are doing and use common sense when operating a power tool. Do not use a power tool while you are tired or under the influence of drugs, alcohol or medication.** A moment of inattention while operating power tools may result in serious personal injury.

10. **Use safety equipment. Always wear eye protection.** Safety equipment such as dust mask, non-skid safety shoes, hard hat, or hearing protection used for appropriate conditions will reduce personal injuries.

11. **Avoid accidental starting. Ensure the switch is in the off-position before plugging in.** Carrying tools with your finger on the switch or plugging in power tools that have the switch on invites accidents.

12. **Remove any adjusting key or wrench before turning the power tool on.** A wrench or a key left attached to a rotating part of the power tool may result in personal injury.

13. **Do not overreach. Keep proper footing and balance at all times.** This enables better control of the power tool in unexpected situations.

14. **Dress properly. Do not wear loose clothing or jewellery. Keep your hair, clothing and gloves away from moving parts.** Loose clothes, jewellery, or long hair can be caught in moving parts.

15. **If devices are provided for the connection of dust extraction and collection facilities, ensure these are connected and properly used.** Use of these devices can reduce dust-related hazards.

page 2

POWER TOOL USE AND CARE

16. **Do not force the power tool. Use the correct power tool for your application.** The correct power tool will do the job better and safer at the rate for which it was designed.

17. **Do not use the power tool if the switch does not turn it on and off.** Any power tool that cannot be controlled with the switch is dangerous and must be repaired.

18. **Disconnect the plug from the power source and/or the battery pack from the power tool before making any adjustments, changing accessories, or storing power tools.** Such preventive safety measures reduce the risk of starting the tool accidentally.

19. **Store idle power tools out of the reach of children and do not allow persons unfamiliar with the power tools or these instructions to operate power tools.** Power tools are dangerous in the hands of untrained users.

20. **Maintain power tools. Check for misalignment or binding of moving parts, breakage of parts and any other condition that may affect the power tool's operation. If damaged, have the power tool repaired before use.** Many accidents are caused by poorly maintained power tools.

21. **Keep cutting tools sharp and clean.** Properly maintained cutting tools with sharp cutting edges are less likely to bind and are easier to control.

22. **Use the power tool, accessories and tool bits etc., in accordance with these instructions and in the manner intended for the particular type of power tool, taking into account the working conditions and the work to be performed.** Use of the power tool for operations different from those intended could result in a hazardous situation.

SERVICE

23. **Have your power tool serviced by a qualified repair person using only identical replacement parts.** This will ensure that the safety of the power tool is maintained.

SPECIFIC SAFETY RULES

1. **Use auxiliary handles supplied with the tool.** Loss of control can cause personal injury.
2. **Wear ear protectors with impact drills.** Exposure to noise can cause hearing loss.
3. **Hold power tools by insulated gripping surfaces when performing an operation where the cutting tool may contact hidden wiring or its own cord.** Contact with a "live" wire will make exposed metal parts of the tool "live" and shock the operator.
4. **Maintain labels and nameplates.** These carry important information. If unreadable or missing, contact a *MILWAUKEE* service facility for a free replacement.
5. **WARNING!** Some dust created by power sanding, sawing, grinding, drilling, and other construction activities contains chemicals known to cause cancer, birth defects or other reproductive harm. Some examples of these chemicals are:
   - lead from lead-based paint
   - crystalline silica from bricks and cement and other masonry products, and
   - arsenic and chromium from chemically-treated lumber.

   Your risk from these exposures varies, depending on how often you do this type of work. To reduce your exposure to these chemicals: work in a well ventilated area, and work with approved safety equipment, such as those dust masks that are specifically designed to filter out microscopic particles.

Symbology

Symbol	Meaning	Symbol	Meaning
(UL)	Underwriters Laboratories, Inc.	V~	Volts Alternating Current
(CSA)	Canadian Standards Association	n_0 xxxx min.$^{-1}$	No Load Revolutions per Minute (RPM)
NOM-ANCE	Mexican Approvals Marking	A	Amperes

Specifications

Cat. No.	Speed	No Load RPM	Blows per minute	Wood					Steel	Concrete	
				Flat boring bit	Auger bit	Ship auger bit	Selfeed bit	Hole saw	Twist drill	Carbide tipped bit	Screw fasteners
5370-1 5371-20 5376-1	low	0 - 1 000	0 - 20 000	1-1/2"	7/8"	7/8"	1-1/8"	3-1/4"	1/2"	5/8"	1/4"
	high	0 - 2 500	0 - 50 000	3/4"	*	*	*	*	3/8"	9/16"	3/16"
5374-1	speed control	0 - 2 500	0 - 50 000	3/4"	Nr	Nr	Nr	1-3/4"	3/8"	9/16"	Nr

* Set gear shift to low speed setting when boring holes in wood.
Nr = Not recommended

FUNCTIONAL DESCRIPTION

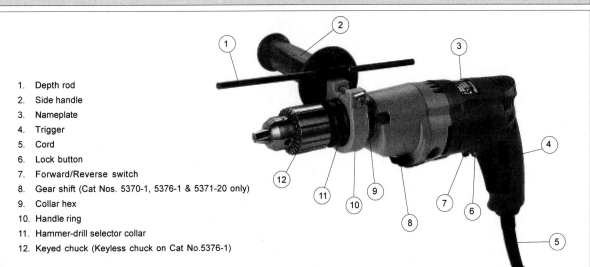

1. Depth rod
2. Side handle
3. Nameplate
4. Trigger
5. Cord
6. Lock button
7. Forward/Reverse switch
8. Gear shift (Cat Nos. 5370-1, 5376-1 & 5371-20 only)
9. Collar hex
10. Handle ring
11. Hammer-drill selector collar
12. Keyed chuck (Keyless chuck on Cat No.5376-1)

page 3

GROUNDING

 WARNING!

Improperly connecting the grounding wire can result in the risk of electric shock. Check with a qualified electrician if you are in doubt as to whether the outlet is properly grounded. Do not modify the plug provided with the tool. Never remove the grounding prong from the plug. Do not use the tool if the cord or plug is damaged. If damaged, have it repaired by a *MILWAUKEE* service facility before use. If the plug will not fit the outlet, have a proper outlet installed by a qualified electrician.

Grounded Tools:
Tools with Three Prong Plugs

Tools marked "Grounding Required" have a three wire cord and three prong grounding plug. The plug must be connected to a properly grounded outlet (See Figure A). If the tool should electrically malfunction or break down, grounding provides a low resistance path to carry electricity away from the user, reducing the risk of electric shock.

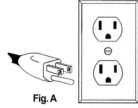

Fig. A

The grounding prong in the plug is connected through the green wire inside the cord to the grounding system in the tool. The green wire in the cord must be the only wire connected to the tool's grounding system and must never be attached to an electrically "live" terminal.

Your tool must be plugged into an appropriate outlet, properly installed and grounded in accordance with all codes and ordinances. The plug and outlet should look like those in Figure A.

Double Insulated Tools:
Tools with Two Prong Plugs

Tools marked "Double Insulated" do not require grounding. They have a special double insulation system which satisfies OSHA requirements and complies with the applicable standards of Underwriters Laboratories, Inc., the Canadian Standard Association and the National Electrical Code. Double Insulated tools may be used in either of the 120 volt outlets shown in Figures B and C.

Fig. B   Fig. C

EXTENSION CORDS

Grounded tools require a three wire extension cord. Double insulated tools can use either a two or three wire extension cord. As the distance from the supply outlet increases, you must use a heavier gauge extension cord. Using extension cords with inadequately sized wire causes a serious drop in voltage, resulting in loss of power and possible tool damage. Refer to the table shown to determine the required minimum wire size.

The smaller the gauge number of the wire, the greater the capacity of the cord. For example, a 14 gauge cord can carry a higher current than a 16 gauge cord. When using more than one extension cord to make up the total length, be sure each cord contains at least the minimum wire size required. If you are using one extension cord for more than one tool, add the nameplate amperes and use the sum to determine the required minimum wire size.

Guidelines for Using Extension Cords

- If you are using an extension cord outdoors, be sure it is marked with the suffix "W-A" ("W" in Canada) to indicate that it is acceptable for outdoor use.
- Be sure your extension cord is properly wired and in good electrical condition. Always replace a damaged extension cord or have it repaired by a qualified person before using it.
- Protect your extension cords from sharp objects, excessive heat and damp or wet areas.

Recommended Minimum Wire Gauge for Extension Cords*

Nameplate Amperes	Extension Cord Length					
	25'	50'	75'	100'	150'	200'
0 - 5	16	16	16	14	12	12
5.1 - 8	16	16	14	12	10	--
8.1 - 12	14	14	12	10	--	--
12.1 - 15	12	12	10	10	--	--
15.1 - 20	10	10	10	--	--	--

* Based on limiting the line voltage drop to five volts at 150% of the rated amperes.

READ AND SAVE ALL INSTRUCTIONS FOR FUTURE USE.

TOOL ASSEMBLY

 WARNING!

To reduce the risk of injury, always unplug tool before attaching or removing accessories or making adjustments. Use only specifically recommended accessories. Others may be hazardous.

Removing and Replacing Quik-Lok® Cords (Fig. 1)

MILWAUKEE's exclusive Quik-Lok® Cords provide instant field replacement or substitution.

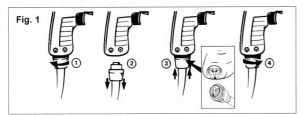

1. To remove the Quik-Lok® Cord, turn the cord nut 1/4 turn to the left and pull it out.
2. To replace the Quik-Lok® Cord, align the connector keyways and push the connector in as far as it will go. Turn the cord nut 1/4 turn to the right to lock.

Attaching the Side Handle

MILWAUKEE Magnum Hammer-Drills are furnished with a side handle to provide an insulated grasping surface and improved control of the tool. A handle ring, which fits behind the hammer/drill selector collar, locks the handle and depth rod in place. To change the setting of the depth rod, loosen the handle slightly and slide the depth rod to the desired position. Always tighten the side handle before operation. Be sure the handle ring is flat against the collar hex and away from the chuck and selector collar.

When using large bits or exerting substantial pressure, position the side handle 180° from the switch handle. This provides a "T" alignment to balance the tipping effect of the force applied to each handle.

Adjusting the Side Handle Position (Fig. 2)

1. Loosen the side handle by unscrewing the handle grip slightly.
2. Rotate the side handle to the desired position.
3. Hold the side handle in the desired position and flat against the collar hex while tightening the handle grip securely.

Setting the Depth Gauge (Fig. 3)

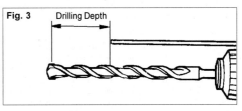

1. Loosen the depth gauge by unscrewing the side handle grip slightly.
2. Slide the depth gauge rod backward or forward until it is set for the desired depth.

   NOTE: The drilling depth is the distance between the tip of the bit and the tip of the depth gauge rod.

3. Hold the side handle in the desired position and flat against the collar hex while tightening the handle grip securely.

 WARNING!

To reduce the risk of personal injury and damage to the tool, hold and brace the tool securely. Brace tools with side handles as shown. If the bit binds, the tool will be forced in the opposite direction. Bits may bind if they are misaligned or when breaking through a hole. Wood boring bits can also bind if they run into nails or knots.

page 5

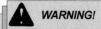

WARNING!

To prevent personal injury, always remove the chuck key from the chuck after each use.

Installing Bits into Keyed Chucks (Fig. 4)

Be sure that the shank of the bit and the chuck jaws are clean. Dirt particles may cause the bit to line up improperly. Do not use bits larger than the maximum recommended capacity of the drill because gear damage or motor overloading may result. For best performance, be sure that the bits are properly sharpened before use.

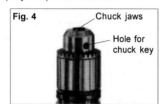

Fig. 4 — Chuck jaws, Hole for chuck key

1. Unplug the tool.
2. Open the chuck jaws wide enough to insert a bit. Allow the bit to strike the bottom of the chuck. Center the bit in the chuck jaws and tighten the jaws by hand to align the bit.
3. Place the chuck key into each of the three holes in the chuck, turning it clockwise to tighten the chuck securely.

   NOTE: Never use a wrench or means other than a chuck key to tighten or loosen the chuck.

4. To remove the bit, insert the chuck key into one of the holes in the chuck and turn it counterclockwise.

Installing Bits into Keyless Chucks (Fig. 5)

For best performance, always use sharp, clean bits and be sure the chuck jaws are clean. Dirt particles may cause the bit to line up improperly. Do not use bits larger than the maximum recommended capacity of the drill because gear damage or motor overloading may result.

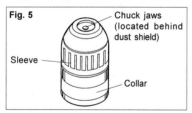

Fig. 5 — Chuck jaws (located behind dust shield), Sleeve, Collar

1. Unplug the tool.
2. To **open** the chuck jaws, turn the sleeve in the direction marked RELEASE.
3. Allow the bit to strike the bottom of the chuck and center the bit in the chuck jaws.
4. To **close** the chuck jaws, hold the collar while turning the sleeve in the direction marked GRIP. Tighten securely.
5. To **remove** the bit, hold the collar while turning the sleeve in the direction marked RELEASE.

OPERATION

WARNING!

To reduce the risk of injury, wear safety goggles or glasses with side shields. Unplug the tool before changing accessories or making adjustments.

Selecting Action (Fig. 6)

MILWAUKEE Hammer-Drills are designed to operate in either a "drill only" mode or a "drilling with hammering action" mode.

Fig. 6

1. To select **Drilling Action**, pull the selector collar toward gear case collar and rotate counter-clockwise until selector collar locks in place.
2. To select **Hammer-Drilling Action**, pull selector collar toward gear case collar and rotate clockwise until selector collar locks in place.

   NOTE: Constant pressure on bit must be maintained to engage hammering mechanism. When pressure on bit is released, hammering action will stop.

Using the Forward/Reverse Switch (Fig. 7)

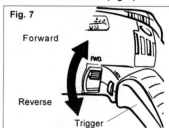

Fig. 7 — Forward, Reverse, Trigger

The forward/reverse switch can only be adjusted when the trigger is not pressed. Always allow the motor to come to a complete stop before using the forward/reverse switch.

1. For **forward** (clockwise) rotation, push the forward/reverse switch to FWD as shown.
2. For **reverse** (counterclockwise) rotation, push the forward/reverse switch to REV as shown. Although an interlock prevents reversing the tool while the motor is running, allow the motor to come to a full stop before reversing.

NOTE: When hammer-drilling, use the tool in forward rotation (clockwise) only.

page 6

Locking Trigger Switch (Fig. 8)

The lock button holds trigger in the ON position for continuous full speed use.

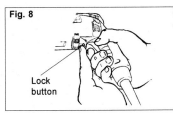
Fig. 8
Lock button

1. To **lock** the trigger switch, push in the lock button while pulling the trigger. Then release the trigger.
2. To **unlock** the trigger switch, pull the trigger and release. The lock button will pop out.

Selecting Speed

The speed can be changed when the tool is at a complete stop or running under no load.

1. For **Low** speed (up to 1 000 rpm), turn the speed selector to position 1.
2. For **High** speed (up to 2 500 rpm), turn the speed selector to position 2.

Starting, Stopping & Controlling Speed

Place the drill bit on the work surface and apply firm pressure before starting. A center punch may be used, in steel, to make starting easier. Start the drill slowly to permit maximum control and to prevent the bit from wandering. When the hole has been properly started, increase the speed until maximum cutting efficiency is reached.

Increasing the speed of the drill permits increasing the amount of pressure applied. However, too much pressure will slow the drill bit and retard drilling efficiency. Too little pressure will cause the bit to slide over the work and cause excessive friction which will dull the point of the bit.

See "Specifications" for RPM and blows per minute information.

1. To **start** the tool, pull the trigger.
2. To **stop** the tool, release the trigger.
3. To **vary** the speed, increase or decrease pressure on trigger. The further the trigger is pulled, the greater the speed.

Stalling

If the drill slows due to loading, increase the speed intil it operates properly. If stalling occurs, reverse the motor and remove the bit from the work and start again.

Operating

Position the tool, grasp the handles firmly and pull the trigger. Always hold the tool securely using both handles and maintain control. This tool has been designed to achieve top performance with only moderate pressure. Let the tool do the work.

If the speed begins to drop off when drilling deep holes, pull the bit partially out of the hole while the tool is running to help clear dust. Do not use water to settle the dust since it will clog the bit flutes and tend to make the bit bind in the hole.

APPLICATIONS

⚠ WARNING!

To reduce the risk of electric shock, check work area for hidden pipes and wires before drilling.

Drilling in Wood, Composition Materials and Plastic

When drilling in wood, composition materials and plastic, select the drill operating mode. Start the drill slowly, gradually increasing speed as you drill. Select low speeds for plastics with a low melting point.

Drilling in Metal

When drilling in metal, select the drill operating mode. Use high speed steel twist drills or hole saws. Use a center punch to start the hole. Lubricate drill bits with cutting oil when drilling in iron or steel. Use a coolant when drilling in nonferrous metals such as copper, brass or aluminum. Back the material to prevent binding and distortion on breakthrough.

Drilling in Masonry

When drilling in masonry, select the hammer-drill operating mode. Use high speed carbide-tipped bits. Drilling soft masonry materials such as cinder block requires little pressure. Hard materials like concrete require more pressure. A smooth, even flow of dust indicates the proper drilling rate. Do not let the bit spin in the hole without cutting. Do not use water to settle dust or to cool bit. Do not attempt to drill through steel reinforcing rods. Both actions will damage the carbide.

page 7

MAINTENANCE

 WARNING!

To reduce the risk of injury, always unplug your tool before performing any maintenance. Never disassemble the tool or try to do any rewiring on the tool's electrical system. Contact a *MILWAUKEE* service facility for ALL repairs.

Maintaining Tools

Keep your tool in good repair by adopting a regular maintenance program. Before use, examine the general condition of your tool. Inspect guards, switches, tool cord set and extension cord for damage. Check for loose screws, misalignment, binding of moving parts, improper mounting, broken parts and any other condition that may affect its safe operation. If abnormal noise or vibration occurs, turn the tool off immediately and have the problem corrected before further use. Do not use a damaged tool. Tag damaged tools "DO NOT USE" until repaired (see "Repairs").

Under normal conditions, relubrication is not necessary until the motor brushes need to be replaced. After six months to one year, depending on use, return your tool to the nearest *MILWAUKEE* service facility for the following:

- Lubrication
- Brush inspection and replacement
- Mechanical inspection and cleaning (gears, spindles, bearings, housing, etc.)
- Electrical inspection (switch, cord, armature, etc.)
- Testing to assure proper mechanical and electrical operation

 WARNING!

To reduce the risk of injury, electric shock and damage to the tool, never immerse your tool in liquid or allow a liquid to flow inside the tool.

Cleaning

Clean dust and debris from vents. Keep the tool handles clean, dry and free of oil or grease. Use only mild soap and a damp cloth to clean your tool since certain cleaning agents and solvents are harmful to plastics and other insulated parts. Some of these include: gasoline, turpentine, lacquer thinner, paint thinner, chlorinated cleaning solvents, ammonia and household detergents containing ammonia. Never use flammable or combustible solvents around tools.

Repairs

If your tool is damaged, return the entire tool to the nearest service center.

FIVE YEAR TOOL LIMITED WARRANTY

Every *MILWAUKEE* tool is tested before leaving the factory and is warranted to be free from defects in material and workmanship. *MILWAUKEE* will repair or replace (at *MILWAUKEE*'s discretion), without charge, any tool (including battery chargers) which examination proves to be defective in material or workmanship from five (5) years after the date of purchase. Return the tool and a copy of the purchase receipt or other proof of purchase to a *MILWAUKEE* Factory Service/Sales Support Branch location or *MILWAUKEE* Authorized Service Station, freight prepaid and insured. This warranty does not cover damage from repairs made or attempted by other than *MILWAUKEE* authorized personnel, abuse, normal wear and tear, lack of maintenance, or accidents.

Battery Packs, Flashlights, and Radios are warranted for one (1) year from the date of purchase.

THE REPAIR AND REPLACEMENT REMEDIES DESCRIBED HEREIN ARE EXCLUSIVE. IN NO EVENT SHALL *MILWAUKEE* BE LIABLE FOR ANY INCIDENTAL, SPECIAL, OR CONSEQUENTIAL DAMAGES, INCLUDING LOSS OF PROFITS.

THIS WARRANTY IS EXCLUSIVE AND IN LIEU OF ALL OTHER WARRANTIES, OR CONDITIONS, WRITTEN OR ORAL, EXPRESSED OR IMPLIED FOR MERCHANTABLILITY OR FITNESS FOR PARTICULAR USE OR PURPOSE.

This warranty gives you specific legal rights. You may also have other rights that vary from state to state and province to province. In those states that do not allow the exclusion of implied warranties or limitation of incidental or consequential damages, the above limitations or exclusions may not apply to you. This warranty applies to the United States, Canada, and Mexico only.

page 8

ACCESSORIES

WARNING!

To reduce the risk of injury, always unplug the tool before attaching or removing accessories. Use only specifically recommended accessories. Others may be hazardous.

For a complete listing of accessories refer to your *MILWAUKEE* Electric Tool catalog or go on-line to www.milwaukeetool.com. To obtain a catalog, contact your local distributor or a service center.

Percussion Carbide-Tipped Bits

These carbide-tipped bits are specially designed for drilling through concrete and masonry. They are made with round shanks for use with Hammer-Drills. Alloy steel shanks and bodies provide durability and long life. The wide spiral with shallow oval flutes removes dust quickly to assure maximum drilling efficiency.

Bit Diameter	Catalog No.	Overall Lenth
1/8"	48-20-6800	3"
3/16"	48-20-6805	4"
3/16"	48-20-6806	6"
1/4"	48-20-6810	4"
1/4"	48-20-6811	6"
5/16"	48-20-6815	4"
5/16"	48-20-6816	6"
3/8"	48-20-6820	4"
3/8"	48-20-6821	6"
3/8"	48-20-6823	13-1/2"
7/16"	48-20-6825	6"
1/2"	48-20-6830	6"
1/2"	48-20-6833	13-1/2"
9/16"	48-20-6835	6"
5/8"	48-20-6840	6"
5/8"	48-20-6843	13-1/2"

Steel Carrying Case
Cat. No. 48-55-0711

Impact Resistant Carrying Case
Cat. No. 48-55-5378

8' Quik-Lok Cord
Cat. No. 48-76-4008

25' Quik-Lok Cord
Cat. No. 48-76-4025

Chuck Key
Cat. No. 48-66-3280

Key Holder
Cat. No. 48-66-4040

1/2" Keyed Chuck
Cat. No. 48-66-1365

Keyless Chuck
Cat. No. 48-66-0600

page 9

LESSON 8

REVIEW – NUMBERS, FRACTIONS AND DECIMALS

Mathematics involves the numerical relationships between things. We use mathematics in many ways in our daily lives — some applications are easy, some are not. This group of lessons will help make the mathematical problems that appear in plumbing work easier to analyze and solve.

ELEMENTARY LOGIC USEFUL FOR SOLVING PROBLEMS

Use the following steps to solve mathematical problems:

1. Study the problem fully and <u>determine the question which is to be answered.</u>

2. Write down the known information and what is asked for as the answer.

3. Determine which mathematical operation(s) is (are) needed to solve the problem.

4. Make an estimate of the answer *(about 3 feet, approximately 4500 BTU, etc.).*

5. Perform the detailed operation.

6. Compare the calculated answer with the estimate.

7. Label the answer with the units involved (dollars, feet, pounds, BTU/hr, etc.).

8. Check your work. It may be helpful to perform the calculation in reverse (the reciprocal operation). This process is always recommended to assure accuracy.

By regularly using these steps you will increase your skills and build self-confidence.

TERMINOLOGY

Terminology means name. All items have names. Numbers are the *stock-in-trade* of mathematics. Numbers can be identified in several ways, as are pointed out in the examples that follow:

INTEGERS

Integers are the ten digits 0, 1, 2, 3, 4, 5, 6, 7, 8, 9. They are the "building blocks" of the numbering system in use in most of the world. Their invention is generally credited to the Phoenecians in ancient times. This number system is often referred to as the "Arabic System," because the Phoenicians were one of the groups of the ancient Arabic world. By combining just ten integer symbols with *the value of position*, the Phoenicians invented a powerful system that can express any number that you can imagine!

WRITTEN REPRESENTATION OF NUMBERS

The Arabic System of numbering is the one most widely used throughout the world.
The ten Arabic numerals —

0, 1, 2, 3, 4, 5, 6, 7, 8, 9

are used to form any number by combining the integers and their positions to express any value.

TABLE 8-A
POSITION NAMES

Etc	Millions	Hundred Thousands	Ten Thousands	Thousands	Hundreds	Tens	Units	Decimals
	0 to 9,000,000	0 to 900,000	0 to 90,000	0 to 9,000	0 to 900	0 to 90	0 to 9	.

Table 8-A illustrates the position concept. The headings are in order of magnitude, starting at the right and proceeding to the left, with each position increasing by a factor of ten.

Thus, starting from the left of the decimal, each column contains digits from 0 to 9, but the *meaning* of each column is as shown below:

1. The first column indicates numbers from 0 to 9

2. The second column indicates numbers from 0 to 9 tens

3. The third column indicates numbers from 0 to 9 hundreds

4. And so on

This system has no limit, because no matter how big (or small) a number you write, it is always possible to write a bigger (or smaller) number.

Translating Table 8-A into common language, we can express any number in terms of this system. For example, if we see the number

843

we think — eight hundred plus forty plus three

We read this number as — eight hundred forty three, leaving out the word plus.

You try it — Write out **9,462**:

_____ _____ plus _____ _____ plus_____ plus _____

which would read (when the "plus" is dropped),

_____ _____ _____ _____ _____ _____

Likewise, if someone reads you a number, you can write it out using this system. Thus,

five million six hundred sixty three

Plumbing Apprentice Student Workbook Year One
Fourth Edition

would be

5,000,663

Since no hundred thousands, ten thousands, or thousands were mentioned, none was listed.

Express <u>five hundred six thousand four hundred fifty three</u> in number form:

_____

Convert the following numbers to word form:

692 _____

5,264 _____

301 _____

18,442 _____

Convert the following words to numerals (digits):

four thousand fifty three _____

thirteen million six _____

seventy eight thousand four hundred two _____

LABELING

Labeling an answer is essential to show a complete answer. The label identifies the units in which the answer is given. Without this label, the answer is incomplete.

Answer this question:

How long is a piece of threaded black pipe as it is shipped from the supply house?

We know this length to be 21 feet; but if your answer is

21

it is only partly correct. Labeling the answer gives the full answer:

21 feet

Label all answers, whether it be in inches, feet, yards, lengths of pipe, etc.

MIXED NUMBERS

Mixed numbers are whole numbers and a part of a whole number. The part of the whole number may be expressed in fraction or decimal form. Examples are the following:

$24\ 1/2$ inches $187\ 3/4$ feet $1\ 1/2"$ pipe   5.2 miles   14.25 pounds   1.5 days

DECIMALS

Decimals also refer to part of the whole and are expressed using multiples of 10 as the **implied** denominator. **Deci** (from Latin) means tenth. Examples of a whole number with a decimal include:

1.5 inches   2.6 feet   16.75 pounds

FRACTIONS

If we need to describe a part of a complete thing, we must use fractions (or decimals, which are a special form of fractions). If a pipe is ten feet long and you cut off five feet, you cut the pipe in half, because both five foot pieces represent one half of the original length.

A fraction is a ratio of two whole numbers.

Fractions refer to part of a whole unit. Examples of a whole number with a fraction include:

$1\ 1/2$ $5\ 1/4$ $11\ 3/4$ $3\ 7/8$

A plumber needs to take measurements for rough-in and finish work, repair work, and service work. Most of the measurements and concepts used in plumbing work involve fractions, especially if we are going to be accurate in our work. The majority of fraction work in the plumbing trade will involve measurements made with a ruler or with gauges.

Consider the fraction "one half." It is written thus:

<u>1 (Numerator)</u>
2 (Denominator)

The number above the horizontal line is the *numerator*.

The number below the horizontal line is the *denominator*.

The denominator expresses the total number of equal parts into which a thing is divided. The numerator expresses the quantity of the equal parts that we are dealing with.

Thus, if a pie is cut into eight equal pieces, and four are eaten, $4/8$ of the pie are gone and $4/8$ remain.

Or, if a concrete slab of 64 square feet has 8 square feet removed, the fraction removed is $8/64$.

$$\frac{8 \text{ square feet}}{64 \text{ square feet}} \quad \frac{\text{(numerator)}}{\text{(denominator)}} \quad \frac{\text{amount removed}}{\text{total units in slab}}$$

REDUCING FRACTIONS

A basic guide followed when working with fractions which makes the process easier is the reducing rule. This guide states that a fraction is usually best expressed in its *lowest terms*.

We can multiply or divide the numerator and denominator by the same number and not change the *value* of the fraction. It is usually best to reduce the numerator and denominator to the smallest numbers possible. Thus, the $4/8$ pie discussed above can be expressed as ½ pie, and the $8/64$ concrete slab can be stated as $1/8$ concrete slab. In order to reduce the fraction, we divide the numerator and denominator by the largest number that divides each of them exactly. For example, to reduce

$$\frac{4}{8}$$

to the smallest numbers possible, we divide both the 4 and the 8 by 4 to get

$$\frac{4 \div 4}{8 \div 4} = \frac{1}{2}$$

Reduce the following fractions to their simplest terms:

$$\frac{7}{42} \qquad \frac{20}{30} \qquad \frac{12}{16} \qquad \frac{15}{20} \qquad \frac{4}{18} \qquad \frac{8}{12}$$

IMPROPER FRACTIONS

If the numerator of a fraction is larger than the denominator, the fraction is called an *improper* fraction. To simplify, divide the numerator by the denominator to get a whole number and a remainder fraction. Most of us get a better mental picture of the value of an improper fraction when it is simplified, but an improper fraction is *not* incorrect!

Examples of improper fractions include the following:

$$\frac{20}{10} \qquad \frac{34}{17} \qquad \frac{27}{3} \qquad \frac{8}{4}$$

For example: $\frac{20}{8} = 2 + \frac{4}{8} = 2 + \frac{1}{2}$, *also written as* $2\ 1/2$

also $\frac{12}{9} = 1 + \frac{3}{9} = 1 + \frac{1}{3}$, *also written as* $1\ 1/3$

PROBLEMS

Simplify these fractions.

$$\frac{24}{7} \qquad \frac{10}{6} \qquad \frac{30}{12} \qquad \frac{22}{16} \qquad \frac{16}{6}$$

CONCEPTS

Fractions are expressed as equal parts of the whole in the following form: $\frac{a}{b}$

where "**a**" is the numerator or the part considered, and "**b**" is the denominator or number of pieces *in the whole section.*

Fractions normally are best-expressed in simplest terms. This is achieved by multiplying (or dividing) the numerator and denominator by the same number. This does not change the value of the fraction.

A fraction is in its lowest terms when the numerator and denominator cannot be divided any further by a common number.

An improper fraction is one in which the numerator is larger than the denominator. Usually, this fraction should be put in the form of a mixed number and the remaining fraction is best expressed in lowest terms.

LESSON 9

MATHEMATICAL OPERATIONS — FRACTIONS

In order to become proficient at mathematics, practice and review are necessary. In the first section on mathematics (Lesson 8), we covered the following terms:

WHOLE NUMBER: A whole number is a number which contains no fractions or decimals.

Examples include: 6 inches   4 feet   32 tons

FRACTION: A fraction is part of a whole number, expressed with the part over the whole amount.

Examples include: a/b $^1/_2$ foot $^{13}/_{16}$ inches

MIXED NUMBERS: A mixed number is a whole number and a part of a whole number, with the part expressed in fraction or decimal form.

Examples include: $2\,^3/_4$ $8^1/_8$ $2^1/_2$

Fractions may appear differently, depending upon the method used to reproduce written material. For ease in printing and typesetting, you will see fractions written several ways in this book. We may write the fraction *one-half* in words or as shown below:

$$½ \quad ^1/_2 \quad \frac{1}{2}$$

These fractions are identical; they are simply written differently.

OPERATIONS WITH FRACTIONS - ADDITION

You will quite often need to perform operations on whole numbers and fractions when taking measurements. The operations involve addition, subtraction, multiplication, and division.

First, we will consider the methods used for addition:

To add fractions, take the following steps:

1. Express all fractions in terms of the same denominator.

2. Add all numerators and set over the common denominator.

3. Simplify this result, if possible.

For example, if we add ½ and ¼, a common denominator of 4 and 2 is 4. The number ¼ can stay the same for this addition, and the number ½ will have to change to a fraction with a denominator of 4. If the denominator is multiplied by 2 to get 4, then the numerator must also be multiplied by 2. Therefore,

$$\frac{1}{2} = \frac{1(2)}{2(2)} = \frac{2}{4}$$

Once a common denominator is established, the numerators are added. The denominators are not added and remain the same.

$$\frac{1}{4} + \frac{1}{2} = \frac{1}{4} + \frac{2}{4}$$

Combining,

$$\frac{1}{4} + \frac{2}{4} = \frac{3}{4}$$

Thus, addition of fractions requires that all fractions have a common denominator before adding the numerators. To convert fractions to the common denominator, we must divide the common denominator by each of the fraction denominators, and multiply the numerator by the results of that division.

Another example: Add the following: $\frac{3}{5} + \frac{1}{4} + \frac{2}{3} + \frac{1}{2} =$

Remember, a common denominator is a number that is exactly divisible by all the denominators. One way to find a common denominator is to multiply the denominators together until you have a number that can be divided evenly by each denominator.

5, 4, and 2 will divide into 20 evenly, but 3 will not.   **5 x 4 = 20**

If we multiply 20 by 3, we will obtain a number that is divisible by 3.
Therefore, 5, 4, 2, and 3 will all divide into 60 evenly.   **20 x 3 = 60**
Therefore, 60 is a common denominator.

To find the new numerators, we multiply the numerator by the number of times that denominator will divide into the common denominator of 60.

5 *Goes into* **60 12 times**   12 x 3 = 36   Therefore, $\frac{3}{5} = \frac{36}{60}$

4 *Goes into* **60 15 times**   15 x 1 = 15   Therefore, $\frac{1}{4} = \frac{15}{60}$

3 *Goes into* **60 20 times**   20 x 2 = 40   Therefore, $\frac{2}{3} = \frac{40}{60}$

2 *Goes into* **60 30 times**   30 x 1 = 1   Therefore, $\frac{1}{2} = \frac{30}{60}$

Thus, the equivalent statement (and partial solution) of the original problem is the following:

$$\frac{36}{60} + \frac{15}{60} + \frac{40}{60} + \frac{30}{60} = \frac{121}{60}$$

We now have an improper fraction which is best converted to a mixed number.

60 goes into 121 2 times with 1 left over

Therefore, the solution to the above addition problem is   2 and $\frac{1}{60}$ or $2\frac{1}{60}$

Note that the smallest denominator that is a common denominator yields the smallest numerators, but it is not essential to use the *lowest* common denominator.

ADDITION OF MIXED NUMBERS

To add mixed numbers, use the following steps:

1. Add the fractions (following the above procedure if necessary)

2. Simplify

3. Add the whole numbers

EXAMPLE

Add $2\frac{3}{4} + 1\frac{3}{8} + 3\frac{1}{2}$

Analysis shows that 8 is the lowest common denominator (the smallest number all of the denominators will divide into evenly).

Begin by converting ¾ (of 2¾) into a fraction with a denominator of 8.

4 *Goes into* 8 2 times   2 x 3 = 6 Therefore, $\frac{3}{4} = \frac{6}{8}$

Since $3/8$ (of $1\,3/8$) is already in the correct form, we leave it as is.

½ (of 3½) converts as follows: $\frac{1}{2} = \frac{4}{8}$

2 *Goes into* 8 4 times   4 x 1 = 4 Therefore,

$2\frac{3}{4} = 2\frac{6}{8}$ $1\frac{3}{8} = 1\frac{3}{8}$ $3\frac{1}{2} = 3\frac{4}{8}$

Our conversions now appear as follows:

Add the Fractions:

$$\frac{6}{8} + \frac{3}{8} + \frac{4}{8} = \frac{13}{8}$$

Simplify

$$\frac{13}{8} = 1 + \frac{5}{8} = 1\frac{5}{8}$$

Add the whole numbers

$$2 + 1 + 3 + 1\frac{5}{8} = 7\frac{5}{8}$$

Since we had no label in our original problem, we need not label our answer.

EXERCISE: Add

$$2\frac{1}{8} + 4\frac{3}{4} + 3\frac{5}{32} + 4\frac{3}{16}$$

Show that final answer is

$$14\frac{7}{32}$$

SUBTRACTION OF MIXED NUMBERS

Subtraction of mixed numbers follows the same principles as addition of mixed numbers.

A common denominator must be established before subtraction of fractions or mixed numbers can be performed.

For example,

$$2\frac{3}{8} - 1\frac{1}{4} = ?$$

We must first arrange the fractions to have the same denominator. The lowest common denominator is 8. Following the steps above, our problem after conversion appears as:

$$2\frac{3}{8} - 1\frac{2}{8} = ?$$

Then, subtract the fractions,

$$\frac{3}{8} - \frac{2}{8} = \frac{1}{8}$$

Subtract the whole numbers,

$$2 - 1 = 1$$

Write the result and check work.

$$2\frac{3}{8} - 1\frac{1}{4} = 1\frac{1}{8}$$

<u>Another example</u>: Subtract $8^1/_3$ from $14^3/_5$

Written out, this problem appears as follows:

$$14\frac{3}{5} - 8\frac{1}{3} = ?$$

Plumbing Apprentice Student Workbook Year One
Fourth Edition

The lowest common denominator is 15.
Convert the fractions. $\dfrac{3}{5} = \dfrac{9}{15}$ $\dfrac{1}{3} = \dfrac{5}{15}$

Rewrite the problem

$$14\dfrac{9}{15}$$
$$-8\dfrac{5}{15}$$

Subtract the numerators and place the result over the common denominator.

$$14\dfrac{9}{15}$$
$$-8\dfrac{5}{15}$$
$$\dfrac{4}{15}$$

Subtract the whole numbers.

$$14\dfrac{9}{15}$$
$$-8\dfrac{5}{15}$$
$$6\dfrac{4}{15}$$

Another example (showing how to solve the problem if the second fraction is larger than the first):

Subtract $1\,^{7}/_{8}$ from $2\,^{5}/_{8}$. Write the problem, making sure the denominators are the same.

$$2\dfrac{5}{8}$$
$$-1\dfrac{7}{8}$$

Borrow one unit from the whole number. $2\,^{5}/_{8} = 1 + {}^{5}/_{8} + {}^{8}/_{8} = 1 + {}^{13}/_{8}$

Rewrite the problem and subtract the fraction.

$$1\dfrac{13}{8}$$
$$-1\dfrac{7}{8}$$
$$\dfrac{6}{8}$$

Subtract the whole numbers.

$$1\dfrac{13}{8}$$
$$-1\dfrac{7}{8}$$
$$0\dfrac{6}{8}$$

Since it is not necessary to write the zero, we would normally write this answer simply

$$2\frac{5}{8} - 1\frac{7}{8} = \frac{6}{8}$$

Since $^6/_8$ is not reduced to its lowest terms, we should reduce the fraction in order to have the best form of the answer.

Thus, our final answer is

$$2\frac{5}{8} - 1\frac{7}{8} = \frac{3}{4}$$

MULTIPLICATION OF FRACTIONS

To multiply fractions, follow these steps:

1. Multiply the numerators.
2. Multiply the denominators.
3. Write as a fraction and reduce if possible.

For example, $\frac{3}{4} \times \frac{5}{8} = ?$

Multiply the numerators. $3 \times 5 = 15$

Multiply the Denominators $4 \times 8 = 32$

Write as a fraction and reduce if possible. $\frac{15}{32}$

MULTIPLICATION OF MIXED NUMBERS

Multiplication of a mixed number and a fraction is similar to multiplication of two fractions.

For example, $3\frac{1}{2} \times \frac{3}{4} = ?$

Convert the mixed number to an improper fraction.

$$3\frac{1}{2} = ? \qquad 3 = \frac{3}{1} \qquad \frac{3}{1} = \frac{6}{2}$$

Thus, $3 + \frac{1}{2} = \frac{6}{2} + \frac{1}{2} = \frac{7}{2}$

Rewrite problem. $\frac{7}{2} \times \frac{3}{4}$

Multiply the numerators $7 \times 3 = 21$

Multiply the denominators $2 \times 4 = 8$

Rewrite the problem $\qquad \dfrac{7}{2} \times \dfrac{3}{4} = \dfrac{21}{8}$

Write as a fraction or mixed number and reduce if possible. $\qquad \dfrac{21}{8} = 2\dfrac{5}{8}$

MULTIPLICATION OF FRACTIONS, MIXED NUMBERS, AND WHOLE NUMBERS

Multiplication of fractions, mixed numbers, and whole numbers is performed like any fractional multiplication.

For example, $\qquad 2 \times 2\dfrac{1}{8} \times \dfrac{3}{4} =$

Change all mixed numbers to fractions $\qquad 2\dfrac{1}{8} = \dfrac{17}{8}$

Put all whole numbers over a denominator of 1. $\qquad 2 = \dfrac{2}{1}$

Rewrite the problem. $\qquad \dfrac{2}{1} \times \dfrac{17}{8} \times \dfrac{3}{4} =$

Multiply the numerators $\qquad \dfrac{2}{1} \times \dfrac{17}{8} \times \dfrac{3}{4} = \dfrac{102}{}$

Multiply the denominators $\qquad \dfrac{2}{1} \times \dfrac{17}{8} \times \dfrac{3}{4} = \dfrac{102}{32}$

Write as fraction or mixed number and reduce if possible

$$\dfrac{102}{32} = \dfrac{51}{16} = 3\dfrac{3}{16}$$

Thus, $\qquad 2 \times 2\dfrac{1}{8} \times \dfrac{3}{4} = 3\dfrac{3}{16}$

To simplify the multiplication of fractions, it is possible to use "cancellation" to reduce the high values of numbers.

For example, $\qquad \dfrac{3}{4} \times \dfrac{4}{5} \times \dfrac{15}{32} =$ $\qquad$ Cancel common multiples $\qquad \dfrac{3}{\underset{1}{\cancel{4}}} \times \dfrac{\cancel{4}}{\underset{1}{\cancel{5}}} \times \dfrac{\overset{3}{\cancel{15}}}{32} =$

Rewrite problem $\qquad \dfrac{3}{1} \times \dfrac{1}{1} \times \dfrac{3}{32} =$

Multiply the numerators $\qquad \dfrac{3}{1} \times \dfrac{1}{1} \times \dfrac{3}{32} = \dfrac{9}{}$

Multiply the denominators $\qquad \dfrac{3}{1} \times \dfrac{1}{1} \times \dfrac{3}{32} = \dfrac{9}{32}$

For the final result: $\qquad \dfrac{9}{32}$

DIVISION OF FRACTIONS

Division of fractions is performed by inverting the divisor and multiplying. The term that follows the division symbol is the divisor. To invert is to **turn upside down**. In other words, make the numerator the denominator and vice-versa.

EXAMPLE $\qquad 2\frac{3}{4} \div \frac{8}{7} = ?\qquad$ (the divisor is $8/7$)

Convert the mixed number to fractional form. $\qquad 2\frac{3}{4} = \frac{8}{4} + \frac{3}{4} = \frac{11}{4}$

Rewrite the problem $\qquad \frac{11}{4} \div \frac{8}{7} =$

Invert the divisor and multiply $\qquad \frac{11}{4} \times \frac{7}{8} =$

Multiply the numerators $\qquad \frac{11}{4} \times \frac{7}{8} = \frac{77}{?}$

Multiply the denominators $\qquad \frac{11}{4} \times \frac{7}{8} = \frac{77}{32}$

Write as a fraction or mixed number and reduce if possible

$$\frac{11}{4} \times \frac{7}{8} = \frac{77}{32} = 2\frac{13}{32}$$

EXERCISE

$$\frac{67}{8} \div \frac{13}{4} =$$

Show that the final answer is $\quad 2\frac{15}{26}$

DIVIDING A WHOLE NUMBER BY A FRACTION

To divide a whole number by a fraction, we make the whole number into a fraction by placing it over the number one (1) and then proceeding as described above.

For example, $\qquad 15 \div 1\frac{3}{5} = ?$

We can put the whole number in fraction form by simply adding a denominator of 1. This does not change the value of the number. Change all mixed numbers to fractions also.

Therefore, $\qquad 15 = \frac{15}{1} \qquad$ and $\qquad 1\frac{3}{5} = \frac{5}{5} + \frac{3}{5} = \frac{8}{5}$

Rewrite the problem $\qquad\qquad\qquad\qquad\dfrac{15}{1} \div \dfrac{8}{5} =$

Invert divisor and multiply numerators. $\quad \dfrac{15}{1} \times \dfrac{5}{8} = \qquad \dfrac{15}{1} \times \dfrac{5}{8} = \dfrac{75}{}$

Multiply denominators. $\qquad \dfrac{15}{1} \times \dfrac{5}{8} = \dfrac{75}{8}$

Write as a fraction or mixed number and reduce if possible.

$$\dfrac{75}{8} = \dfrac{72}{8} + \dfrac{3}{8} = 9\dfrac{3}{8}$$

Therefore, $\quad 15 \div 1\dfrac{3}{5} = 9\dfrac{3}{8}$

<u>This method also works if the divisor is a whole number.</u>

CONCEPTS

1. **Adding (or subtracting) fractions,** all fractions must be put in common denominator form before adding (or subtracting) the numerators.

2. When fractions must be made larger for subtraction, borrow a unit from the whole number and add it to the fractional amount.

3. **Multiplying fractions**, multiply the numerators, multiply the denominators, put in fraction form and reduce to lowest terms.

4. **Dividing fractions**, invert the divisor and multiply.

5. Before multiplying and/or dividing mixed numbers, change the mixed number into an improper fraction.

6. When multiplying or dividing by a whole number, put the whole number in fraction form by placing it over one (1).

7. Use the cancellation method whenever possible to simplify the multiplication of fractions.

LESSON 10

MATHEMATICAL OPERATIONS — DECIMALS AND FRACTIONS

We have considered numbers, fractions, mixed numbers, and decimals. In the previous lesson, we learned how to perform operations with fractions. Many problems appear that require operating with decimals or combinations of fractions and decimals.

One way to solve such problems is to convert the decimals to fractions and then perform the operations as described earlier.

To convert a decimal to a fraction, write the decimal as the numerator of the fraction. The denominator will be a **1** followed by as many zeros as there are digits to the right of the decimal point in the numerator. The fraction can then be reduced to lowest terms. The required operations can then be performed according to the methods of fractions.

For example, convert 0.8125 to a fraction.

Write the decimal as a numerator. **0.8125**

The denominator will be 1, followed by as many zeros as the number of places to the right of the decimal point in the numerator.

0.8125
1.0000

Move the decimal points to the right

8125.
10000.

Reduce to lowest terms. If you do not recognize the largest divisor (and who would see 625 hiding in 8125 and 10,000?), simplify by whatever common factors you do see, and keep simplifying until the problem is reduced as far as possible. Remember, you simplify by dividing the numerator and denominator by the same number (must divide evenly).

8125 *divides by* 25 = **325**   **325** *divided by* 25 = **13**
10000 *divides by* 25 = **400**   **400** *divided by* 25 = **16**

Thus, $0.8125 = \dfrac{13}{16}$

<u>Another example:</u> convert **10.625** to a mixed number in fraction form.

Write as a fraction.   **0.625**
                        1.000

Move the decimal points so that you are dealing with whole numbers.   **625**
                                                                       1000

Reduce to lowest terms.   **625** *divides by* 125 = **5**
                          1000 *divides by* 125 = **8**

Therefore, the mixed number in fraction form is $10\,\tfrac{5}{8}$.

ADDING DECIMAL-FRACTION COMBINATIONS

Addition of decimal-fraction combinations is accomplished by changing the decimal to a fraction, finding the common denominator, and adding as shown in the previous Lesson.

Add $\quad 3.175 + 14.625 + 2\frac{3}{8}$

Begin by setting up the decimal amounts (.175 and .625) as fractions. $\quad \frac{.175}{1.000} \quad \frac{.625}{1.000}$

Move the decimal points. $\quad \frac{175}{1000} \quad \frac{625}{1000}$

Reduce the fractions. $\quad \frac{175 \text{ divided by } 25}{1000 \text{ divided by } 25} = \frac{7}{40} \quad \frac{625 \text{ divided by } 125}{1000 \text{ divided by } 125} = \frac{5}{8}$

Rewrite the problem so that you know what numbers you are now dealing with:

$$3\frac{7}{40}$$
$$+14\frac{5}{8}$$
$$+2\frac{3}{8}$$

The fractions must now be stated in terms of a common denominator (40, in this case)

$$3\frac{7}{40} = 3\frac{7}{40}$$
$$14\frac{5}{8} = 14\frac{25}{40}$$
$$2\frac{3}{8} = 2\frac{15}{40}$$

Rewrite the problem for clarity

$$3\frac{7}{40}$$
$$+14\frac{25}{40}$$
$$+2\frac{15}{40}$$

Add the fractions.

$$3\frac{7}{40}$$
$$+14\frac{25}{40}$$
$$+2\frac{15}{40}$$
$$\overline{\quad\frac{47}{40}}$$

Add the whole numbers.

$$3\frac{7}{40}$$
$$+14\frac{25}{40}$$
$$+2\frac{15}{40}$$
$$\overline{19\frac{47}{40}}$$

This is a whole number with an improper fraction.

The fraction should be reduced to its lowest terms.

$$\frac{47}{40} = 1\frac{7}{40}$$

Complete the addition to finish.

$$19$$
$$+1\frac{7}{40}$$
$$\overline{20\frac{7}{40}}$$

EXAMPLE: Add

$$602.2315 + 211\frac{7}{16} + 4\frac{3}{32}$$

Set up the decimal amount (**.2315**) as a fraction $\quad \frac{.2315}{1.0000} \quad$ Move the decimal points $\quad \frac{2315}{10000}$

Reduce the fraction. $\quad \frac{2315 \text{ divides by } 5}{10000 \text{ divides by } 5} = \frac{463}{2000}$

Plumbing Apprentice Student Workbook Year One
Fourth Edition

Rewrite the problem so that you understand what numbers you are now dealing with.

$$602\text{-}463/2000 + 211\text{-}7/16 + 4\text{-}3/32 = ?$$

The fractions must now be stated in terms of a common denominator. Starting with 2000, divide this value by 16 and 32 to determine if the result is even. As the division by 32 is not even, double the 2000 and perform the same task. Since all values — 2000, 16, and 32 — divide equally into 4000, 4000 is the common denominator.

Thus, $602\frac{463}{2000} = 602\frac{926}{4000}$ Rewrite the problem for clarity $602\frac{926}{4000}$

$211\frac{7}{16} = 211\frac{1750}{4000}$ $+211\frac{1750}{4000}$

$4\frac{3}{32} = 4\frac{375}{4000}$ $+ 4\frac{375}{4000}$

Add the Fractions, $602\frac{926}{4000}$ Add the whole numbers. $602\frac{926}{4000}$

$+211\frac{1750}{4000}$ $+211\frac{1750}{4000}$

$+ 4\frac{375}{4000}$ $+ 4\frac{375}{4000}$

$\frac{3051}{4000}$ $817\frac{3051}{4000}$

This answer is in its lowest terms

SUBTRACTION OF DECIMAL-FRACTION COMBINATIONS

Subtraction of decimal-fraction combinations can also be accomplished by changing the decimal to a fraction, finding a common denominator and subtracting as described above for addition.

EXAMPLE   Subtract **0.175 from $^3/_4$**   Set up the problem $\frac{3}{4} - 0.175 =$

Convert the decimal to a fraction as shown above $\frac{3}{4} - \frac{175}{1000}$
and rewrite the problem for clarity.

Reduce the fraction. $^3/_4$ is in its lowest form but $^{175}/_{1000}$ must be $\frac{175 \text{ divided by } 25}{1000 \text{ divided by } 25} = \frac{7}{40}$

Rewrite the problem. $\frac{3}{4} - \frac{7}{40}$

Find a common denominator. In this case it is 40. 4 goes into 40 10 times $\frac{30}{40} - \frac{7}{40} = \frac{23}{40}$
so we multiply both the numerator and denominator of the fraction $^3/_4$ by 10.

Another example, subtract **23.272 from 68 $^7/_{16}$**.   Set up the problem $68\,^7/_{16} - 23.372$

Convert the decimal to a fraction,   or   Reduce the fraction to its lowest form

$\frac{.272}{1.000}$ $\frac{272}{1000}$ $\frac{272 \text{ divided by } 8}{1000 \text{ divided by } 8} = \frac{34}{125}$

Rewrite the problem

$$68 \frac{7}{16} - 23 \frac{34}{125}$$

Find a common denominator

$$16 \times 125 = 2000$$

Rewrite the problem

$$68 \frac{875}{2000}$$
$$- 23 \frac{544}{2000}$$
$$\frac{331}{2000}$$

Express the numerators in terms of the common denominator

$$16 \times 125 = 875$$
$$34 \times 16 = 544$$

Subtract the whole number

$$68 \frac{875}{2000}$$
$$- 23 \frac{544}{2000}$$
$$45 \frac{331}{2000}$$

Rewrite the problem

$$68 \frac{875}{2000}$$
$$- 23 \frac{544}{2000}$$

The answer is complete at this point since the fraction is in its lowest form.

MULTIPLICATION OF FRACTIONS AND DECIMALS

Multiply .025 by $^{16}/_{35}$.

Set up the problem and simplify

$$\frac{\cancel{25}^5}{\cancel{1000}_{125}} \times \frac{\cancel{16}^2}{\cancel{35}_7} =$$

and rewrite the problem.

Simplify

$$\frac{1}{25} \times \frac{2}{7} =$$

The answer is in lowest terms.

$$\frac{\cancel{5}^1}{\cancel{125}_{25}} \times \frac{2}{7} =$$

$$\frac{1}{25} \times \frac{2}{7} = \frac{2}{175}$$

EXAMPLE Multiply 1.25 by $^6/_{25}$.

Set up the problem.

$$1.25 \times \frac{6}{25} =$$

Convert to fraction form.

$$1.25 = 1 + \frac{25}{100} = \frac{100}{100} + \frac{25}{100} = \frac{125}{100}$$

Rewrite the problem

$$\frac{125}{100} \times \frac{6}{25} =$$

Simplify

$$\frac{\cancel{125}^{\cancel{5}^1}}{\cancel{100}_{\cancel{50}_{10}}} \times \frac{\cancel{6}^3}{\cancel{25}_1} =$$

Rewrite the problem for clarity.

$$\frac{1}{10} \times \frac{3}{1} =$$

Multiply.

$$\frac{1}{10} \times \frac{3}{1} = \frac{3}{10}$$

The result, $^3/_{10}$, is in its simplest form

DIVISION OF FRACTIONS AND DECIMALS

EXAMPLES   Divide 0.875 by $^1/_8$.

Set up the problem $\quad$ Write as a fraction $\quad$ Invert divisor and simplify

$$0.875 \div \frac{1}{8} = \qquad \frac{875}{1000} \div \frac{1}{8} = \qquad \frac{\cancel{875}^{7}}{\cancel{1000}_{125}} \times \frac{\cancel{8}^{1}}{1} =$$

Multiply out $\quad \frac{7}{1} \times \frac{1}{1} = \frac{7}{1}$

The answer is simply 7

Multiply **2.125** by **⁹/₁₆**. $\qquad$ Set up the problem. $\qquad 2.125 \times \frac{9}{16} =$

Convert the decimal to a fraction by starting to the right of the decimal point and simplify.

$$\frac{.125}{1.000} = \frac{125}{1000} \qquad \text{Rewrite the problem.} \qquad 2\frac{125}{1000} \times \frac{9}{16} =$$

Convert mixed number to an improper fraction. $\quad \frac{2}{1} + \frac{125}{1000} = \frac{2000}{1000} + \frac{125}{1000} = \frac{2125}{1000}$

Rewrite problem. $\quad \frac{2125}{1000} \times \frac{9}{16} =$

Invert the divisor and simplify

$$\frac{\cancel{2125}^{17}}{\cancel{1000}_{8}} \times \frac{16}{9} = \qquad \frac{17}{\cancel{8}_{1}} \times \frac{\cancel{16}^{2}}{9} =$$

Rewrite the problem and simplify again if possible.

$$\frac{17}{1} \times \frac{2}{9} = \frac{34}{9}$$

This is an improper fraction and is best converted to a mixed number, for the best answer. $\quad 3\frac{7}{9}$

<u>Note that when fractions are reduced to their lowest terms, you work with smaller numbers and the chances for error are also reduced.</u>

In the examples above, note that fractions containing large numbers are more manageable when reduced as soon as possible whenever they appear in your calculations.

CONCEPTS

To convert a decimal to a fraction observe the following rules:

1. The numerator is the number to the right of the decimal point.

2. The denominator is 1 plus the same number of zeros as there are places to the right of the decimal point.

3. Reduce the fraction to the lowest terms to minimize chance of error.

FRACTIONS TO DECIMALS

We have demonstrated how to convert decimals to fractions in order to perform operations in problems where some numbers are fractions and some are decimals. In most cases, such operations can be more easily done if we express all numbers as decimals.

Use a hand-held calculators to make these operations easier whenever you encounter them.

CONVERTING A FRACTION TO A DECIMAL

To convert a fraction to a decimal, simply divide the numerator by the denominator.

For example, convert $7/15$ to decimal form. Set up the problem. 15)7

Place the decimal point after the numerator, 15)7. then in the quotient above 15)7.

Add sufficient zeros after the decimal point in the numerator for the desired quotient range. 15)7.000

Perform the division and check work

```
      .466
15)7.000
   6 0
   1 00
     90
    100
```

Therefore, $7/15$ = 0.4666. Note that this number will never come out even and that the 6 will continue to repeat. Therefore, the answer would be 0.467 (to three places).

Another example,

Convert $22/7$ to decimal form. Carry out to three decimal places.

Set up the problem. 7)22 Place the decimal point after the numerator. 7)22. and in the quotient. 7)22.

Add sufficient zeros after the decimal point in the numerator for the desired quotient range 7)22.0000

```
    3.1428
7)22.0000
  21
   10
    7
   30
   28
    20
    14
```

Since the instruction was to carry out the operation to three decimals, the calculation was extended to the fourth place to see if it is necessary to "round up" or "round down" in the third place.

Since the last digit is an "8," the required answer is: $22/7$ = 3.143

ADDITION

Addition is done by expressing all numbers in a problem in decimal form, keeping the decimal points lined up, and adding.

EXAMPLE: Add $5/8$, 6.125, and 14.175

First, convert the fraction to a decimal. Set up the problem. 8)5

Add a decimal point after the numerator.	Place the decimal point in the quotient.	Add sufficient zeros after the decimal point in the numerator for the desired quotient range.

8)5̄.	8)5̇.̄ 8)5̇.̄0̄0̄0̄	.625
		8)5.000
		   48
		     20
		     16
		       40
		       40

Rewrite your problem	Add the numbers	Bring down the decimal
$\frac{5}{8} = 0.625$ + 6.125 = + 6.125 +14.175 = + 14.175	0.625 6.125 +14.175 20 925	0.625 6.125 +14.175 20.925

SUBTRACTION

Subtraction is handled similarly to addition.

EXAMPLE: subtract **1.025 from 24 3/4**   Set up the problem. $24\frac{3}{4} - 1.025 =$

Convert 24 3/4 to a decimal form.

    .75
4)3.00
  2 8
    20
    20

Therefore, $\frac{3}{4} = .75$

Rewrite the problem

24.75
- 1.025

Subtract and check.

24.750
- 1.025
23.725

EXERCISE: Subtract 8.125 from 25 15/16. Express in decimal form. show that the answer is 17.8125.

MULTIPLICATION

Multiplication of fractions and decimals is performed after the conversion of fractions to decimals.

<u>For example</u>, multiply **5/16 by 8.125.**   Set up the problem $\frac{5}{16} \times 8.125 =$

Convert the fraction.   **5 *divided by* 16 = 0.3125**

Perform the operation

```
    0.3125      4 places to the right of the decimal
  x 8.125      + 3 places to the right of the decimal
   15625
    6250
    3125
   25000
  2.5390625    7 places to the right of the decimal
```

EXERCISE: Multiply $5/8$ by 6.35.   Verify that the correct answer is **3.96875**

DIVISION

As in the other operations, division is performed after converting the fractions to decimals.

EXAMPLE: Divide **15 $3/4$** by **0.125**.   Convert $3/4$ to a decimal   **3 divided by 4 = 0.75**

Set up the problem.   Move the decimal point in the divisor.   Move the decimal point in the dividend and locate in the quotient.

$0.125 \overline{)15.75}$ $125. \overline{)15\ 75}$ $125. \overline{)15\ 750.}$

Carry out the operation and check.

```
          126.
    125)15 750.
        12 5
         3 25
         2 50
           750
           750
```

EXERCISE: Divide **8.125** by **2 $7/16$**. Express in decimal form.
(The correct answer is 3.33...)

COMMENTS ON FRACTION AND DECIMAL CALCULATIONS

Examine problems like these to see whether it is easier to be solved by converting fractions to decimals, or vice versa. Either way, you won't go wrong, it just might be easier to perform with some sets of numbers as fractions, and with some others as decimals.

With metric measurements, you are more likely to see decimal numbers than fractions. It is important, therefore, that you develop skill in the use of decimals and the conversion of fractions to decimals so that you can be the best possible worker in this field, anywhere in the world!

LESSON 11

MEASURING TAPES, FOLDING RULES, AND SCALE RULERS

FOLDING RULES

Plumbers must make many measurements in the course of their work. The most often used length-measuring device used to be the folding rule (also called spring-joint or zig-zag rule). Today most mechanics use steel tape rulers. Folding rules are usually six or eight feet long when fully extended. These rules are marked on both sides, as described below.

The folding rule is accurate, inexpensive, and compact. It requires little maintenance, and, unlike tapes, is stiff when extended. As with all tools, you must follow certain procedures to ensure accuracy.

High-quality rules have the marks *scratched*, or *etched*, into the surface of the rule. In that way, the mark is still visible even if the paint marking is rubbed off.

STEEL TAPES

Steel tapes are available in convenient cases in various lengths and widths. Probably the most popular is 25' long, 1" wide. These are marked off in the same units as folding rules.

Steel tapes are made with a fairly deep curve across the steel ribbon which helps to keep them somewhat stiff when extended.

READING MEASURMENTS

Look at a six foot ruler or steel tape. Note that the measurements are divided using a series of calibrated lines.

1. Examine the zero-to-one inch increment of the ruler closely.

2. There are fifteen lines between zero and 1, 17 including zero and 1. Therefore, the space between each pair of lines equals one-sixteenth of an inch.

3. Two of these spaces equal two-sixteenths, or one-eighth of an inch.

4. Four spaces equal four-sixteenths (one quarter) inch, and so on.

5. The small line markings appear progressively larger at the eighth, quarter and half inch marks respectively.

When lengths are measured using a ruler, they will usually be a mixed number.

Since these values are in fraction or mixed number form, all operations which can be performed using fractions and mixed numbers can be used with these values.

CARE AND MAINTENANCE

Folding rulers should have the joints lightly lubricated every six months or so. Steel tapes do not need maintenance, except to be kept clean. Steel tapes are designed to minimize the risk of being kinked, but if they are kinked, they probably will have to be replaced.

EXAMPLE

Three pieces of pipe measured $2\frac{1}{2}$", $1\frac{3}{4}$", and $8\frac{1}{8}$" in length. What is the total length of these pieces?

Set up the problem

$$2\frac{1"}{2}$$
$$+1\frac{3"}{4}$$
$$+8\frac{1"}{8}$$

Express fractions with a common denominator.

$$2\frac{4"}{8}$$
$$+1\frac{6"}{8}$$
$$+8\frac{1"}{8}$$

Add the fractions

$$2\frac{4"}{8}$$
$$+1\frac{6"}{8}$$
$$+8\frac{1"}{8}$$
$$\frac{11"}{8}$$

Simplify the fraction

$$\frac{11"}{8} = 1\frac{3"}{8}$$

Add the whole numbers

$$2"$$
$$+\ 1"$$
$$+\ 8"$$
$$+\ 1\frac{3"}{8}$$
$$12\frac{3"}{8}$$

Lengths greater than 12" often will be expressed in feet and inches.

Therefore, 12-3/8" = 1' 0-3/8"

EXAMPLE

What is 182 1/2" in terms of feet?

Divide by 12 (12" in 1')

$$\begin{array}{r} 15 \\ 12\overline{)182\tfrac{1}{2}} \\ \underline{12} \\ 62 \\ \underline{60} \\ 2\tfrac{1}{2} \end{array}$$

ANSWER

(To avoid confusion, be sure to label the answer.)

$$182\frac{1"}{2} = 15'\ 2\frac{1"}{2}$$

COMMENT

It is difficult to provide a definite rule as to whether you should use feet and inches or keep all measurements in inches. Either way can be *correct* — it is a matter of reducing the chance for error. Remember that every time you perform an operation, there is a chance for error. Therefore, use the method that requires the fewest operations. Usually, this will also be the quickest!

Note that metric measurements will largely remove this issue — length measurements will be in meters and millimeters, and the millimeter is simply a decimal value of meters!

EXAMPLE

Lay out three plumbing fixtures according to the following spacing: Side wall to center of first: $32\frac{1}{2}"$, to center of second: 41", to partition: $19\frac{1}{4}"$, thickness of partition: $4\frac{1}{4}"$, to center of third: $16\frac{1}{4}"$, and to face of opposite sidewall: $26\frac{5}{8}"$. What total length is required for this layout?

As there are fewer operations required to add the values as given, we will not convert each one to feet and inches before adding.

Set up the problem

$$32\frac{1}{2}"$$
$$+\ 41"$$
$$+\ 19\frac{1}{4}"$$
$$+\ 4\frac{1}{4}"$$
$$+\ 16\frac{1}{4}"$$
$$+\ 26\frac{5}{8}"$$

Find the common denominator for the fractions and add

$$32\frac{4}{8}"$$
$$+\ 41"$$
$$+\ 19\frac{2}{8}"$$
$$+\ 4\frac{2}{8}"$$
$$+\ 16\frac{2}{8}"$$
$$+\ 26\frac{5}{8}"$$
$$\frac{15"}{8}$$

Simplify

$$\frac{15"}{8} = 1\frac{7}{8}"$$

Add the whole numbers: $32" + 41" + 19" + 4" + 16" + 26" = 138"$

Add the fractions of inches: $138" + 15/8 = 138" + 1\frac{7}{8}" = 139\frac{7}{8}"$

Convert to feet and inches by dividing by 12. $139\frac{7}{8}" = 11'\ 7\frac{7}{8}"$

EXAMPLE

Express the fractions with a common denominator and add the fractions.
A building has three rooms across the front. Each outside wall is 15", partition $6\text{-}\frac{1}{4}"$, room widths of $15'\ 8\text{-}\frac{1}{4}"$, $22'\ 7\text{-}\frac{1}{2}"$, and $9'\ 10"$.

What is the total building width?
Set up the problem for fewest operations.

$$\begin{aligned}&0'\ 15"\\ +\ &15'\ 8\frac{1}{4}"\\ +\ &0'\ 6\frac{1}{4}"\\ +\ &22'\ 7\frac{2}{4}"\\ +\ &0'\ 6\frac{1}{4}"\\ +\ &9'\ 10"\\ +\ &0'\ 15"\\ &\frac{5"}{4}\end{aligned}$$

$$\begin{aligned}&0'\ 15"\\ +\ &15'\ 8"\\ +\ &0'\ 6"\\ +\ &22'\ 7"\\ +\ &0'\ 6"\\ +\ &9'\ 10"\\ +\ &15"\\ +\ &1\frac{1}{4}"\\ \hline &68\frac{1}{4}"\end{aligned}$$

Convert the sum-of-inches to feet-and-inches.

$68\frac{1}{4}" = 5'8\frac{1}{4}"$

Rewrite the problem and add:

$0' + 15' + 0' + 22' + 0' + 9' + 0' + 5' 8 - \frac{1}{4}"$

$= 51' 8 - \frac{1}{4}" = $ Width of Building

SCALE RULERS AND GAUGES

SCALE RULERS

Scale rulers are measuring devices calibrated so that you can measure a length on a drawing (blueprint) and read the full-sized dimension directly on the scale. The use of such scale rulers reduces the chance for error. Scale rulers are available in lengths of 6", 12", 18", and 24". Other lengths are available for special purposes.

Scale rulers are available for many specialties — Architects scale, Mechanical Engineers scale, Civil Engineers scale (decimal), and the Metric scale. The most commonly used of these in construction is the Architects scale. Rulers are flat or triangular in shape (see Figure 11-A) with different scales on each edge of each face.

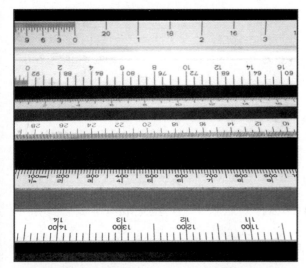

Figure 11-A – Engineers and Architects Scale

Some of the scale markings on the Architects scale are as follows:

$\frac{1}{16}" = 1'\text{-}0"$, $3/32" = 1'\text{-}0"$, $\frac{1}{8}" = 1'0"$, $\frac{3}{16}" = 1'\text{-}0"$, $\frac{1}{4}" = 1'0"$, $\frac{3}{8}" = 1'\text{-}0"$, $\frac{1}{2}" = 1'\text{-}0"$, etc.

By making a drawing to scale, the designer makes a diagram of the item being represented (a tool, a building, the cross-section of a wall, etc.) at less than full-size. In this way, economical drawings can be prepared for a project and most job problems discovered and resolved before the actual work is started.

Most buildings are depicted on drawings to a scale of $\frac{1}{4}" = 1'0"$ or $\frac{1}{8}" = 1'0"$. It is common practice to show details of a job to larger scale, often $\frac{3}{4}" = 1'0"$ or $1" = 1'0"$.

In order to produce the architect's or engineer's design, the mechanic must measure the drawings and convert to full size. Using the scale ruler is the easiest way to do this. This is also the way most jobs are **taken off** project drawings. You can also **scale** drawings with rulers calibrated full size (for example, your six-foot folding rule). If the scale is $\frac{1}{4}" = 1'0"$, every inch you measure on the drawing represents 4'. If the scale is $\frac{1}{8}" = 1'0"$, every inch you measure represents 8'.

EXAMPLE

What is the actual size of a building that measures 20"x46" on the drawing if the scale is $\frac{1}{8}" = 1'0"$?

20" x 8 = 160', and 46" x 8 = 368'

Therefore, The building is 160' x 368'.

EXAMPLE

A pipe line on a drawing measures 2½"long. The drawing is marked 1/4" = 1'0".

How long is the pipe?

$2\frac{1}{2}" \div \frac{1}{4} = 2\frac{1}{2} \times \frac{4}{1} =$

$\frac{5}{2} \times \frac{4}{1} = \frac{20}{2} = \frac{20}{2} = 10$

The pipe is 10' long.

EXERCISE

Scale is $1/4$" = 1'0". A building measures 36" x $50\frac{1}{2}$" on the drawing. What are the dimensions of the building? (You should get 144' x 202'.)

REVIEW

In the discussion above, we saw that scale rulers are used to produce drawings that are proportionately reduced in size from the object being represented in the drawing. We also saw that the plumber most often will use the $1/4$" = 1'0" or $1/8$" = 1'0" scales.

The designer of a project will mark the drawing with a note such as:

SCALE $\frac{1"}{4}$ = 1'0" OR SCALE $\frac{1"}{8}$ = 1'0"

Using this information, the installer will select the proper scale on the scale ruler so that all readings will be correct, full-sized values.

If a scale ruler is not available, you can still determine the length of a measurement taken from the drawing — simply divide the actual drawing measurement in inches by the scale of the drawing (i.e., multiply by the scale inverted). For the two common scales mentioned, this simply means multiplying by 4 or 8.

PROBLEMS

SCALE $1/4$" = 1'0"   Drawing measurements are: Find the full-sized lengths.

$6\frac{1"}{4}$ = _____

$2\frac{3"}{4}$ = _____

$41\frac{3"}{4}$ = _____

SCALE $1/8$" = 1'0"

$3\frac{3"}{8}$ = _____

$\frac{5"}{8}$ = _____

$2\frac{3"}{8}$ = _____

SCALE ¹/₄" = 1'0"
Full-sized lengths are

3'

10'

1'

Find the length on the drawing to represent these lengths:

_____

_____

_____

GAUGES

Gauges are measuring tools. You can connect a gauge to the item being tested. A gauge uses a particular means (dial face, mercury column, water column, etc.) to give us a reading that indicates the condition being observed. Thus, a thermometer usually has a mercury column which shows temperature; a pressure gauge usually has a dial face with a moving needle to show pressure. This general class of gauges are mechanical analog types, because a needle deflection, a change in a mercury column, or the level of water in a manometer are produced by direct mechanical means, and they must be observed and interpreted to mean a certain numerical value.

Other gauge types are known as digital, where the condition being measured is shown in a numerical readout on the instrument. Most such gauges can also be referred to as electronic because it almost always requires an electronic network to produce the digital reading.

Thermometers, pressure gauges, and flow meters are examples of gauges commonly used in plumbing. The face of the gauge should show the units of the calibration (pounds per square inch, degrees F, pounds per hour, etc.). For the analog types, the major subdivisions of the calibration should be marked also so that the indication can be read quickly and accurately. Digital types can be read more quickly (and they often respond more quickly to change in the observed condition) than the analog versions.

Gauges are used to observe the operating condition of the things we work with. We usually have a value in mind that we expect to observe, and we check with the gauge to see how close to that norm the system is performing.

The principle use of a gauge, therefore, is to indicate whether the reading is in a normal or an abnormal range. Should the reading be in the abnormal range, we must take appropriate action. The gauge we use, therefore, must be accurate, applied properly, and read correctly.

EXAMPLES

1. We want to limit hot water temperature delivered to the building plumbing fixtures to 130°F. We read a thermometer in the hot water line to determine whether or not we have to take some action (relight pilot, change thermostat, search for leak, or leave everything alone).

2. If we know that a sewage pump must develop a discharge pressure of 16 psi to lift the sewage to the piping leaving the building, a pressure gauge at the pump outlet will indicate whether or not we have to take some action (clean strainer, repair check valve, repair pump, etc.).

It follows then, that it is important that the gauges we use are accurate, that they are applied properly, and that they are read correctly.

As your lessons continue, we will continue the study of gauges and their applications.

Temperature Products
5" Industrial Glass Tube Thermometers

DESCRIPTION

The U.S. Gauge 5" industrial thermometer is a small, compact instrument of sturdy steel construction. It combines the advantages of easy readability, small size, and adaptability to a variety of installations requiring an economical thermometer. Careful consideration must be given to the material of the bulb shield. Failure of the bulb shield by corrosion will allow the medium being measured to escape from its container.

Model T-571 is a lower mount (LM) design with a steel bulb shield, while Model D-57B is lower mount with a brass bulb shield. Model T-572 is a lower back mount (LBM) design and has a steel bulb shield; Model D-57B has a brass bulb shield.

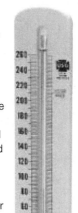

SPECIFICATIONS

SCALE LENGTH: 5"

CASE: Drawn steel case shell, finished in white baked enamel; case fits directly on the metal bulb shield forming an integral assembly

INDICATION: Magnifying glass tube filled with red liquid; black temperature scale printed on white case shell

BULB SHIELD: Steel zinc plated or brass

T-571, Low connection only, scale may be oriented by rotation of the bulb within its shield; insertion length is 1-3/8"

CONNECTION:
T-571: 1/2-14 NPT (LM); T-572: 1/2-14 NPT (LBM)

SPEC NUMBER SELECTION CHART

5" Glass Tube Thermometers – Plain Case

Model No.	Bulb Shield	Connection	Spec No.
T-571	Steel	1/2-14 NPT LM	5640
D-57B	Brass	1/2-14 NPT LM	38829
T-572	Steel	1/2-14 NPT LBM	19154
D-52B	Brass	1/2-14 NPT LBM	38830

© 2002, by AMETEK, Inc. All rights reserved. 15M1102A (160126) Specifications are subject to change without notice. Visit our Web sites for the most up-to-date information.

AMETEK®

ISO 9001 REGISTERED MANUFACTURER

For Gauges/Thermometers:
U.S. GAUGE
820 Pennsylvania Blvd.
Feasterville, PA 19053 U.S.A.
Tel: (215) 355-6900
Fax: (215) 354-1802
www.ametekusg.com
Customer Service Tel: (863) 534-1504
Customer Service Fax: (863) 533-7465

For Electronic Products:
PMT PRODUCTS
820 Pennsylvania Blvd.
Feasterville, PA 19053 U.S.A.
Tel: (215) 355-6900
Fax: (215) 354-1800
www.ametekusg.com

For Diaphragm Seals:
M&G PRODUCTS
8600 Somerset Drive
Largo, FL 33773 U.S.A.
Tel: (727) 536-7831
Fax: (727) 539-6882
www.ametek.com/tci

Figure 11-B – AMETEK US Gauge 5" Industrial Glass Tube Thermometers

General Purpose Gauges
Series P-500 Low Cost Utility Gauges

DESCRIPTION

U.S. Gauge Series P-500 utility gauges provide economical, reliable service in a wide variety of applications including pumps, compressors, and other equipment.

Series P-500 utility gauges come in English and metric ranges with accuracy of either ±3-2-3% or ±1.6% full scale.

Gauges are available in 1-1/2", 2", 2-1/2", 3-1/2", and 4-1/2" steel cases with a choice of English or metric connections in center-back and lower mount configurations.

SPECIFICATIONS

RANGES: 30" Hg VAC through 0-5000 psi

ACCURACY: ±3-2-3% (Grade B) or ±1.6% of span

BOURDON TUBE: Phosphor bronze

DIAL: ABS (aluminum optional)

POINTER: Aluminum

CASE: Painted steel

CONNECTION:
Brass, 1/8-27 NPT for 1-1/2" and 2"; 1/4-18 NPT for 2" through 4-1/2"; G1/8B, G1/4B, R1/8, R1/4, center-back or lower mount

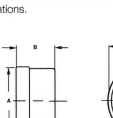 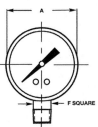 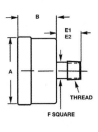 

INCHES (MM)

MODEL NUMBER	DIAL SIZE	UNIT	A	B	C	E1	E2	F	L
P-500	1-1/2"	inches	1.64	.92	.34	.62	–	.44	.55
		mm	41.66	23.37	8.66	15.75	–	11.18	13.97
P-500	2"	inches	2.16	1.15	.40	.84	.90	.55	.87
		mm	54.86	29.21	10.16	21.34	22.86	13.97	22.10
P-505	2-1/2" LM	inches	2.73	1.15	.39	.48	.25	.82	–
		mm	69.34	29.21	9.91	12.19	6.35	20.83	–
P-505	2-1/2" CBM	inches	2.49	1.21	1.22	–	.28	.84	–
		mm	63.25	30.73	30.99	–	7.11	21.34	–
P-505	3-1/2" LM	inches	3.53	1.17	.40	–	.24	.83	–
		mm	89.66	29.72	10.16	–	6.10	21.08	–
P-505	4-1/2" LM	inches	4.80	1.12	.38	–	.45	1.10	–
		mm	121.92	28.45	9.65	–	11.43	27.94	–

* Length E1 refers to 1/8 NPT connections; E2 refers to 1/4 NPT connections
** The E dimension is the minimum length of the wrench square from the case bottom

© 2002, by AMETEK, Inc. All rights reserved. 15M1102A (160126) Specifications are subject to change without notice. Visit our Web sites for the most up-to-date information.

ISO 9001 REGISTERED MANUFACTURER

For Gauges/Thermometers:
U.S. GAUGE
820 Pennsylvania Blvd.
Feasterville, PA 19053 U.S.A.
Tel: (215) 355-6900
Fax: (215) 354-1802
www.ametekusg.com
Customer Service Tel: (863) 534-1504
Customer Service Fax: (863) 533-7465

For Electronic Products:
PMT PRODUCTS
820 Pennsylvania Blvd.
Feasterville, PA 19053 U.S.A.
Tel: (215) 355-6900
Fax: (215) 354-1800
www.ametekusg.com

For Diaphragm Seals:
M&G PRODUCTS
8600 Somerset Drive
Largo, FL 33773 U.S.A.
Tel: (727) 536-7831
Fax: (727) 539-6882
www.ametek.com/tci

Figure 11-C – Series P-500 Low Cost Utility Gauge Ametek U. S. Gauge

LESSON 12

GOALS OF PLUMBING, WATER SOURCES, WASTE DISPOSAL

Plumbing provides a system of piping and equipment installations which deliver potable water (that is, water safe to drink for human beings) to plumbing fixtures, and removes wastes from these fixtures. Plumbing must keep the incoming water safe and uncontaminated, and it must remove wastes in a safe and sanitary way to maintain the health of all of us.

The plumbing industry also includes installations such as sewage plants and water supply treatment plants. These facilities must be installed and operated properly to provide us with these needed services, as well as to minimize the impact of our activities on the environment.

The skilled mechanic in this field needs to be aware of the hazards that must be avoided to maintain the integrity of plumbing systems.

WATER SUPPLY HAZARDS

Water supply hazards threaten the safety of the potable water. All these hazards have a common element:

In some way, there is the possibility that an unacceptable material may be introduced into the water supply system.

CROSS - CONNECTION

In nearly all cases of contamination, the unacceptable material moves into the potable water system through a *cross-connection*.

A cross-connection is a physical connection or arrangement between two separate piping systems. Cross-connections can occur in various ways. In your training to become a plumber, you must learn how to recognize the many forms of possible cross-connections.

Entry of any contaminant into the potable water system is called backflow. Backflow is usually divided into three modes:

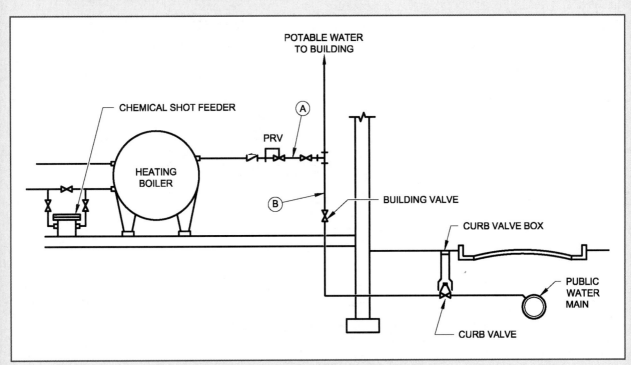

Figure 12-A – Backflow Caused by Back-Pressure

GRAVITY BACKFLOW

Gravity backflow contamination occurs when an open potable water supply pipe receives uncertain fluids from a source that is above the potable line: for example an open water line in a trench in the ground is contaminated by trench water.

BACK PRESSURE BACKFLOW (Figure 12-A)

Back pressure backflow occurs through a cross-connection when the pressure in the contaminated system is greater than the pressure in the potable system (which is above atmospheric pressure), thereby forcing the contaminated fluid into the potable fluid piping.

BACK - SIPHONAGE BACKFLOW (Figure 12-B)

Back-siphonage backflow occurs through a cross-connection when a vacuum is formed in the potable system which draws contaminated fluids from the questionable source, which is at atmospheric pressure. Gravity, backpressure, or back-siphonage backflow usually occurs in some unusual or emergency circumstance. The plumber must learn how to recognize the cross-connection arrangement in many disguises and to realize the proper protection procedure for each form.

OTHER HAZARDS

A pollution source may also be present within the potable water system because of soil or other fouling material in a new system, through contaminants which may enter an elevated gravity tank, or the deterioration of inadequate system materials (tank lining, gaskets, thread sealant, etc.). Standards from the National Sanitation Foundation address testing requirements for direct and indirect additives to prevent the contamination of the water supply from pipes, valves, fittings, tanks, and other devices.

PRINCIPAL WASTE HAZARDS

Three major hazards are frequently encountered in waste systems:

1. Improper drainage as the result of drain and waste line leakage; or lines installed without proper slope, use of regular fittings rather than drainage types (which are designed to develop streamline flow and to minimize stoppages), undersized lines, or lines with defective joints which may leak, hold solids, and/or develop frequent stoppages.

2. Sewer gas — **the mixture of vapors, odors, and gases found in a sewer** — can be hazardous and is generally offensive.

3. Waste products backing up in plumbing fixtures — as a result of pressurized sewers or from other waste piping elements — introduce serious hazards into the building or other structural entity.

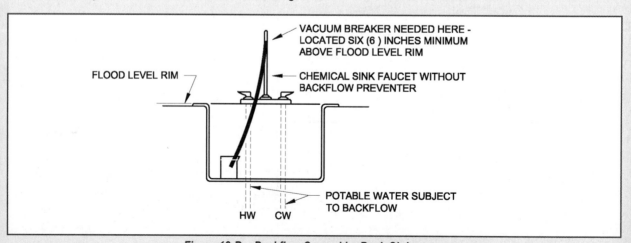

Figure 12-B – Backflow Caused by Back-Siphonage

In addition to using quality materials in the proper way for a long-lasting installation, the plumbing system must be designed to operate properly. Proper operation means that waste materials are conveyed away as they are generated and that noxious gases are kept contained in the waste system, not released inside the building.

There are many types of drainage and venting materials available to the plumber. Knowledge of the techniques of workmanship for each type is the **stock-in-trade** of the skilled mechanic.

Waste systems also require maintenance to keep the systems operating properly.

During the late 1800's, a system of vent piping was developed which, while open to the outdoors, could keep the sewer gases out of the building. The device that makes this possible is the waste trap. A major effort in the plumbing industry over the years has been to determine the minimum requirements for satisfactory venting of drainage systems, but no improvement on the concept of the waste trap has yet appeared.

Regardless of material used in the waste system, the piping must be adequately sized for the drainage loads to be served; be sloped downward in the direction of flow; be fitted with proper drainage fittings; be equipped with traps at every fixture connected to the waste system (to assure that noxious gases are not released inside the building); and, be connected properly to vent systems to insure the integrity of trap seals.

FIXTURE TRAPS

The fixture trap allows waste materials to pass readily into the drainage system while maintaining a seal so that gases in the waste system are not released into the building. The trap is a **U** shaped fitting which contains a water seal to prevent the passage of air or gas through a pipe without materially affecting the flow of sewage or liquid waste. The trap is effective only when a vertical **seal** of water exists between the crown weir and the top of the dip of the trap. If the trap seal is lost for any reason, the trap is ineffective. Figure 12-C shows the components of a common fixture P-trap.

In Figure 12-C, (a) shows the water in the trap at rest; (b) demonstrates the reversal of water in the trap if negative pressure occurs in the drainage system; (c) shows the spillover as a result of the momentum if the negative pressure is relieved quickly.

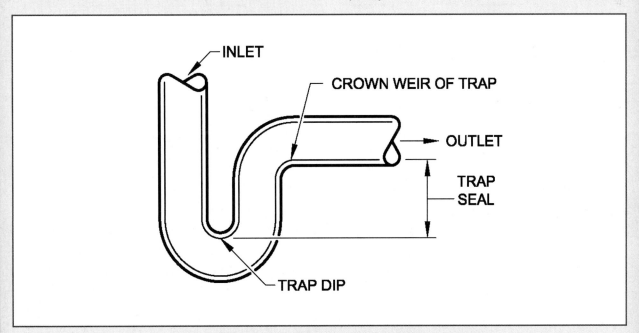

Figure 12-C-a – Elements of a Fixture Trap

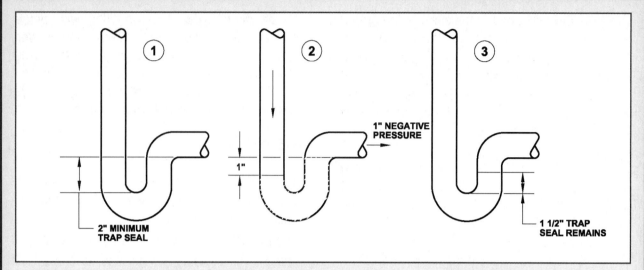

Figure 12-C-b – Trap Seal Reduction from 1" Negative Pressure

NOTES:
(1) Trap at rest with 2" trap seal.
(2) Trap subjected to 1" suction from building drainage piping.
(3) Trap at rest with 1/2" loss of trap seal. Trap will continue to spillover and lose trap seal when subjected to 1" suction until the trap seal is reduced to 1". The 2" initial trap seal permits the trap to withstand 1" suction and still maintain a trap seal of at least 1".

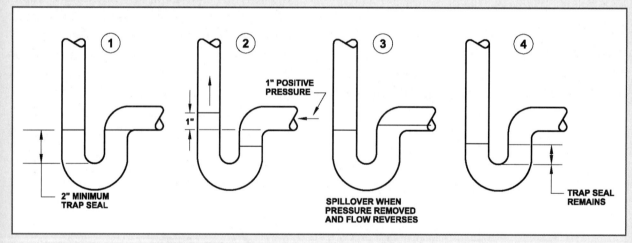

Figure 12-C-c – Trap Seal Reduction from 1" Positive Pressure

(1) Trap at rest with 2" trap seal.
(2) Trap subjected to 1" positive pressure from building drainage piping.
(3) When pressure is removed, some spillover occurs from the momentum of the trap legs equalizing.
(4) Trap will continue to spillover and lose trap seal when subjected to 1" positive pressure until the trap seal is reduced to 1". The 2" initial trap seal permits the trap to withstand 1" positive pressure and still maintain a trap seal of at least 1".

Traps for kitchen sinks, lavatories, laundry tubs, and similar fixtures are designed for a 2" water seal for normal operation. The conventional industry allowance for maximum pressure variations in the waste system is plus or minus 1" of water column. A 1" increase in the waste piping system will only cause the water in the trap seal to rise in the fixture tailpiece, so the water is not lost except by oscillation (see Fig. 12-C, [b & c]). A 1" pressure reduction in the waste piping system could pull 1" of the seal water out of the trap and down the drain.

Thus, the worst case seal loss after a major negative drainage event that produced this -1" water column variation would be 1", so a 1" seal would still be present in the trap. It should be realized that the actual occurrence of this maximum pressure variation is quite rare, but if it should occur, the plumbing traps would still keep out the noxious sewer gases.

Trap seals can be lost in many ways. The acronym SAMOBECC has been coined to aid in remembering them:

Siphonage
Aspiration
Momentum
Oscillation
Back-pressure
Evaporation
Capillary action
Cracked pipe

The details for these modes of seal loss are as follows:

Siphonage is the loss of the trap seal from negative pressure at the trap outlet. It may be induced by water flowing through the trap itself (if it is improperly vented), or by the flow in a waste pipe from other fixtures on the line.

Aspiration is also the loss of the seal from negative pressure at the trap outlet, but the source of the negative pressure is high velocity flow through a fitting, and is due to the geometry of the fitting. A vent on the trap outlet will prevent this occurrence. Reducing the flow through the fitting, or increasing the size of the fitting, will also prevent aspiration.

Momentum is the loss of trap seal by developing a large, high speed flow through a trap and trap arm. The inertia of the moving water maintains the flow long enough to pull the water from the trap before the flow stops. To minimize the chance for this event, codes limit the vertical distance from the fixture outlet to the trap, usually about 24" maximum, and also limit clothes washer standpipes to 48" maximum.

Oscillation is the loss of trap seal by the water in the trap moving back and forth rapidly enough to cause the seal water to run over the crown weir. The process starts as a result of cycling high and low pressures at the trap outlet.

Back-pressure leads to trap seal loss by aiding the above processes. High flow rates approaching the trap outlet produce an increase in pressure, which makes the seal rise on the inlet side of the trap. Then, when the wave passes, a negative pressure develops and the seal water reverses, beginning the oscillation. Also, if the back-pressure is high enough, stack gases will be forced into the room through the seal, splashing seal water into the space.

Capillary action is the loss of the seal due to debris laying in the trap over the crown weir. Water is wicked over the weir where it can gradually be dripped away.

Cracked pipes permit the loss of seal through a broken or failed pipe or fitting. This problem will show up as a leak below the fixture, and can only be remedied by repairing the leak.

FIXTURE AND EQUIPMENT HAZARDS

Potential hazards from fixtures and equipment include burns from hot pipes; hot-water scalding at fixtures; drowning in spas and hot tubs due to hair and body entrapment; equipment damage or explosion from pressure increase of heated water; cross-connections within the fixtures themselves, and exposure to noxious or explosive gases from system malfunctions.

Plumbing codes have minimum requirements for potable water pipe materials and installations, cross-connection protection, drainage and vent materials and installations, sizing methods, fixture and equipment installations, and maintenance requirements. Adhering to the code provisions assures sufficient, safe water supply, and adequate drain and vent systems. Licensing the plumber helps to enforce the basic rule that code provisions must be followed.

Table 12-A lists some of these problems with possible remedies.

THE PLUMBING PROFESSIONAL

The total water supply system includes both water supply plants that prepare and deliver potable water to our buildings or other points of use, and distribution system within the buildings (or other types of facilities).

The total waste system includes the waste system within the building and the connections to the (usually) public sewage disposal plants that process our wastes to render them harmless.

In addition, the plumbing engineer and plumbing installer are obligated to minimize energy required to operate our systems, and to be alert to new methods of reducing energy needs for our systems in the future. One such effort involves the installation of low consumption plumbing fixtures and fix-

Table 12-A
Fixture and Equipment Hazards

HAZARD	PREVENTION
Burns from hot pipes	Isolate and/or insulate exposed piping.
Scalding from hot water	Place appropriate control devices which limit water temperature or system pressures to safe levels.
Drowning	Use appropriately designed suction fittings to reduce the potential for entrapment. Properly supervise bathers. Advise children not to drop their heads below the water.
Excessive pressures and explosions	Place relief valves and blow-off devices on fixtures which heat water.
Cross-connections	Place backflow devices on potentially hazardous fixtures, fittings, or piping arrangements.
Fouling surfaces	Use fixtures only for designed purposes — provide adequate flow to wash fixture.
Noxious gases and fumes	Install piping according to codes. Test piping after installation.
Explosive gases	Install piping according to codes. Test after installation. Use the proper safety control devices to prevent explosions.

ture fittings, as it is estimated that 3% of current energy use in the USA is involved in conveying water and 6% is used for water heating. If less water is used, then less water must be pumped or heated, thereby saving millions of dollars in annual energy use (on a national basis).

Plumbers and pipefitters (persons trained to handle and fabricate piping systems) are also the logical mechanics to work on cross-country pipe lines, on irrigation systems, and process piping installations of all types.

The plumber has many obligations for top-quality performance in the various facets of the industry.

We ask — no, we expect — you to put forth maximum effort to be a true professional in this demanding industry.

WATER SOURCES

Water is one of the most plentiful compounds found in our world. It is present as vapor in the atmosphere; as water in and on the earth; and as ice in the polar regions, on the higher mountains, and in much of the temperate regions in some seasons.

By far the largest amount of water is in the oceans, but the oceans are not a good source of water for human consumption or irrigation because of their mineral content. Scientists have been working for many years on methods for using sea water for human consumption or for irrigation, but to date all the installations for obtaining potable water from the sea are either experimental or very costly to operate. In some parts of the world, however, such plants are in use because there is no other choice available.

All our water eventually comes to us via solar energy from ocean evaporation and rain- and snow-fall. Because of the variations encountered in obtaining and using this water, however, we will discuss the following sources as being substantially different from each other.

SURFACE WATER

"Surface" water is taken from shallow wells, streams, lakes, or reservoirs. This water usually needs filtering to remove coarse solids, settling processes to remove fine suspended solids, and chlorination to eliminate bacteria. Surface water contains a minimum of dissolved solids. The distinction between "surface" water and "ground" is not clear-cut, but the principal difference is in the length of time between when it fell as rain, and when it is withdrawn for our use. Surface water is only in the ground for a short time, and ground water has been there for a while, perhaps as long as years.

GROUND WATER

Ground water from deep below the surface of the ground has been *in transit* many months or years since it fell as rain or snow. Such water is obtained from wells with appropriate pumps. This water is usually clear since its rate of movement is slow, allowing fine materials in suspension plenty of time to settle out. However, the water is high in dissolved solids (i.e., minerals) since it also has time to dissolve rock strata through which it has been moving. It may require chlorination to eliminate bacteria and treatment to remove objectionable minerals.

SURFACE WATER - CISTERN

A special category of surface water is collected from direct rainfall in cisterns — (large tanks or holding areas fed by gutters or downspouts). Water from cisterns usually has very limited use — laundry or irrigation, for example.

GRAY WATER

Water from plumbing fixtures such as sinks, lavatories, and laundry facilities may also be used in some jurisdictions for gray water uses such as irrigation or flushing water closets and urinals. Many practical problems have kept this concept largely unused, at least in the United States.

MUNICIPAL WATER SYSTEMS

The source of water for most large municipal systems is surface water. It is usually less expensive to develop surface reservoirs than to establish large well fields. Large-scale systems then treat water before distribution by some or all of the following methods:

AERATION (ADDING AIR TO WATER)

Air is dissolved into the water by releasing compressed air under the surface of the water, by spraying a thin sheet of water into the air, or by tumbling the water over rocks. The object is to saturate the water with dissolved air which will drive out other gases; and the high oxygen content will tend to convert other adverse items to less harmful forms.

PRECIPITATION (REMOVING SOLIDS FROM WATER)

Foreign objects are caused to cluster together by adding certain chemicals to the water. These solids (precipitates) can then be separated and removed.

SETTLING

By holding the water still for several hours, fine suspended particles are allowed to settle to the bottom so that they can be drawn off.

SAND FILTRATION

The finest suspended particles are removed by passing the water through fine filters.

STERLILIZATION

Chlorine is added to the water to kill any bacteria and to keep bacteria from developing.

MONITORING

All systems include a means of monitoring the water which is distributed to the public. Such monitoring includes total flow quantities, testing for undesirable elements, as well as checking pressure in various places. In the U.S., municipal plant operators are responsible for submitting reports on all these data on a regular basis to the Environmental Protection Agency (EPA), so the protection of the water users is thus assured. Figure 12-D shows a typical arrangement of a public water supply pipe into a building.

PRIVATE SYSTEMS

Private water supplies must be utilized in areas where there is no city or municipal water supply or where the cost of providing such a supply is prohibitive. Private systems, large or small, will have some or all of the large system equipment, depening upon the needs of the private system. Figure 12-E shows the typical arrangement of a private water supply.

The private system water source (usually a well for a single family residence) should be placed uphill from any source of pollution — local codes specify minimum clearances to sources of pollution.

Figure 12-F and Table 12-B present examples of minimum clearances that must be observed for placing a private well. Since the geological nature of the area will affect what are safe distances, the values shown in the figure and table should be considered as minimums — verify the requirements for the area in which the well is to be placed.

Wells can be dug, bored, driven, or drilled. The method used depends on the depth of the water table and the nature of the earth to be penetrated. Because of their depth, drilled wells are least susceptible to drought. Also, owing to their depth and their seal at the top of the well casing, they are also least susceptible to contamination.

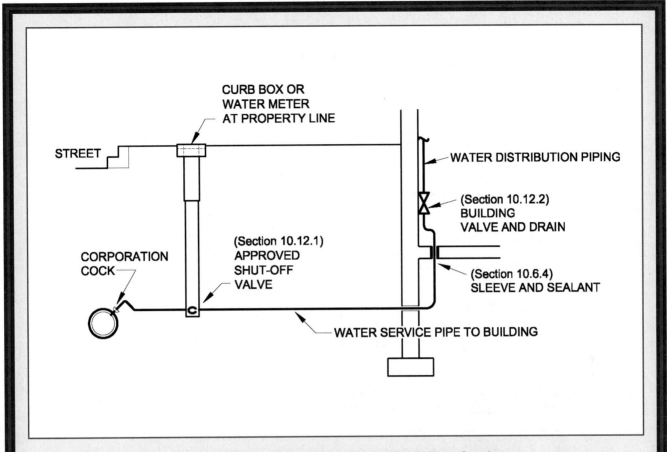

Figure 12-D – The Water Service Pipe in a Public Water Supply

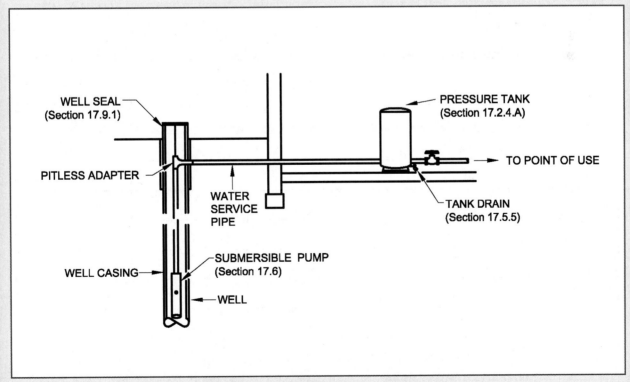

Figure 12-E – A Typical Private Water Supply System

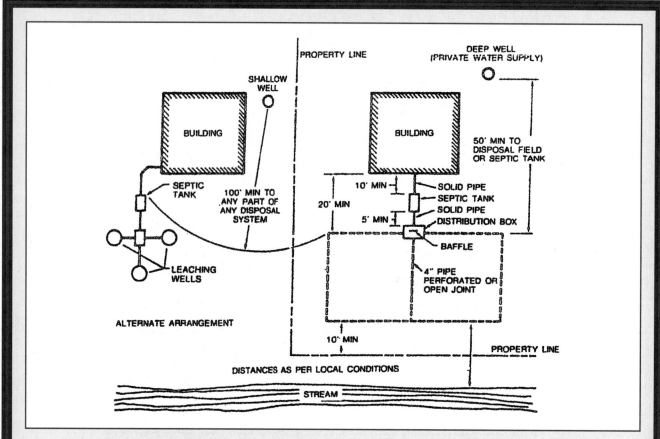

Figure 12-F – Minimum Distances for Location of Components of a Private Sewage Disposal System

Table 12-B Minimum Distances in Feet for Location of Components of A Private Water Supply System			
	Shallow Well	Deep Well	Single Suction Line
Building Sewer Other than Cast Iron	50	50	50
Building Sewer Cast Iron	10	10	10
Septic Tank	100	50	50
Distribution Box	100	50	50
Disposal Field	100	50	50
Seepage Pit	100	50	50
Dry Well	100	50	50

NOTE: Every effort and precaution shall be exercised to assure that sewage and septic by-products shall not contaminate any water supply. All piping for both water supply and drainage systems shall be tested to insure its integrity, as a form of added protection.

Once a supply of water is established, a pump is used to deliver water from the well to the building plumbing system. Pumps are rated in terms of flow (usually in gallons per minute), pressure, horsepower, suction pipe diameter, and required electrical characteristics. The pump is selected so that its capabilities match, or slightly exceed, the requirements of the load served by the well.

When a new well is drilled and before it is used for human consumption, it is usual procedure to have an initial bacteria test performed by a laboratory. This test will also indicate any other characteristics of the water being tested, such as **hardness**. An annual bacteriological test is also recommended for such systems.

WATER SOFTENING

Water obtained from deep wells contains dissolved chemicals that originate in the deep rock strata where the water is *in residence* for months or years. Most of these chemicals are calcium and magnesium carbonates. When this water is used without any treatment, the calcium and magnesium ions combine with the active components of soaps, thus requiring more soap to accomplish the cleaning. These ions also develop a precipitate (usually called **scum**) that reduces the cleaning ability of the soap.

The compounds also precipitate-out against heated surfaces, producing what is commonly called lime scale in water heaters and boilers. Heavy deposits of this scale lead to early failure of such equipment.

A partial solution to these problems is provided by a water softener. A softener contains chemicals that exchange calcium and magnesium ions for sodium ions so that after the water has passed through the softener, the water contains dissolved sodium salts (principally carbonates). Such water does not interfere with the action of soaps, and the **lime scale** is greatly reduced when this treated water is heated.

After a period of use, the softener chemical must be regenerated by placing a very strong salt (sodium chloride) solution in the softener, and then washing it out.

The water softening process is expensive so that, in most cases, only water that is to be heated is softened.

NEUTRALIZING

When well waters are aggressive, that is to say, when the water tends to have an acidic nature, the water must be neutralized. Acidity of water is measured in terms of a pH scale. The scale has a range from 0-14 with neutral water reading 7.0. The closer to the 0 reading, the more aggressive (acidic) the water will be. Conversely, if the water reads between 7 and 14, the water is alkaline.

Neutralizing usually involves the passing of water through a tank which contains a chemical which reacts with the aggressive compounds in the water. Like the softener, the reaction results in compounds which are periodically backwashed from the neutralizer either automatically or manually. The neutralizer will require periodic replenishment of the chemicals to continue its performance.

You can see that a great deal of work and equipment is needed to bring to us a very low-cost, vital product — potable water!

Figures 12-G and 12-H show proper installations to prevent cross-connections.

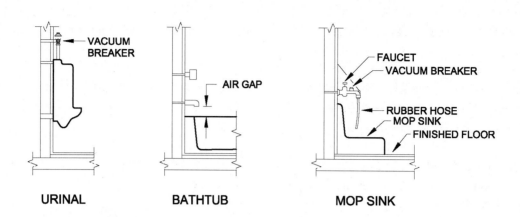

NOTES:
1. Individual outlet protection is required by this Code. It protects the potable water distribution piping from being contaminated by a cross connection within the property.

Figure 12-G – Cross Connection Control by Individual Outlet Protection

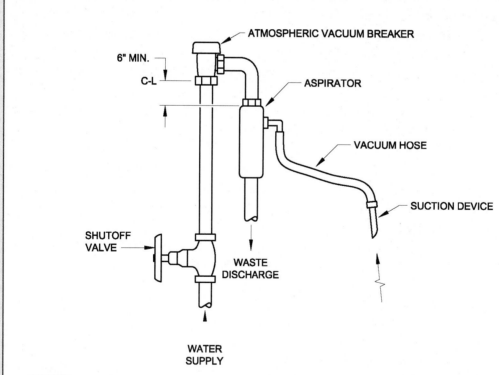

NOTES:
1. Atmospheric vacuum breakers are not rated for periods of more than 12 hours under continuous water pressure.

Figure 12-H – The Potable Water Supply to an Aspirating Device Protected by An Atmospheric Vacuum Breaker

LESSON 13

SEWAGE DISPOSAL

People have always tended to simply drop things they are through with — whether a fast-food sandwich box today, or a broken bowl or bones from a meal in prehistoric times. If there are very few people in a very large space, this system works fairly well. However, as population grows in a town of fixed physical limits, random discard of our leftovers creates serious problems and hazards.

If rotting garbage and human excrement were not offensive to us, mankind would not have survived to the nineteenth century when the scientists of the times proved how deadly these materials can be. In ancient times, waste disposal was a matter of getting rid of offensive materials. It was also desirable to get rid of rain water, because mud greatly impedes any travel, whether on foot, horseback, or wagon. Thus, the earliest sewer systems were for storm drainage; offensive waste materials were simply washed along. These systems ended at the nearest convenient part of the natural watershed, usually a stream or river; or, for sea coast towns, at the ocean. As sanitary sewers developed, they were terminated in the same places. Now, as shown in Figure 13-A, sanitary sewers and storm drains are preferred to be installed and maintained as separate systems.

Only in the last century have we come to realize that we cannot simply dump our leftovers and wastes into the natural watersheds (which frequently are the source of our drinking water) and survive. In the United States, soon after World War II, and starting with major population centers and progressing to smaller cities, sewage treatment facilities have been installed so that the nation's natural waterways are being *cleaned up*. At the other end of the

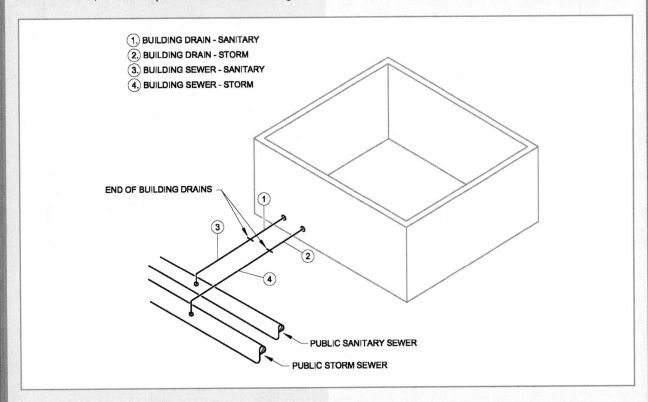

Figure 13-A – Separate Sanitary and Stormwater Building Drains and Sewers

size range, private sewage systems for an individual home are installed and operated so that they do not pollute the environment.

Some parts of the world are ahead of the U.S. and many parts are far behind. Without doubt, some students of the plumbing craft will be offered work in various parts of the world to help other people to further their desire to improve their lives.

As plumbers, we must be aware of our place in this overall system, and be proud of our work in carrying out the trust that the public places in us.

PUBLIC SYSTEMS

A modern public or municipal sewage system consists of the sewer piping and the treatment plant. Most sewers are gravity flow because it is more economical, but in some cities the terrain may require lift stations and (sometimes) pressurized sewer mains (also called force-mains) to pass high obstructions.

The size of the piping to the plant is dependent upon the probable volume of waste material flowing from domestic and industrial users and upon available slope (or fall) to the treatment plant. It is usually considered the area of expertise of Civil Engineers to design all but the smallest public sewer systems, and that branch of engineering has developed capacity ratings for sewer pipes based on type of material, available slope, and pipe diameter.

The piping layout of the municipal system is from building sewer to the *lateral line*. The lateral line is the line from the private owner's property line to the local main. The *local main* is the line to the trunk sewers.

Trunk lines serve sections of a city and connect many local mains. The trunk lines then connect to *interceptors*, which are the large diameter pipes transporting the sewage to the treatment plant. These lines are called interceptors because they intercept the trunk lines at or near the former river discharge points.

If there is an area in a system where a pipe is lower than the elevation required for gravity flow, the sewage must be pumped up to the higher line at a lift station. A lift station consists of a receiving tank, pumps, and liquid-level sensors to control the pumps. When the waste materials reach a certain level in the receiving tank, the pumps are automatically activated to pump the sewage into the higher elevated pipe.

Sewer segments are installed in straight lines with manholes located in the system wherever there is a change of direction, pipe size, slope, or where one line connects to another. Manholes are large enough for easy inspection or to accommodate

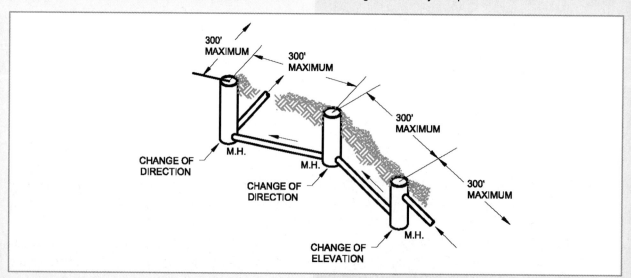

Figure 13-B – Location and Spacing of Manholes

cleaning equipment. They are placed in long runs for cleaning and inspection. Figure 13-B shows location of manholes and cleanouts as required by many plumbing codes.

SEWAGE TREATMENT PLANT

The sewage treatment plant receives all the wastes for treatment. The sewage is usually treated by one of two methods — the trickling filter process or the activated sludge process.

The sewage travels through the following stages in the trickling filter process:

Stage 1 **SCREEN** — Removes cans, branches, and other coarse materials in the incoming material

Stage 2 **GRIT CHAMBER** — Removes gravel- and sand-sized solid materials

Stage 3 **PRIMARY SETTLING TANK** — Holds materials until the wastes settle to the bottom (sludge) and the water (effluent) remains on the top

Stage 4 **DIGESTER** — Sludge is sent to a digester where it is retained for approximately one month while bacteria reduce the volume of solids by converting most of it into harmless products in solution. In thirty days the remaining sludge is removed from the digester and transported to the sludge drying tank. When the sludge is removed from the drying tank, it is disposed of.

Stage 5 **TRICKLING BED** — The effluent is drawn off to a trickling bed which *aerates* or provides oxygen. Oxygen causes further chemical and bacterial action to convert the particles in the waters to harmless chemicals.

Stage 6 **DISPOSAL** — Once aeration is completed, this water is discharged into a stream, which usually is not the source of potable water for the city.

ACTIVATED SLUDGE VARIATION

The activated sludge process uses the screen, grit chamber and primary settling tank, but the succeeding steps are these:

Stage 1 **SCREEN**

Stage 2 **GRIT CHAMBER**

Stage 3 **PRIMARY SETTLING TANK**

Stage 4 **OXYGENATED TANK** — The sewage passes into a huge oxygenated tank where the bacterial action is continued on both the sludge and the effluent. After 5-6 hours in this tank, the sludge and effluent are separated and pumped to final processing.

Stage 5 **BACTERIAL SUPPLEMENTS** — Extra bacterial growths are sometimes added to this system to "eat" away at the sludge.

SYSTEMS OF BACTERIAL GROWTH

The action in a treatment system is bacterial, either aerobic or anaerobic. Because the sewage reaches the plant 99½% water and ½% dissolved and suspended solids, the bacterial action is usually on the verge of being diluted out of being.

Aerobic bacteria dissolve materials with the presence of oxygen. This process is usually quicker than the anaerobic method, and does not produce offensive odors. *This process will take place with any water-borne sewage if oxygen is present.*

Anaerobic bacterial action does not use oxygen. This process is slower than the one discussed above, and usually does produce significant odors.

A waterway that contains organic materials will be undergoing aerobic sewage reduction. The source of this oxygen is mainly the dissolved oxygen in the water. The waterway (stream, river, pond, or lake) is rated in terms of "B.O.D." *(biological oxygen demand)*. From the point of view of a sewage treatment plant, water having a high B.O.D. number requires a higher supply of oxygen in the system to keep the aerobic bacteria active. This replenishing of the oxygen is usually accomplished by discharging compressed air below the surface of the water.

Remember, once the oxygen is depleted the anaerobic bacteria take over and odor development begins.

From the point of view of the recreational user of a stream or lake, the high B.O.D. means that the dissolved oxygen in the water will be reduced (perhaps even to near zero) and the water will not support aquatic life, nor will it be safe for swimming or other such uses.

PUBLIC TREATMENT PLANT CHALLENGES

Municipal systems may be overloaded so that the treatment plant does not operate properly. As indicated above, storm and sanitary drains are being separated to help alleviate this condition. Too much water dilutes the bacterial action and delays the cleansing process.

Recently, however, there has been a rethinking about keeping storm water away from the treatment plant. The idea is that typical storm water content is not the pure rainwater that we wish it were, but rather contaminated water, even though the makeup is different from what is in waste water from the activities of man and animals. Storm water that falls on paved areas (the typical source of water that does not seep into the ground) will contain rubber (from tires), all manner of petroleum products, various salts, sand and dirt, and some animal waste products.

The leaders in the field of sewage treatment do not believe that this is good material to put into our rivers or lakes!

The normal sewage plant bacterial process can be **poisoned** by chemicals such as petroleum products and industrial wastes or other unusual waste products. Special effort should be made to prevent cooling water or storm water from entering the sewer without verifying the acceptability of such waste to the treatment plant authority. You need to be aware that even small amounts of certain chemicals can be very bad for the typical sewage plant process. Check with the operators when in doubt.

PRIVATE SYSTEMS

Private sewage systems serve small buildings with a limited number of fixtures and people. They are similar to the municipal system in that bacterial action is the means of reducing waste. Figure 13-

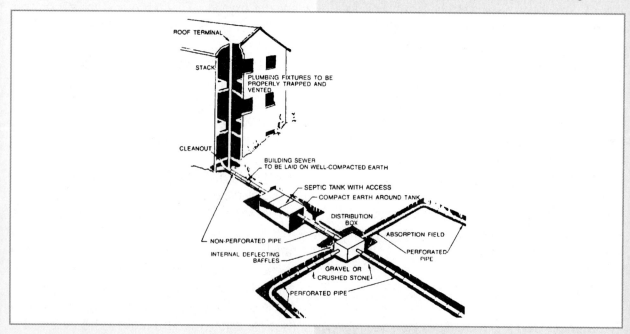

Figure 13-C – Typical Private Sewage Disposal System

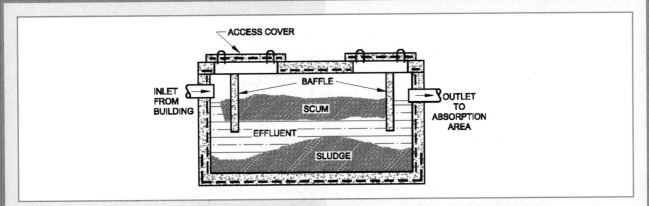

Figure 13-D – A Typical Septic Tank

C shows a typical private disposal system.

The three main parts of a private system are the following:

SEPTIC TANK — The septic tank is similar to the settling tank in the municipal system. The bacterial action breaks down the waste and causes the sludge and effluent to separate in the tank. Figure 13-D shows a concrete septic tank. The digested sludge in the septic tank builds up until it must be removed by pumping.

DISTRIBUTION BOX — The distribution box receives the effluent from the septic tank and distributes the watery material to the leaching fields. Figure 13-E shows a typical distribution box. Note that the three exit pipes must be at the same elevation to assure equal flow to all parts of the disposal field.

LEACHING FIELD — The leaching field is a system of pipes or structure which distributes the flow of water into the soil. The effluent water dissipates

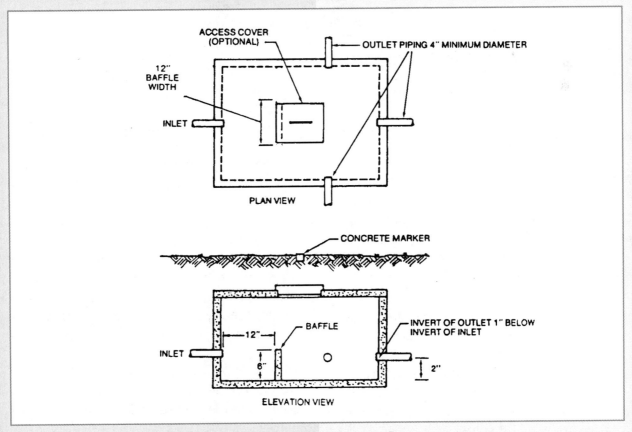

Figure 13-E – Typical Distribution Box

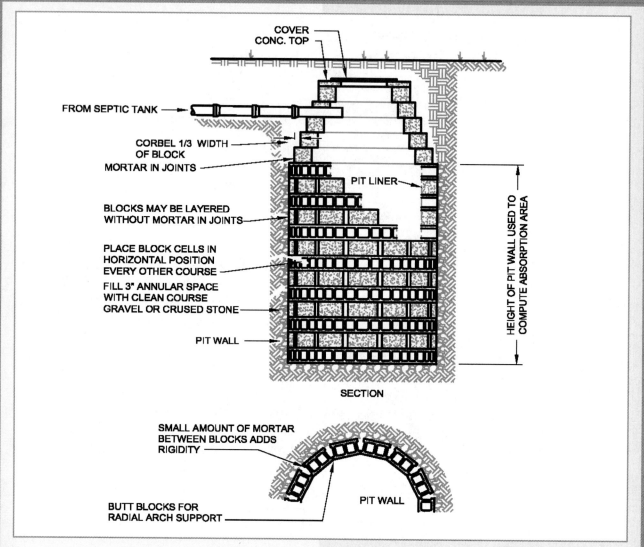

Figure 13-F – A Leaching or Seepage Well or Pit

through absorption and evaporation. Figure 13-F shows a leaching well.

Some private systems are equipped with an aerating fan or air pump. The addition of air to the septic process enables the aerobic bacteria to remain active, thereby improving the waste decomposition.

PERCOLATION TEST

The key to the private septic system is the percolation of effluent into the ground. Therefore, a **perc** test is performed to determine if the soil around a building site is suitable to absorb the effluent material. The test involves digging a test hole(s) in the ground, adding water to the hole(s), and observing the rate of loss (or **percolation**) into the earth.

The **perc** test result is used to determine how large a leaching field is needed to handle the sewage load of a building

A perc test is required prior to building the sewage system, and usually prior to receiving a construction permit. In some areas, your local governing body will inform you of requirements for a leaching field without the necessity of a percolation test. This information is based on soil types in the local area.

Another term used in private systems is the **cesspool**. A cesspool is a sewage receiver pit. It is usually small in diameter but relatively deep. A cesspool differs from a septic tank in that a septic tank is water tight and the effluent is piped away. A cesspool is not water tight so the effluent seeps into the earth directly.

A *drywell* is a fabricated tank placed in the ground to receive septic tank effluent or storm water. The sides and bottom of the drywell permit water to pass into the earth.

DANGERS OF WASTE PRODUCTS

During the Middle Ages, bacterial diseases were constantly present in the general population. Persons who were not in robust health — the very young and those who were middle-aged or older — were especially at risk from human and animal waste products. In those earlier times, industrial and agricultural wastes were much less plentiful than they are now, but only because the population numbers were much smaller.

As discussed above, wastes were removed because they were considered offensive. Only in the past two centuries have we come to understand how serious a health hazard waste products of all types can be.

Similar bacterial actions that pose hazards to us must be encouraged in the sewage treatment plant — these activities convert the waste materials to harmless products (or, at least, much less dangerous ones). Bacterial action reduce wastes to a nearly inert residue (called sludge) plus water, carbon dioxide, and other gases. A major challenge to the sewage treatment industry has been to develop means to reduce the hazardous levels to ever smaller amounts, and to deal with non-typical products that they receive in the waste stream from their "customers." An ongoing problem is the increasing sensitivity of the monitoring equipment that the laws require them to use, which means that they must achieve greater levels of processing, and that the research scientists keep lowering the values of safe exposure to many of the standard components of typical waste materials.

CAUSES OF CONTAMINATION

Historically speaking, contamination has been caused by the factors and circumstances that are described in the sample stories that follow:

INADEQUATE INSTALLATION

In 1932, people all over the country were getting sick with intestinal disorders. Finally, a pattern was

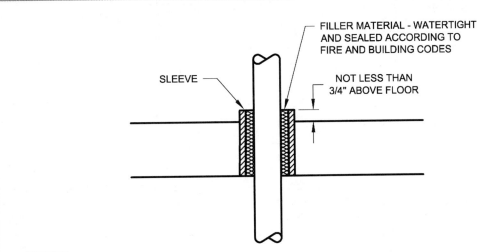

NOTES:
1. The 3/4" extended sleeve prevents spillage on the floor above from leaking down into the food handling area and contaminating the food products.

Figure 13-G – A Pipe Penetration of a Floor Above a Food Handling Area

noted that these people had all attended the World's Fair in Chicago, and *they stayed at one of two hotels!* It was discovered that there was a kitchen employee in one hotel who was a carrier of an intestinal malady, but he didn't get sick with it himself. Because of low city water pressure, it frequently happened that the hotel had negative water pressure, especially. on the higher floors, which would pull water out of any fixture that was full of water and that had a low fill spout. In those days, tub filler spouts were piped through the face of the tub so that it would be submerged in the bath water, depending on how much water the tenant had put in the tub. In addition, at least some fixtures in the kitchen had similar arrangements. Thus, there was frequent cycling of "used water" into the (presumed) potable water piping and the system became thoroughly contaminated. Typical tenants, who stayed only a short time, would not get sick until they returned home — hence the mystery!

As a result of this major health detective story, all plumbing fixtures must have their fill spout discharge opening located a minimum distance *above* the overflow level of the fixture. This distance is called an "air gap," and there are specific ways to determine what it has to be. For typical residential fixtures, this dimension is usually about 1" of vertical separation. Where the air gap arrangement is not possible, very stringent rules apply to the fill opening and approach piping for such unusual fixtures.

MATERIAL FAILURE

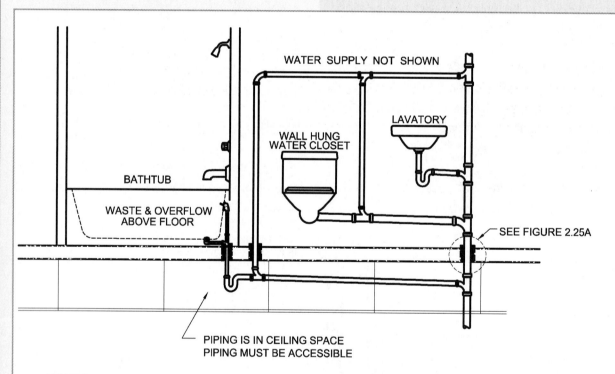

NOTES:
1. All pipe penetrations through the floor above must have extended pipe sleeves. See Figure 13-G The number of floor penetrations should be minimized.
2. Floor-outlet water closets are not permitted above a food handling area.
3. Bathtubs and showers must have above-the-floor waste outlets with piping extended to a point where an extended pipe sleeve can be installed.
4. Piping run in a ceiling space above a food handling area must be fully accessible.

Figure 13-H – Plumbing Fixtures Above a Food Handling Area

In New York (1946) twenty cases of typhoid fever were traced to a leaking lead closet bend which dripped into a fruit bin in a market. The user of the water closet was a typhoid carrier. The faulty lead waste pipe material had worn through but a secondary cause of the problem was the placement of a plumbing fixture over a food storage container.

We have learned proper methods of installing plumbing fixtures where food handling facilities are present. Figure 13-G shows the floor penetration procedures and Figure 13-H shows the additional protection required for such installations.

Figure 13-H shows a typical installation of bathroom fixtures over a food-preparation area.

IMPROPER INSTALLATION PRACTICES

The Holy Cross College football team was poisoned (in 1972) from contaminants entering ground hydrants and polluting the water supply. Proper backflow preventers were not installed, so the team members were not protected from the foreign material. The contamination that poisoned the Holy Cross football team originated in a lawn pit faucet that had been exposed to urine and defecation from youngsters.

POOR MAINTENANCE PROCEDURES

A contaminated rag was tied around a leaking pipe joint in a dairy (in 1945). The leaking water seeped through the rag and dripped into milk cans. The contaminated milk caused nine deaths and over 400 illnesses.

INADEQUATE DESIGN OF WASTE TREATMENT SYSTEMS

A few years ago, Cincinnati, Ohio, was discharging much waste into the Ohio River. Cities downstream experienced severe problems with purifying the water for their municipal systems. As soon as the sewage plant was upgraded, the river cleaned itself, aquatic life improved, and the water supplies of downstream towns were no longer threatened.

HISTORICAL RECORD

Proper plumbing is a major factor in infectious disease control. The records for water-borne diseases in the United States show dramatic improvement in the past century:

Diarrhea — enteritis and dysenteries
    1900   100,000 annual deaths
    1944   fewer than 15,000 deaths

Typhoid fever

    1900   23,000 annual deaths
    1944   less than 600 deaths

On the reverse side, from 1961 through 1970, there were 128 known water-borne disease outbreaks, resulting in over 46,000 illnesses and 20 deaths. More recently, the SARS pathogen that started in China was made possible by inadequate plumbing in major buildings.

A review of Centers for Disease Control and Prevention (CDC) information shows that there are only a few reported incidents of water borne disease outbreaks in the USA over each of recent years (some data are for the 1980's, others are for the 1990's).

Two major "strings" are attached to the data: first, not all outbreaks are reported, and second, the number of persons made ill in each outbreak is very large in many cases (one outbreak caused 13,000 illnesses!).

Also, most of the cases had to do with defective materials or methods at the source of the water, rather than with plumbing, although there were plumbing cases reported.

These hazards impose a substantial responsibility on us to practice personal sanitary habits to assure that wastes enter the system designed to handle them. Children must be taught good hygienic practices and to use bathroom facilities properly.

The plumber must know the risks that waste materials pose and the procedures necessary to guard against these risks. However, as good citizens, we need to use proper sanitary procedures and train our children to do the same.

HAZARDS TO THE SERVICE TECHNICIAN

The increasing popularity of drain cleaning chemicals for use by the homeowner has introduced another hazard to plumbing service personnel. Such chemicals can burn skin, eyes, and breathing passages. Use protective gear to isolate yourself from these hazardous solutions.

Human waste is laden with a variety of potential diseases. Waste may contain bacteria which may expose the plumber to diseases such as hepatitis, diarrhea, and influenza. Although disease organisms such as AIDS are very vulnerable outside the human body, the plumber should wear protective gear when servicing existing drains, especially when working in medical buildings.

Sewer gases can also be harmful to public health. Methane gas is a byproduct of human waste decomposition. This gas will burn, and will deplete oxygen in sewer manholes. The plumber should always ventilate the manhole prior to entry. Breathing apparatus should be worn when entering any confined space such as a manhole. A safety harness on the plumber and an observer outside the manhole are necessary safeguards whenever a plumber has to work in such places.

Be sure to follow current safety practices whenever entering any confined space!

LESSON 14

INTRODUCTION TO GASES

GENERAL

We encounter matter in three possible states or conditions. The three conditions depend on the molecular activity (that is, the energy level) of the molecules that make up the substance. These three states and their energy levels are listed below:

STATE OF MATTER	ENERGY LEVEL
Solid	Low Molecular Activity
Liquid	Intermediate Molecular Activity
Gas	High Molecular Activity

A gas is a fluid that expands in all directions to fill its container. Some substances are in a gaseous state at normal temperature and pressure. Some substances have to be heated to become a gas (for example, water). Some substances are changed chemically to release a gas (calcium carbide plus water yields acetylene). Commercially useful gases are extracted from wells, from the air, or are manufactured in chemical plants.

Natural gas is a fuel gas found with petroleum deposits deep in the ground. Principally methane, natural gas is the lightest member of a large family of hydrocarbons which includes butane, propane, and the fuels we call **gasoline, diesel fuel,** and heating oils. These hydrocarbons are the product of long-time changes in organic material that grew on the earth millions of years ago.

We can consider gases from two points-of-view —

1) those that are useful in our work, and

2) those that are *accidentally* or *incidentally* encountered as we are doing other things

GASES USEFUL IN OUR WORK

The gases that we use in our work, or for which we provide piping systems, include the following:

- Air
- Acetylene
- Ammonia
- Argon
- Carbon Dioxide
- Chlorine
- Fuel Gases
- Nitrogen
- Oxygen
- Refrigerants
- Steam

GASES ACCIDENTALLY ENCOUNTERED

Gases encountered more or less *accidentally* include the following:

- Methane
- Carbon Dioxide
- Air dissolved in liquids
- Hydrogen sulfide
- Carbon Monoxide
- Water Vapor

Note that some gases appear on both lists. Gases such as chlorine, sulfur dioxide, carbon monoxide, and ammonia are chemically hazardous when encountered; others could be hazardous because of temperature (steam); potential for explosion (fuel gases); or simply because of energy stored at high pressure.

We are concerned with flow characteristics, ways that hazardous gases can be encountered, and levels of toxicity for various gases. Piping systems in larger buildings are color-coded so that the contents of the pipe may be readily identified. The color-code is based upon the use and hazard-level of the gas in the pipe.

You should realize that many gases are formed as a byproduct of some necessary operation. Among such operations are using flames, processing sewage, metal refining, working with plastic solvents, and recovering contaminated or environmentally-damaging refrigerants.

We can classify commercially useful gases as **naturally-occurring** or **manufactured**. The following lists illustrate the two categories of gases:

NATURALLY- OCCURING GASES

Naturally-occurring fuel gases are **natural gas** and LP gases. Other naturally occurring gases are the components of air:

- Argon
- Carbon dioxide
- Helium
- Nitrogen
- Oxygen
- Various others

MANUFACTURED GASES

Manufactured gases are generated by chemical process.

- Acetylene
- Ammonia
- Carbon monoxide
- Refrigerants
- And Many Others
- Coal gas
- Hydrogen
- Chlorine

GASES USED IN PLUMBING

All parts of air are useful to the plumber — oxygen for high-temperature flames, nitrogen for testing and for shielding the inside of piping during brazing operations, carbon dioxide and the rarer gases used mostly for weld shielding.

Gases most familiar in the plumbing industry are divided into two categories: 1) fuel gases, and 2) other gases.

FUEL GASES

Fuel gases are flammable and care must be taken to contain them. Leaks are serious issues; fire and explosion are a great risk. Fuel gases include the following, with comments on each:

1. **Oxygen**
   Not a fuel itself, but combines with fuel gases to produce heat

2. **Acetylene**
   Burns with high-temperature flame, ideal for working with steel and brazing operations, used in portable torches Fuel with widest combustion range – very flammable and dangerous

3. **Natural Gas**
   High heat content, lowest-cost fuel, most convenient at fixed locations (served by utility mains), used for comfort heating. Lighter than air and will collect in high pockets; lately being marketed as a fuel for vehicles. Natural gas is found in many parts of the world. Many countries have little or no use for the product, so a major industry is arising that involves shipping liquid natural gas to the United States

4. **LP Gases**
   (various mixtures of butane and propane) Lower flame temperature than acetylene, higher heat content than acetylene, most convenient fuel for portable torches, also being marketed as a motor fuel

OTHER GASES

Other gases include the following, with comments on each:

1. **Ammonia**
   Used in large refrigeration installations, used as a feedstock for many chemical processes, agricultural fertilizer

2. **Chlorine**
   Used to purify water at treatment plants and swimming pools, used as a feedstock for many chemical processes

NOTE: Ammonia and chlorine are deadly gases. They will produce life-threatening burns if they

contact your person in any way. They will destroy your lungs if inhaled. Only experienced workers, using proper safety equipment, should install or operate systems involving these gases.

3. **Carbon Dioxide**
   Used to form the shielding atmosphere around the work during welding operations, used to put the *fizz* in soft drinks, used for industrial cooling, used in some types of fire extinguishers. In its solid state (dry ice), it is used for various cold-holding applications

NOTE: Gaseous carbon dioxide is hazardous only if it has displaced the air in a confined space. Solid carbon dioxide (dry ice) is hazardous because of its extremely cold temperature. Severe skin damage will occur if dry ice contacts your body.

4. **Carbon Monoxide**
   Used as fuel in some heavy industrial applications, present whenever there is incomplete combustion of hydrocarbons

NOTE: **Carbon monoxide is poisonous in very low concentrations. Its presence should be suspected with any combustion process. Even the best burners will leave some of the fuel only partly burned. Because any internal-combustion engine is burning a hydrocarbon, do not operate any engine except in a well-ventilated area or if the engine exhaust is piped outdoors.**

Most gases are supplied to us in pressurized containers (cylinders). Extreme caution should be used when working with these cylinders as a bumped, tipped, or otherwise jolted cylinder may explode. Transport caps should be used when moving cylinders. Always chain cylinders to a substantial building member or holding device. Periodically check valves and piping for leaks using a soap solution. **NEVER** test for fuel gas leaks using an open flame if a fuel gas leak is suspected, do not operate any electrical equipment as a spark may ignite escaped gases.

PROPERTIES OF GASES

To compare gases, and to understand how they function, the following properties of gases need to be considered:

DENSITY

The weight per unit of volume of a gas

MOLECULAR WEIGHT

The atomic weight of the gas molecule

SPECIFIC GRAVITY

The ratio of the density of a gas to the density of air

GAS LAWS

The mathematical relationships that describe gas behavior help us understand what a gas will do when conditions are changed. The first equation describes volume and pressure changes with temperature remaining constant.

The general gas law states that pressure **P** times the volume **V** equals the temperature **T** times a constant **R** times the number of moles n of that gas or

$$PV = nRT$$

Where volume is in ft^3, pressure is in lb/in^2 absolute, and temperature is in degrees Rankine. Note that Rankine temperature is the following:

$$T_{Rankine} = T_{Fahrenheit} + 460$$

Simply stated, pressure, volume, temperature, and the gases' physical properties are all related proportionately.

Boyle's Law states that the volume of a gas is inversely proportional to pressure, or

$$\frac{V_1}{V_2} = \frac{P_2}{P_1}$$

Where V_1 = Initial volume
V_2 = Final volume
P_1 = intial pressure (absolute units)
P_2 = final pressure (absolute units)

This equation can be changed (as any equation can be), to make use of the information we already have to find a missing element. Thus, the equation is also correct in the following form:

$$\frac{V_1 \times P_1}{P_2} = V_2$$

EXAMPLE

A sample of gas occupies a volume of 1000 cubic feet at 685 psig. If the pressure is increased to 785 psig, what volume will the gas occupy? (Remember, atmospheric pressure is approximately 15 psig.) Set up the equation:

$$\frac{1000}{V_2} = \frac{800}{700} \qquad 1000 = \frac{800(V_2)}{700} \qquad \text{Therefore,} \qquad \frac{1000 \times 700}{800} = V_2$$

$$V_2 = 875 \text{ cu ft}$$

Of equal importance to us is the relationship between pressure and temperature (volume remaining constant). The following equation demonstrates how gas will behave when temperature changes.

$$\frac{P_1}{P_2} = \frac{T_1}{T_2}$$

Where P_1 = intial pressure (absolute units)
P_2 = final pressure (absolute units)
T_1 = intial temperature (absolute units)
T_2 = final temperature (absolute units)

Remember, atmospheric pressure is approximately 15 psig and absolute zero is -460 F.

EXAMPLE

An oxygen cylinder is stored at 80°F with a pressure of 2685 psig. What will the pressure be if direct sunlight warms this cylinder to 200°F?

P_1 = 2685 + 15 = 2700 *psia*
T_1 = 80 + 460 = 540 *Rankine*
P_2 = ?
T_2 = 200 + 460 = 660 *Rankine*

$$\frac{2700}{P_2} = \frac{540}{660}$$

$$\frac{2700(660)}{540} = P_2$$

$$P_2 = 3300 \text{ psia}$$

Therefore, the pressure read on a gauge would be   3300 - 15 = 3285 *psig*

LESSON 15

MECHANICAL PROPERTIES OF MATERIALS AND STRUCTURES

INTRODUCTION

There are many ways to consider materials and their properties — chemical characteristics, electrical properties, and thermal behavior are of interest in many cases. This lesson will explore the mechanical properties of materials, and the consequences of these properties when materials are used as components of structures.

The intent is to enable you, as a plumbing journeyman, to be aware of the limitations inherent in the structure and structural elements that you must consider when installing plumbing work in a building. With an awareness of these limitations, you will be able to place your work as required by your assignments, and still keep the building structure sound and intact.

The mechanical properties that must be understood in this context are the responses of the particular material under consideration to three types of loads:

Compression
Tension
Shear

Companion sets of properties that interest us are the ease of installing and the ability of the material to accept and hold fasteners — screws, bolts, nails, etc.; and the ease of cutting, shaping, or otherwise modifying the building material to be able to fit our plumbing materials into a building structure.

THE THREE LOADS

Compression

A compression load is one the tends to push the material together, to crush it, or to make it smaller. Since all building materials have weight, and since buildings are usually piles of materials heaped together (ouch!, says the architect), compressive loads are common within building structural elements.

Tension

A tensile load is one that tends to pull a material apart. If we pull a coaster wagon with a rope, there is a tensile load in the rope. For reasons that we shall see later, there is almost as much tension loading in building structures as there is compression loading.

Shear

Shear loads act to make the parts of a material slide past each other. This loading is more difficult for most

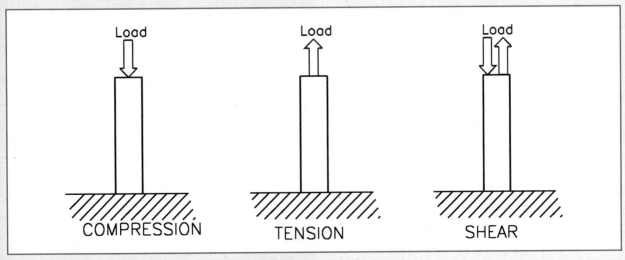

Figure 15-A – Three Categories of Loads

of us to envision, and it frequently leads to building member failures that "we do not see coming."

The individual figures shown in Figure 15-A illustrate the three categories of loads.

STRESS LINES

To help us visualize what is happening in a structural member under stress, we add the stress lines shown in Figure 15-B. Each line represents a certain stress magnitude, and the stress per unit area is the total stress divided by the cross-sectional area.

cross-sectional shape of the structural member changes. As the stress lines "tighten up" to pass by the area of cross-section reduction, the local intensity of the stress increases. Figure 15-C shows this modification. As the shape of the members get more complicated, the stress lines are likely to become concentrated, and the material is more likely to fail. This result fits with our experience — namely, that if we cut holes, or otherwise reduce the size of the member, we increase the chance of collapse.

The members shown in Figure 15-C are shown with smooth, round holes to reduce the cross-section. If

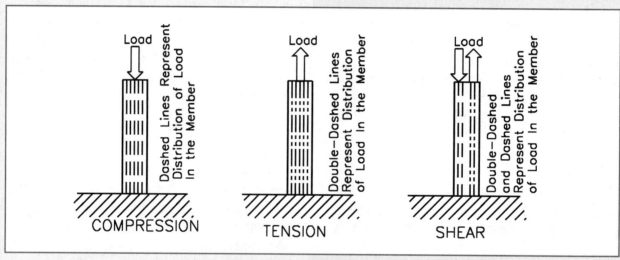

Figure 15-B – Structural Member Under Stress

STRESS LINES IN MEMBERS THAT HAVE VARYING CROSS-SECTIONS

Notice what happens to the stress lines when the

the openings are irregular or jagged, the stress concentrations are more severe, and failure is more likely as shown in Figure 15-D. Thus, you should always make cuts (when and where permitted) in

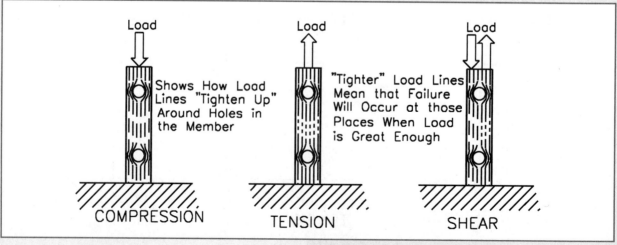

Figure 15-C – Stress Lines "Tighten Up"

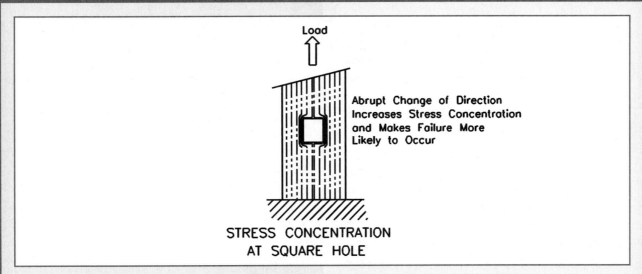

Figure 15-D – Stress Concentration at Square Hole

structural members with sharp tools, and do everything you can to keep the openings free of such irregularities.

BEAMS AND COLUMNS

Beams and columns are the typical forms that building members take in any practical structure. The following descriptions will cover beams and columns, what they are intended to do, how they resist the applied loads, and how they can fail.

BEAMS

Figure 15-E shows a typical beam. Most beams are supported at the ends, and loads are applied between the supports. The beam resists failing by bending, with the side toward the load trying to shorten (that is, it is in compression), and the side opposite the load trying to lengthen (that is, it is in tension). For a beam of symmetrical cross-section, the beam will bend along the center plane of the beam.

The failure mode of a beam is largely dependent upon the material — steel beams fail by collapse of the top member in compression, whereas wood and concrete beams usually fail in 45° shear lines!

Typical beams are installed so that the strongest direction is the one to which the load is applied, and much weaker in the direction perpendicular to the direction of intended loading. Therefore, if the load is applied just slightly off-center, the beam can fail by rolling into its weaker direction. To guard against this type of failure, beams are tied together when possible or otherwise braced so that they cannot roll sideways.

Beams are usually visualized as being horizontal,

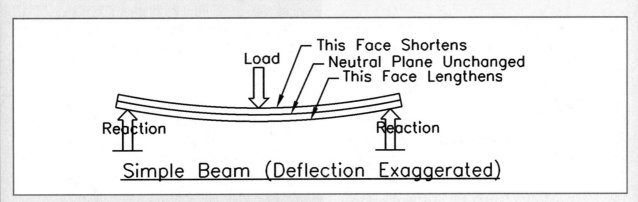

Figure 15-E – Simple Beam (Deflection Exaggerated)

with the load(s) applied above, but the beam and load can be in any configuration. For example, a basement wall is a vertical beam resisting outside earth loads that are trying to move the wall inward.

The most common beams — from earliest times — are made of wood and they are rectangular in cross-section. Typical beams of this sort used today are floor joists and roof rafters. Such beams, when kept dry (that is, inside buildings) have a very long life when loaded within their strength limits.

Standard lumber shapes of fir, spruce, or yellow pine of 2x8 nominal size are applied (for typical residential floor loading) at 16" on center and up to 12'-6" of span from face-to-face of the supports. 2x10 joists can be installed up to 15' span. The allowable load increases if the joist centers are reduced, but the span cannot be increased significantly because it is limited not only by the plain strength of the lumber, but also by the "firmness" or rigidity desired in a floor. In other words, with longer spans, the floor would seem to be "spongy" when you walked on it.

When longer spans or greater strength are required, it is most economic to change materials. Industry has developed trusses made of lumber components that can be made to fit much longer spans and still have the required rigidity. A very economical design uses the approximate equivalent of a 2x4's on the top and bottom members, with a sheet of pressed-board in between to hold the top and bottom members apart. The manufacturers provide directions for the size and permitted locations of openings that can be cut into the pressed-board web.

For very heavy loads or very long spans, steel beams and steel bar joists are available for any application.

Reinforced concrete beams are excellent choices for many applications, but they are seen more often on heavy construction outdoors. Bridges very often are made with reinforced concrete beams. Concrete is reinforced with steel rod to take up tension loads; concrete is very strong in compression but weak in tension.

As in most things — the initial cost, ease (or difficulty) of working with the product, meeting space requirements (whether very restricted or not), delivery schedule, probable life required of the material — all combine to dictate the choice.

While the plumber seldom has any input as to the building material selection, you must be aware of the limitations of working with any of these beam types.

COLUMNS

Columns are loaded in compression — that is, the load is attempting to shorten the member. They are usually divided into three broad classes: **Posts**, which have a slenderness ratio of 10 or less; **Columns**, which have a slenderness ratio up to 120; and **Struts**, which have a *slenderness ratio* of more than 120. Slenderness ratio is (approximately) the unsupported length of the member divided by half the width (in consistent units). As an example, a pipe column that is 10' long and 6" in diameter has a slenderness ratio calculated as follows:

10' = 120" Half the diameter = 6"/2 = 3"
Ratio = 120"/3" = 40

Notice that the slenderness ratio has no units, as it is feet/feet, inches/inches, etc.

For a column with circular cross-section, the slenderness ratio is the same in all directions, but for steel shapes or most lumber shapes, the slenderness ratio is different in different directions. Thus, it is normal for columns to be tied together in their weaker directions to guard against failure in those directions.

Columns fail by buckling. Therefore, it is vitally important that columns be absolutely straight and loaded symmetrically. Posts fail by crushing the material, but true posts are seldom found in buildings. Struts are not used in any vital member because their ability to withstand failure is unpre-

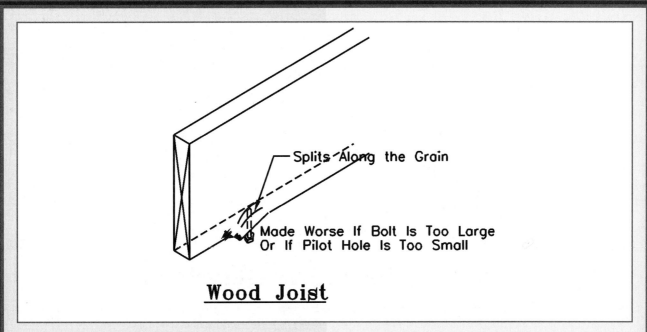

Figure 15-F – Wood Joist Failure

dictable.

ATTACHING TO STRUCTURAL MEMBERS

Nearly every plumbing job will require that attachments have to made to the structure. It is necessary that you realize the limitations of the materials available to support the plumbing work, otherwise failures will occur.

As a general rule, try to attach to the entire member, not just a piece of it. Also try to have the load on the structural part compressive, rather than tensile. Figures 15-F and 15-G show cases where local failure caused the failure of the plumber's attachment. Local failure is likely to occur in lumber where screws are placed in the edge because the wood fibers pull apart!

A common problem is the need to carry heavy loads under steel sheet roofing. The support for the roof material is "Z" beams which are formed from sheet

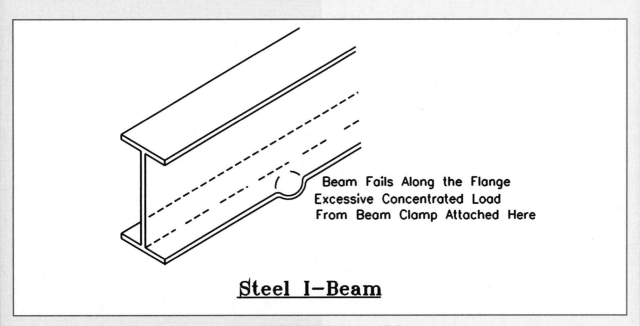

Figure 15-G – Steel I-Beam Local Failure

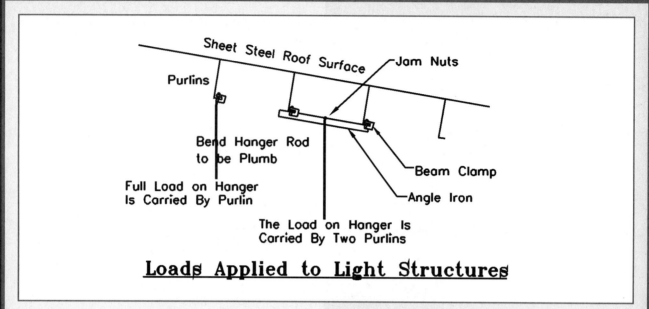

Figure 15-H – Loads Applied to Light Structures

metal. These members are called purlins, and they are sufficiently strong to carry the roof load, but to attach at a single point is a problem! Figure 15-H shows one way to distribute the concentrated load over two purlins. Notice that the load carried by each purlin is only one-half the load in the pipe hanger.

DYNAMIC DOWNLOADS

The actual load seen by the building is much greater if the loads are dynamic, that is, the piping moves in response to shock waves due to pumps starting and stopping, fluid hammer due to rapid closing of valves, or even items like the opening and closing of large, heavy doors.

LESSON 16

CUTTING, DRILLING, AND NAIL PROTECTION FOR BUILDING STRUCTURAL ELEMENTS

Plumbing work almost always involves modifying the building components to install our plumbing work—piping as well as fixtures and appurtenances. This lesson explores the requirements and limitations of what can and cannot be done to fit our work into the building.

What follows are general rules of cuts, notches, and holes that can be placed in the building elements. These limitations come from the 2000 International Residential Code.

The very first limitation, however, is the ruling of the general contractor foreman or the building architect. These people must approve any cuts in the building before you can proceed!

The illustrations in this lesson show the many variations of required openings in the building materials. If at all possible, do not notch any building member if a drilled hole can be used to accommodate our work.

NAIL PROTECTION

Plumbing piping, gas piping, and electrical cables must be protected from punctures by nails, sheetrock screws, or other fasteners that are introduced into studs, joists, or rafters.

To increase the probability of success, the trades people are encouraged to use fasteners that are long enough, but not too long! It doesn't make the job better to use 3" sheetrock screws when all the required strength is achieved with 1" penetration beyond the sheetrock into the wood stud or ceiling joist.

The second line of defense is to place piping or electrical work in the middle of the wall or ceiling structure. If minimum length (but adequate) fasteners are used, this will greatly increase the chances of success.

The last line of defense is placing striker plates on the face of studs or joists over the routing of the line to be protected. The striker plate is a piece of 16 gauge steel that is as wide as the stud or joist, and that is long enough to extend 4" beyond the pipe, tubing, or cable (in each direction).

The figures that accompany this discussion show the basic rules. Be sure to consult your local codes before modifying any structural element.

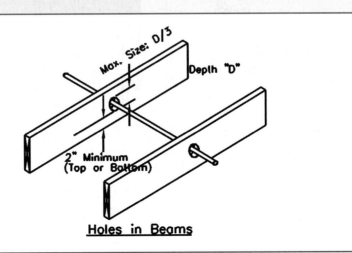

Figure 16-A – Holes in Beams

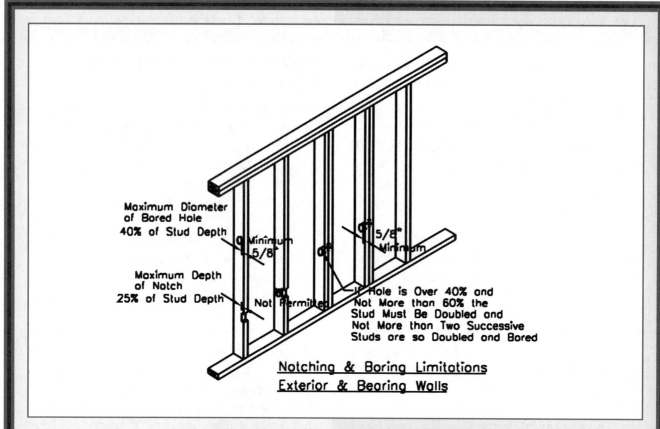

Figure 16-B – Notching & Boring Limitations – Exterior & Bearing Walls

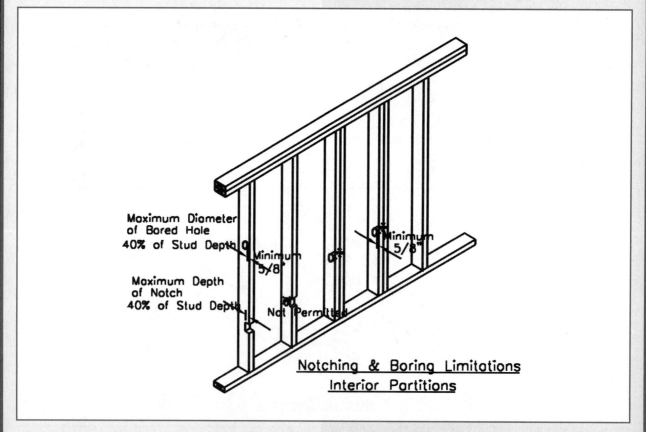

Figure 16-C – Notching & Boring Limitations – Interior Partitions

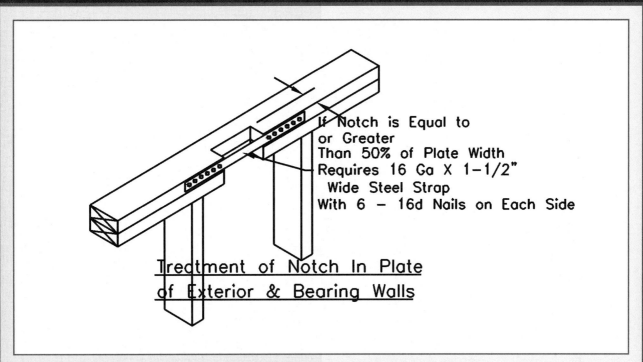

Figure 16-D – Treatment of Notch in Plate of Exterior & Bearing Walls

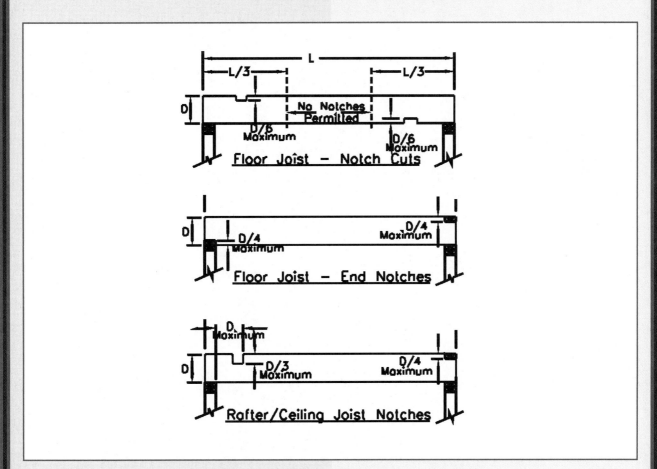

Figure 16-E –
Floor Joist – Notch Cuts
Floor Joist – End Notches
Rafter/Ceiling Joist Notches

LESSON 17

PIPING MATERIALS USED IN PLUMBING WORK – PRESSURE

Plumbing work almost always involves installing or servicing piping systems. Industry experience keeps broadening the range of piping material available to accomplish the needed functions in plumbing systems. This lesson explores many of the requirements, characteristics, and performance of the large variety of pressure pipe selections available to the plumber in executing his or her work.

Before you can select piping materials for the various plumbing systems in a typical building, several factors must be considered, including all of the following (not necessarily in this order):

1. Cost of the piping material and matching fittings

2. Suitability for the job — anticipated life of the facility may change the choices

3. Code requirements

4. Ease of assembly

5. Weight

6. Durability

This lesson will begin the process of material review by considering several piping types for pressure applications. Comments regarding advantages and disadvantages are included about each material described. Local code regulations will inform you about application and acceptability of these products.

COPPER TUBE

Copper tube has both pressure and DWV applications. In this lesson we will describe the manufacturing methods of producing copper tube, as well as the pressure uses of this material.

1. Hot-piercing process

2. Extrusion process

3. Cup and draw process

4. Drawing process

5. Tubing roll process

6. Welded seam

Of all material, copper tube has the widest range of plumbing uses. Types K and L are used for water distribution, water services, and many commercial and industrial applications. Type M is used for water distribution and heat-transfer systems (that is, hot water heating and chilled water cooling). Type DWV is used for drain, waste, and vent. Type ACR is used for refrigerant piping for air conditioning and refrigeration. Most tubing is color-coded for quick identification; however, if the color coding is not present, the tube is also stamped for identification.

Copper is a dense, strong, ductile metal that is resistant to attacks by many chemicals. Consequently, a relatively thin-wall copper tube can provide an excellent piping material that is light weight, strong, and long-lasting.

NOTE: Copper is referred to as tubing, not pipe. Although copper pipe is available by special order, the term pipe implies that the wall thickness of the copper would permit it to be threaded.

TYPE K Type K is the heaviest-wall tubing, and is available in hard-drawn temper in diameters from ¼" to 12" and in lengths up to 20'. It is available in soft-temper coils up to 2" in tube size and up to 100' long. Type K tubing color code is a green stripe and lettering.

TYPE L Type L is the medium thickness wall tubing, available in the same range of outside diameters and soft and hard configurations as Type K. Type L color code is a blue stripe and lettering.

TYPE M Type M is a thinner wall tubing available in hard temper only in 20' lengths. Its color code is a red stripe and lettering.

DWV DWV (drainage-waste-vent) is the thinnest wall tubing; it is suitable for drain, waste, and vent only. It is available in diameters from 1½" to 8" diameter in 20' lengths. DWV tubing color code is a yellow stripe and lettering.

ACR ACR is approximately the same as Type L tubing. It is shipped from the factory already cleaned and dried and with sealed ends. It is used for refrigeration or for medical gases. The tubing is available in 20' hard-temper lengths or 50' soft-temper coils.

TYPE G Type G copper tubing is intended for use in gas distribution. It is made in 20' hard temper lengths and 50' coils.

ADVANTAGES

The advantages of copper are the following:

1. Ease of assembly, with many possible methods
2. Light weight
3. Least bulk (it will fit in a minimum of space) of any piping material available
4. Resistance to many corrosion products
5. Wide variety of types for any job requirement
6. Capable of being formed or shaped
7. Non-combustible

DISADVANTAGES

1. A major disadvantage is the fire hazard created by the solder operations using a torch flame for heating. Where flame-joint is not desirable, alternate joining methods are available, such as roll-grooving, flameless soldering, a press-connection joining method with special tooling, or even epoxying.
2. Rapid tubing erosion caused by some water conditions and flow rates
3. Solder joint or fitting failure in heat-cool stress cycles
4. Relative ease of puncture by external forces
5. Noise when water is flowing inside at high velocities
6. High coefficient of expansion/contraction compared with iron or steel products

GLASS PIPE

Glass pipe is made by the extrusion method and comes in 10' lengths. It is used in special industrial or laboratory pressure and waste piping applications. Although it resists most chemicals, it is expensive and requires special skill to assemble.

Some types of plastic pipe such as CPVC, PVC, and PVDF are now being used in place of glass. Some of these products are available in transparent versions so that the internal condition of the line is always available for inspection.

PLASTIC PIPE

Plastic pipe is made of materials from petroleum-based feed stocks. It is manufactured either by extrusion (drawing plastic stock material through a metal mold by high pressure or vacuum methods), molding (shaping the pipe by thermoforming, using heated rollers to form the desired shape), or casting (pouring molten plastic into a metal mold).

These products, which comes in many forms, can be used for a wide range of applications. Plastic pipe materials also vary widely in cost and suitable applications. The range of applications is listed below:

1. Water service and distribution
2. Drain, waste, and vent (see Lesson 18)

3. Heating and cooling

4. Gas distribution – for fuel gases it is limited to outdoors, buried only

5. Various special industrial applications

Several types of plastics have plumbing uses:

1. PE – polyethylene

2. ABS – acrylonitrile butadiene styrene

3. PVC – polyvinyl chloride

4. CPVC – chlorinated polyvinyl chloride

5. PB – polybutylene

6. PP – polypropylene

7. PEX – cross-linked polyethylene

8. PVDF – polyvinylidenefluoride

9. PFA – polyfluoroalcoxy

10. PTFE – polytetrafluoroethylene

These materials are available in several wall thicknesses or Standard Dimension Ratios.

PVC (POLYVINYL CHLORIDE)

PVC is a tough material that is highly resistant to many chemicals that are typical in building waste streams and in industrial or commercial pressure piping installations.

Pipe is manufactured by extrusion processes. Fittings are made by injection molding methods wherein molten plastic is forced into molds. This produces a high density product and a completely homogeneous fitting.

As a general statement, all plastics must be protected from ultra-violet light. This is the most active component of sunlight. PVC is no exception, and pipe that is to be installed outdoors must have some alloying chemical in the mix when it is manufactured to achieve this protection.

Other advantages of PVC include these: it will not spark when struck; it is safe around explosives and flammable vapors; it has not been proven to build up water contaminants.

The working strength is affected by temperature; therefore, pressure ratings are decreased at higher temperatures. At a later lesson in this series, a more detailed discussion of the limitations of PVC pipe will be explored.

PVC is not to be used for compressed air because it may splinter if it is ruptured suddenly.

PVC pipe is available in various sizes in three pipe schedules — 40, 80, 120 — all differing in wall thicknesses and corresponding pressure ratings. It is available in straight lengths or coiled lengths, depending upon usage. The favorite joining method for pressure applications is to solvent weld the fittings and pipe together (see Figure 18-O).

PVC is also available in SDR sizes (Standard Dimension Ratio), which is a ratio between the outside diameter and the wall thickness.

PVC PRESSURE FITTINGS

Pressure fittings available include the following:

- ells — 90° and 45°, regular and street pattern
- tees
- adapters
- reducers
- couplings
- bushings

For pressure applications, joints are usually made with solvent cement methods (see Figure 17-A), or threaded in Schedule 80 or Schedule 120 piping.

CPVC (CHLORINATED POLYVINYL CHLORIDE)

CPVC is used in water distribution systems for hot and cold supply up to 100 pounds maximum working pressure at 180° F. While it could be used for

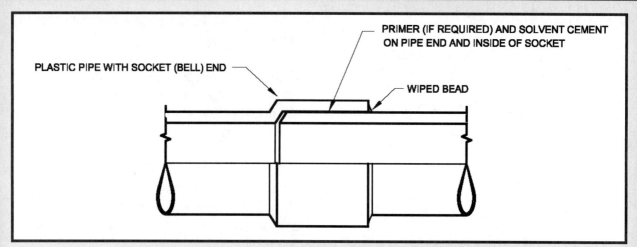

Figure 17-A – A Solvent Cement Joint in Socket (Bell) End Plastic Pressure Pipe

drainage, it is more expensive than PVC or ABS, so it is less likely to be used unless its higher temperature rating is required.

Some pipe manufacturers provide pipe with a socket for joining formed on one end (see Figure 17-A).

PE (POLYETHYLENE)

PE is produced under extremely high temperatures and pressures. It is preferred over other plastics for high pressures because of its burst strength and easy installation. Most applications are outside of buildings.

PE is made in clear form, carbon black dye is added to improve resistance to sunlight. It is usually sold in coiled form.

PE is used extensively for jet wells, water supply systems, farm sprinkler systems, and is also good for salt lines, chemical well lines, and gas-gathering systems. It is workable with mechanical tools and is usually found to be well suited for burial due to its high crush strength. Because PE pipe material is not dimensionally stable (like PVC, CPVC, or ABS), joints are usually made with barbed insert fittings (see Figure 17-B), or by thermal welding (see Figure 17-C).

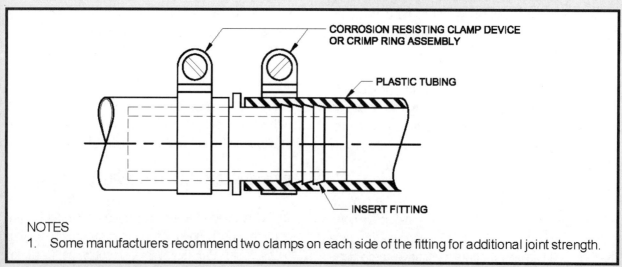

NOTES
1. Some manufacturers recommend two clamps on each side of the fitting for additional joint strength.

Figure 17-B– An Insert Fitting Joint in Plastic Tubing

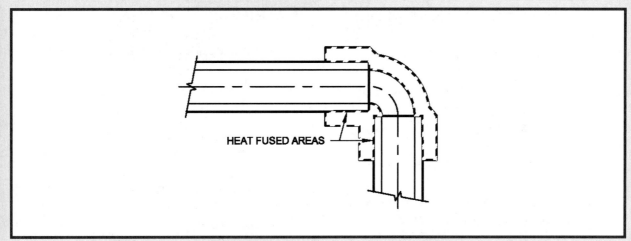

Figure 17-C– A Heat Fused Joint in Plastic Water Piping

PB (POLYBUTYLENE)

NOTE — THIS PRODUCT IS NO LONGER AVAILABLE!

While PB is not available at this time, there is a large amount of it installed that you may encounter in repair and remodeling work. If at all possible, it should be replaced with current pipe or tubing materials.

Polybutylene tubing was popular for water distribution in buildings and gas distribution outside buildings underground. Joints can be made with insert fittings (Figure 17-B) or with heat fusion (Figure 17-C).

PEX

Cross-linked polyethylene is a relatively recent offering to the U.S. plumbing water distribution market, although it has been used for a number of years in Europe. Its pressure rating is higher than other coiled plastic materials. Joints are special insert type. Manifold systems are available with this material.

There will be much more about PEX in the Year Two Apprentice Manual.

PEX-AL-PEX

Polyethylene-aluminum-polyethylene tubing for water distribution is comprised of a sandwiched tubing of cross-linked polyethylene extruded on both sides of a thin aluminum tube. The semi-rigid nature of this tubing allows it to retain its shape when bent with special tools (manually works, too). Brass fittings are used to assemble tubing lengths together.

ADVANTAGES

In general, the advantages of plastics for pressure applications are considered to be the following:

1. Light weight
2. Resistance to many chemicals
3. Ease of installation
4. Excellent flow characteristics
5. Low cost

DISADVANTAGES

The disadvantages of plastics are usually thought to be the following:

1. Plastics may release harmful fumes if exposed to flame
2. Some plastic cements or solvents are very flammable
3. Solvent-cement vapors can be harmful, especially in confined spaces

4. Piping materials can be punctured easily

5. More supporting hangers are required

6. Some types can change from extremely brittle to extremely soft with moderate temperature changes

7. Water-flowing noises are transmitted readily

8. A greater expansion/contraction coefficient than other pipe materials

STEEL PIPE

Steel pipe is manufactured by continuous weld (beginning with steel materials already processed into billets called skelp), ERW (a coil strip that is drawn through forming rolls until rounded, then is fusion welded), or seamless (a round billet is pierced with a rod).

Steel pipe is the heaviest and strongest piping material. Its uses include the following:

1. Drain, waste, and vent

2. Water service and distribution

3. Heating and cooling

4. Snow melting and snow making

5. Radiant panel heating

6. Fire protection

7. Gas distribution

Steel pipe is either galvanized or black. Black pipe has a light coat of black paint applied to protect it in storage.

Steel pipe is available in lengths to 21' (longer lengths on special order), and in a wide variety of wall thicknesses.

ADVANTAGES

The advantages of steel pipe are the following:

1. Durability

2. Great strength

3. Low coefficient of expansion

4. Multiple means of joining

DISADVANTAGES

The disadvantages of steel pipe are the following:

1. Its installation is more difficult and consumes more time than other materials

2. It requires expensive, heavy, specialized tooling

3. It can be corroded by many common liquids

Steel pipe is made to dimensional (outside diameters and wall thickness) standards called Schedules. In the size ranges we normally encounter ($1/_8$" to 4"), standard weight pipe is called Schedule 40 and extra heavy pipe is Schedule 80. Many plastic materials are made to Schedule 40 or Schedule 80 dimensions. (See Table 17-A)

THREADED PRESSURE FITTINGS

There are several types of threaded pressure-rated fittings for use with steel pipe. A brief description of each follows:

Malleable iron

This class is available in standard and extra-strong weights, galvanized or painted black. The fittings are available in a wide variety of ells, 45's, tees, and reducing types. They are low cost and very long-lasting.

Cast-iron

This fitting type is used in pressure piping applications such as heating or cooling systems and fire-protection sprinkler systems. These fittings are available in sizes from 1/4" through 8", and in plain *(black)* or galvanized finishes.

The fittings are available in standard or extra-heavy weights. The standard weight is suitable for working pressures to 175 psig from -20° F to 100° F, commonly called WOG (water-oil-gas) rating. The working pressure at elevated temperatures is 125 psig at saturated steam temperature (353° F). The extra heavy weight is used at pressures to 250 psig saturated steam (406° F), or 400 psig at room temperature. The working pressures are reduced at temperatures above or below the ranges given.

This type material is also selected for piping that could be exposed to heavy mechanical loads, such as drain, blow-down, and feedwater applications for high-pressure steam boilers.

These fittings are very strong and durable. Cast iron fittings will not distort under mechanical or thermal loads — therefore, a pipe joint **made** tight will **stay** tight.

An interesting and useful aspect of the strength and rigidity of cast iron fittings is that they can be cracked (i.e., broken) in place without damaging the pipe in the joint. The skilled mechanic can readily break out a cast-iron fitting to facilitate repair or system modification.

Steel

A line of forged steel fittings is available that includes a wide range of pressure and temperature ratings. These fittings are used for many applications in a variety of commercial and industrial applications.

OTHER PRESSURE FITTINGS

Welding

These fittings are used with black pipe to form very strong systems. They are described in detail in later lessons.

Grooved

These fittings have grooves placed near the fitting end to accept a clamping device to hold the fitting to the pipe. This type is also discussed in detail in later lessons.

Compression

There are several versions of compression fittings to join steel pipe. The many options and size ranges are also described in lessons that follow.

NATIONAL PIPE thread-TAPER

NPT, National Pipe thread-Taper, is a standard which was established in order to insure that the threads on all pipe and fittings would fit together properly. There are product standards for threaded fittings which guarantee consistency of a product size despite the large number of manufacturers making that product.

National Pipe thread-Taper means that the pipe end is tapered to fit the fitting socket which is also tapered at the same angle. The NPT taper is equal to 3/4" per foot (on the diameter). See Table 17-A for details of these threads. Table 17-B lists the dimensions and pressures for steel pipe of various sizes and Schedules (that is, various wall thicknesses).

MALE AND FEMALE

The terms *male* and *female* are used to determine whether a fitting has an external pipe thread or has internal pipe threads. A male thread or spigot can be visually examined as an outside thread or spigot. (*Spigot* and *male* are used synonymously.) A female thread (or hub or socket) must be inspected on the inside.

A simple rule to distinguish a male from a female fitting is that the male end is inserted into the female socket and mechanically secured. Figure 17-D shows a joint with these thread types.

STAINLESS STEEL PIPE

Stainless steel pipe is available for special or unusu-

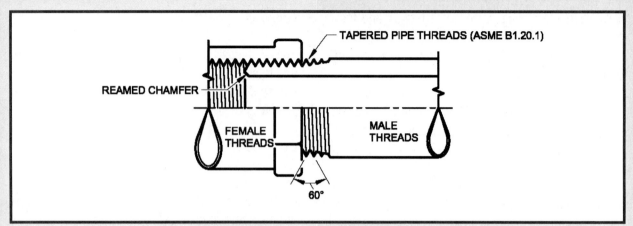

Figure 17-D – A Threaded Pipe Joint

al applications. This material is usually very expensive, but alloys are available that are resistant to most chemical attack. It is available in pressure and drainage grades, and of many "stainless" alloys. Types 304 and Type 316 alloys are among the highest quality stainless steel options. The Type 316 alloy is often installed in chemical waste or below grade applications. Pressure piping is often threaded (if the wall thickness is based on Schedule 40 as a minimum); drainage pipe is joined by o-ring or hubless coupling joints.

CORRUGATED STAINLESS STEEL FUEL GAS TUBING

A relatively inexpensive (for stainless steel) tubing for fuel-gas distribution systems is available. The stainless steel tubing is covered with a bright yellow jacket. It is corrugated for ease in forming bends, to pass around obstructions, and to permit long runs without joints. The best applications are for 2 psi systems, which permit large volume flows with small tubing.

GALVANIC ACTION

An important point which should be examined before the materials lesson is closed is the principle known as galvanic action.

Galvanic action is a chemical reaction between two dissimilar metals (such as iron and copper) when the metals are placed in a solution in which electron transfer can occur. Practically stated, if a copper male adapter is screwed into a female galvanized fitting, a reaction occurs in the presence of fresh water. The copper and iron form a short-circuited battery which eventually leads to corrosion of the iron and degradation of the joint.

Scientifically, the galvanic action principle will be detailed in a later lesson; yet it must be understood that joining two dissimilar metals can lead to problems. Reactions occur only in the presence of fresh water and with two dissimilar metals electrically connected together.

For galvanic action to be a practical problem, three conditions are required:

1. A large amount of copper (or brass) and a small amount of steel must be present

2. The system must contain fresh water

3. The dissimilar metals must be physically connected.

Galvanic action is not a problem under these conditions:

1. In steel pipe systems using brass valves and/or brass supply nipples

2. In drainage systems

3. In closed process systems (heating, chilled

water, etc.)

4. Where insulating unions or couplings are used between dissimilar materials

Table 17-A
Characteristics of Tapered Pipe Threads
National Pipe thread-Taper (NPT)*

Nominal Pipe Size (Inches)	Threads Per Inch	Approximate Length of Thread (Inches)	Approximate Number of Threads to be Cut	Approximate Total Thread Makeup, Hand and Wrench (Inches)
1/8	27	3/8	10	1/4
1/4	18	5/8	11	3/8
3/8	18	5/8	11	3/8
1/2	14	3/4	10	7/16
3/4	14	3/4	10	1/2
1	11 1/2	7/8	10	9/16
1 1/4	11 1/2	1	11	9/16
1 1/2	11 1/2	1	11	9/16
2	11 1/2	1	11	5/8
2 1/2	8	1 1/2	12	7/8
3	8	1 1/2	12	1
3 1/2	8	1 5/8	13	1 1/16
4	8	1 5/8	13	1 1/16
5	8	1 3/4	14	1 3/16
6	8	1 3/4	14	1 3/16

* Consult industry standards and manufacturer's data for additional information. Variations for drilling and tapping may affect these dimensions.

Table 17-B
USS Standard Pipe and Line Pipe — Continuous Butt-welded Steel Pipe
A53 and API 5L Dimensions, Weights, Test Pressure

Size Nominal and (O.D.)	Wall Thickness	Plain End Weight	Nominal T & C Weight		Class	Schedule Number	Applicable Specifications‡		Test Pressure*
			Standard Pipe** Coupling	Line Pipe Coupling			A-53	API 5L	
Inches	Inches	Lbs per Ft	Lbs per Foot						psi
1/8 (.405)	0.68	.24	.24	.25	Std	40	x	x	700
	0.95	.31	......	.32	XS	80	x	x	850
1/4 (.540)	.088	.42	.42	.43	Std	40	x	x	700
	.199	.54	......	.54	XS	80	x	x	850
3/8 (.675)	.091	.57	.57	.57	Std	40	x	x	700
	.126	.74	......	.74	XS	80	x	x	850
1/2 (.840)	.109	.85	.85	.86	Std	40	x	x	700
	.147	1.09	......	1.09	XS	80	x	x	850
3/4 (1.050)	.113	1.13	1.13	1.14	Std	40	x	x	700
	.154	1.47	......	1.48	XS	80	x	x	850
1 (1.315)	.133	1.68	1.68	1.70	Std	40	x	x	700
	.179	2.17	......	2.18	XS	80	x	x	850
1¼ (1.660)	.140	2.27	2.28	2.30	Std	40	x	x	1000
	.191	3.00	......	3.02	XS	80	x	x	1300
1½ (1.900)	.145	2.72	2.73	2.75	Std	40	x	x	1000
	.200	3.63	......	3.66	XS	80	x	x	1300
2 (2.375)	.154	3.65	3.68	3.75	Std	40	x	x	1000
	.218	5.02	......	5.07	XS	80	x	x	1300
2½ (2.875)	.203	5.79	5.82	5.90	Std	40	x	x	1000
	.276	7.66	......	7.73	XS	80	x	x	1300
3 (3.500)	.188	6.63	......	......	....	....	..	x	1000
	.216	7.58	7.62	7.70	Std	40	x	x	1000
	.300	10.25	......	10.33	XS	80	x	x	1300
3½ (4.000)	.188	7.63	......	......	....	....	x	x	1200
	.226	9.11	9.20	9.25	Std	40	x	x	1200
	.318	12.51	......	12.63	XS	80	x	x	1700
4 (4.500)	.156	7.25	......	......	....	....	x	x	1000
	.188	8.64	......	......	....	....	x	x	1200
	.219	10.00	......	......	....	....	x	..	1200
	.237	10.79	10.89	11.00	Std	40	x	x	1200
	.337	14.98	......	15.17	XS	80	x	x	1700

* Test pressures shown are standard USS test pressure which equal or exceed specification test pressures.
** It is standard practice to furnish T & C extra strong pipe with a line pipe coupling.
‡ "X" indicates that size is made to the indicated specifications.
The weight per foot of pipe with threads and couplings is based on a length of 20 ft. including the coupling.

Data from U.S. Steel Information

LESSON 18

PIPING MATERIALS USED IN PLUMBING WORK – DWV

This lesson continues the subject of piping selection, with emphasis on DWV applications.

Cast-iron soil pipe, glass pipe, vitrified clay pipe, concrete pipe, and plastic pipe — with appropriate fittings — are used for drain, waste, and vent systems and storm water systems. As noted below, even aluminum pipe was offered for a few years. The many manufacturers in these industries have improved their products and assembly methods over the years, making for a highly competitive marketplace so that contractors, engineers, and building owners have a very wide range of options to choose from.

ALUMINUM

For a short time, aluminum pipe was available for DWV applications. While it was generally conceded to be satisfactory for its intended uses, the manufacturer was unable to develop enough sales volume to continue making the product.

CAST-IRON SOIL PIPE

Cast-iron soil pipe is made of gray iron with a metallurgical makeup which includes graphite flakes throughout the iron material. The inner and outer surfaces of the pipe (and fittings) contain oxides of iron-graphite flakes, thereby protecting the pipe from further corrosion. Since cast-iron pipe is a carbon-iron alloy product, the pipe is brittle.

Raw materials used in the manufacture of cast-iron soil pipe include pig-iron, scrap-iron and steel, metallurgical coke, and limestone.

The traditional method of manufacturing cast-iron pipe starts with placing all of the raw materials in a large melting furnace (called a cupola) and heating them to 2700° F or above. The molten material is poured into sand or metal molds. This method is called static casting which means that the mold remains stationary while the liquid iron is poured into it.

Sand molds are pre-formed shapes with hollowed areas forming the pipe shapes. A core is placed in the center of the mold to form the hollow area. This shell around the hollow area is the actual pipe.

BASIC CAST-IRON FITTING DESCRIPTION

Bends, which are the most common fittings, are described in terms of fractions of a full circle. For more complicated fittings, the straight path is read first and then the side branch(es). If the branch is tapped, it should be stated. If the fitting is hubless, it should be stated. For example, a 4" tee with a 2" branch is needed in hubless style. It would be called 4 x 2 hubless tee. If the branch is desired to be tapped, it would be *4 x 2 hubless tapped tee*. (The trade name would be a *4"x2" no hub tapped tee*.)

PHYSICAL CHARACTERISTICS OF CAST-IRON SOIL PIPE

Cast-iron pipe and fittings are strong and durable, but the material is brittle and cracked pipe or fittings may occasionally be encountered. This damage can be caused by rough handling in shipping or on the job, although some (usually very few) damaged pieces are shipped from the manufacturer's plant. The careful mechanic will check each piece as the job is installed. A light hammer tap will cause sound pipe or fitting to ring, whereas a cracked product will sound a dull thud.

An additional problem — much harder to check — is a product which has a sand hole from the casting process. Although the tar coating tends to seal these holes, they will show up eventually, sometimes long after the job is put in service.

Pipe and fittings are marked to size, weight class (service or extra heavy), and manufacturer.

Cast-iron soil pipe is available in two product types:

hubbed (that is, bell and spigot) and hubless. Hubbed pipe is available in various configurations and in two weights. Figure 18-A shows two forms of hubbed pipe — single hub and double hub.

The industry is undergoing a change at this point and many manufacturers are replacing double hub with 42" single hub pieces.

The double-hub pipe has a hub on each end, so that short pieces can be cut from these lengths with minimum pipe waste. (See Figure 18-A)

HUBLESS

Hubless pipe and fittings are made in service weight only. Historically, the difficulty of maintaining a constant pipe-wall thickness was overcome by using extra-heavy weight. Improvements in foundry technology since the 1950's and '60's have enabled the producers to make an excellent product with the standard-weight wall. This development results in material saving, reduced shipping costs, and easier installations, so everyone benefits.

Most fitting configurations of hubless and hubbed are similar so we will describe them only once.

Each pipe type is available from 1½" to 15" in diameter, and 5' or 10' in length. All hubbed fittings have hub(s) at the inlet(s) and a plain end at the outlet.

All illustrations in this section are due to the courtesy of the Cast Iron Soil Pipe Institute, from their Cast Iron Soil Pipe Handbook.

CAST-IRON SOIL PIPE FITTINGS

BENDS

Figure 18-B shows a typical bend. Bends are fittings that make a turn. They are described in terms of fractions of a full circle. Usual bends are 1/4 (90° turn), 1/5 (72° turn), 1/6 (60° turn), 1/8 (45° turn), and 1/16 (22½° turn). Long-turn bends with the greatest radii are called sweeps. One-quarter bends and sweeps are available with a foot, which is a projection that looks like a support leg. The sweep is placed at the foot of a vertical stack and the fitting is placed on a support (usually concrete) to maintain the stack in place.

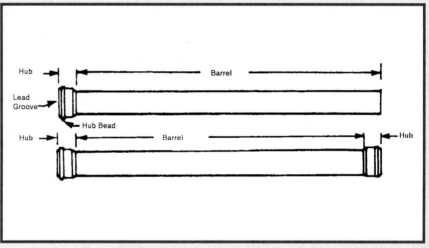

Figure 18-A – Single Hub and Double Hub Cast-Iron Soil Pipe

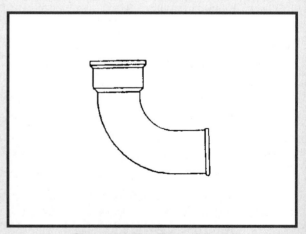

Figure 18-B – Typical Bend

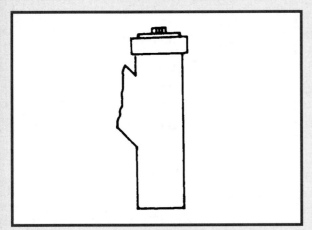

Figure 18-C – Y Branch With Cleanout Plug

WYES

Wyes are shaped only somewhat like the letter Y. The main flow path is straight, and the side branch enters at an angle — usually 45°. These fittings are used to join two lines into one line leaving the fitting. They are made the same size on all three openings, or the side opening is made smaller than the straight-through pathway. Figure 18-C shows a wye fitting that has a tapped cleanout plug in the run position.

SANITARY TEES

Sanitary tees are shaped approximately like the letter T. The side branch enters at 90° to the straight path but with a small radius to enter the main flow. Tee branches are made the same size as straight-through branches as well as the side opening smaller than the straight-through pathway. Figure 18-D shows a hubbed tee.

TEE-WYE

A tee-wye is a combination of a wye and a $1/8$ bend. This fitting directs the flow from the side branch to the outlet end, as shown in Figure 18-E.

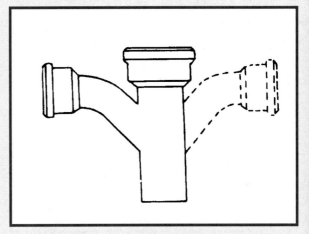

Figure 18-E – Tee Wye and Double Tee Wye

DOUBLES

Wyes, tees, and tee-wyes are also available in double configurations. Generally, these double fittings may be used only in vertical arrangements. Figure 18-D and 18-E also show the double configuration.

TAPPED TEES

Tapped tees are a variation of the tee where the side opening is tapped with threads rather than a hub.

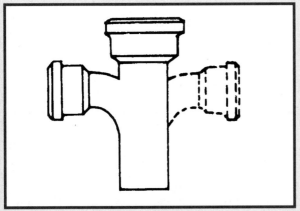

Figure 18-D – Sanitary T Branches, Single and Double

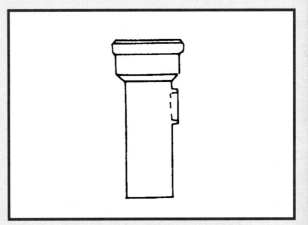

Figure 18-F – Tapped Vent Tee

Tapped tees are made in sanitary tee (for drainage) or straight tee (for venting) configurations. Figure 18-F shows a tapped vent tee.

TEE CLEANOUT

A tee cleanout is a special tee with a side tapping and cleanout plug that facilitates testing or provides access for rodding. See Figure 18-G.

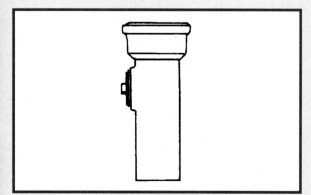

Figure 18-G – Tee Cleanout

OFFSET

An offset is a fitting that moves the flow line laterally without changing direction. The turns within the fitting are at 45° to assure satisfactory flow. See Figure 18-H.

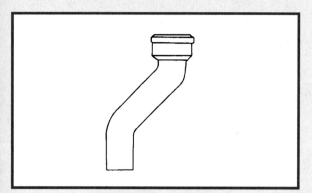

Figure 18-H – Offset

REDUCER

A reducer has an inlet that is smaller than its outlet, as shown in Figure 18-I.

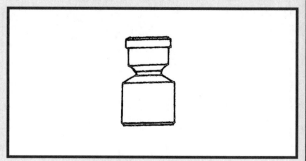

Figure 18-I – Reducer

INCREASER

An increaser has an inlet that is larger than its outlet. See Figure 18-J.

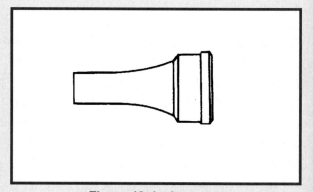

Figure 18-J – Increaser

P TRAP

A P Trap is a trap fitting with a vertical inlet and horizontal outlet, as shown in Figure 18-K.

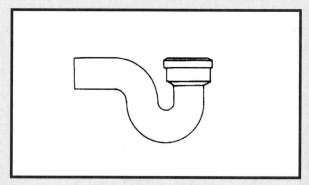

Figure 18-K – P-Trap

RUNNING TRAP

A running trap is a trap with horizontal inlet and outlet. See Figure 18-L.

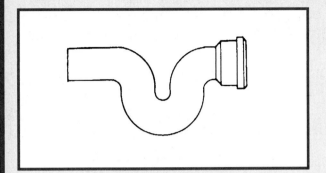

Figure 18-L – Running Trap

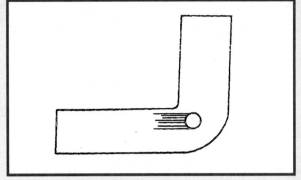

Figure 18-N – Closet Bend With Tapped Side Outlet

FERRULE

Ferrules are used to blank off a hub. These are usually tapped and provided with a cleanout plug. See Figure 18-M.

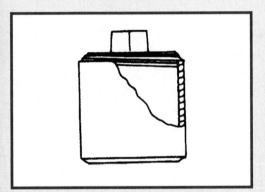

Figure 18-M – Ferrule and Plug

CLOSET BEND

A closet bend is similar to a ¼ bend except there are straight segments on each side of the bend, and the fitting does not have a hub. Closet bends are also available with 1½" or 2" tappings. See Figure 18-N for an example of a tapped closet bend.

READING CAST-IRON FITTINGS

We have learned that bends are expressed in fractions of 360°. The rules for reading wyes, tees, tee-wyes, double tees, etc. are as follows:

How to Determine Right or Left Hand Inlets

Use the following illustrations and descriptions to determine whether a fitting has a right or left hand inlet.

Right Hand Illustrated

- **Closet Bends and P Traps** — Place the inlet upright near you, with the spigot facing away. If the side inlet is on your right, it is a right hand inlet.

Right Hand Illustrated

- **All Branch Fittings** — Place the spigot near you, with the branch facing upwards. If the side inlet is on your right, it is a right hand inlet.
- **All Bends** — Place the fitting upright, with the spigot near you. If the side inlet is on your right, it is a right hand inlet.

Figure 18-N(2) – How to Determine Right or Left Hand Inlets (Charlotte Pipe and Foundry Company)

1. Read the fitting with the straight-through size

2. Read the side inlet size (not necessary if it is the same as straight-through)

3. Indicate whether the inlet is tapped and if so, what size the tap is.

EXAMPLE

If you are installing a system in hubless pipe and need a closet bend with two side inlets of two-inch diameter size, you ask for a 4" x 4" x 2" x 2" double outlet hubless closet bend.

When ordering a 4" tee with a 2" hub branch, you ask for a 4" x 2" tee.

If that tee had to have a tapped branch instead of a hubbed (caulked branch), you would order a 4" x 2" tapped tee.

CAST-IRON THREADED FITTINGS USED WITH GALVANIZED STEEL PIPE

DURHAM FITTINGS

A *fine-grained* cast-iron material is used to manufacture threaded cast-iron drainage fittings. One series of these products, called *Durham* fittings, is intended for drainage applications. The inside diameter of the Durham fitting is held close to the inside diameter of the matching pipe. When assembled, the flow path through the system does not contain enlargements at the fitting, thereby producing (it is alleged) a superior flow-way for gravity drainage applications.

The fittings that make an approximate 90° turn (ells and tees) are tapped to actually give a slope of ¼" per foot. That is, the actual angle between tappings is a little less than 92°. These fittings are available from 1¼" through 2" and from 3" through 6" sizes.

AVAILABLE PATTERNS

Some available patterns are the following:

90° elbows   22½° elbows
45° elbows   11¼° elbows
60° elbows   Sanitary tees
Wyes   Tee-wyes
    many, many special types

These fittings are covered with a paint or coal-tar coating which protects them both in storage and in service.

The completed system of Durham fittings and steel piping produces a stronger system than any other drainage system. This strength is important in larger jobs or high-rises, where total building movement might overstress other materials.

DURIRON PIPE

A special proprietary pipe that is resistant to many aggressive chemicals is known as Duriron. It is similar to cast iron, but it is alloyed with materials that increase the chemical resistance. The material is very hard, and it is generally considered to be somewhat more difficult to work with.

GLASS PIPE AND FITTINGS

Glass pipe and fittings are used in special drainage systems. For moderate pressure piping applications for aggressive chemicals, see Lesson 17. Both drainage and pressure systems are used in chemical plants, food and beverage plants, pharmaceutical plants, textile and paper mills, and laboratories.

Glass pipe and fittings are highly resistant to most corrosive chemicals, have a low thermal coefficient of expansion, and are non-absorbent. The piping can be cleaned vigorously (e.g., with steam) and, being transparent, it can be visually inspected throughout its length.

Glass pipe is available in lengths up to 10' and in all the usual diameters for drain, waste, and vent applications. Operating pressure limitations for glass piping are 60 psig up to 3" size, 35 psig from 3" to 4" size, and 20 psig for over 4" to 6" size.

AVAILABLE FITTINGS

Available fittings include:

90° elbows	60° elbows
Tees	45° elbows
22½° elbows	Crosses
As well as a variety of reducers	

Glass fittings are joined with a gasket and clamp system. These joints are discussed in a later lesson.

Glass piping must be supported adequately with hangers that are plastic-coated or similarly protected, otherwise the glass will be scratched. If the glass is scratched, local stresses can cause a crack and subsequent failure.

Some words of caution about the contents of a glass piping system: such systems are installed for very aggressive chemicals. The discharge of the glass system must be neutralized before it enters the building general-purpose drainage system. Unusual chemicals — aggressive or not — may not be introduced into a municipal or private sewer system. All such chemicals must be made harmless before they are passed into any waste system.

Furthermore, to properly clean or dismantle a system, you must know its contents and proper procedures for safely dealing with them.

VITRIFIED CLAY SEWER PIPE AND CONCRETE PIPE

Vitrified clay sewer pipe and concrete pipe are used for sanitary and storm sewers and for infiltration and exfiltration applications. Clay pipe and concrete pipe applications are limited to installations that are outside of buildings.

USES

Clay and concrete pipe are available unperforated, perforated for infiltration applications and perforated for exfiltration application. Unperforated pipe is used for sewage, industrial wastes, and storm water. Pipe with perforations along the barrel for exfiltration is used for leaching fields for septic tanks or a leaching field for storm water disposal. Pipe used for infiltration is installed for foundation drains, swamp draining, or farm field draining.

Like all buried piping, clay and concrete piping must be laid on a firm, continuous bed. Before the pipe can be installed, unstable or swampy soil must be replaced by a bed of gravel or concrete, which is then covered with a few inches of sand as the final pipe support.

Concrete pipe is somewhat more susceptible than vitrified clay pipe to attack by the chemicals in sewage. Concrete is more often used on the large sewers where it is the lowest cost material.

Concrete pipe formerly was made with two different aggregates — asbestos cement for pressure applications, and gravel for gravity sewers. Asbestos is no longer used because of the serious hazard to long-time workers exposed to asbestos fibers and dusts.

Concrete pipe is reinforced with steel mesh so that it can achieve the required strengths as called for in the product standards. Clay and concrete pipe standards have requirements that assure a long, successful service life for pipe materials that conform to the standard.

General characteristics of both pipes are similar and similar fitting types are available.

Clay pipe is made in diameters of 4" to 42"; concrete pipe is also made in these diameters and in much larger sizes.

Pipe lengths are usually 2', 3', or 5'.

Joints are hub-and-spigot design; joint filler material and configurations vary from one manufacturer to another.

Pipe walls are about ¾" thick in the smallest sizes; thickness increases as the pipe diameter increases.

Clay and concrete pipe, like all materials, are marked as to weight, size (inside diameter), manufacturer, and may include the manufacturer's address. Weight classifications for clay pipe are either **Extra Strength** or **Standard Strength**.

FITTINGS

Relatively few fitting types are made because most sewers are laid in straight segments with size changes, turns, and elevation changes occurring only at manholes. In smaller sizes (4"-8"), ells, tees, wyes, and reducers are available. Short-pattern fittings, tapped tees, etc., are not made in clay and concrete pipe because sewage or storm waste easily clog, or because the pipe material is not suitable (for example, for threading).

ELLS

Ells are made in the following angles:

- 90°
- 45°
- 30°
- 22½°

TEES

Tees are made with a radiused turn on the side branch. This radiused turn provides better flow characteristics and helps sewer cleaning tools follow the direction of the flow.

WYES

Wyes are the preferred fitting whenever the layout will permit them.

STANDARDS REQUIREMENTS

Vitrified clay pipe and concrete pipe are manufactured to meet the requirements of ASTM documents. ASTM is the American Society for Testing and Materials. This organization provides the framework for the development of industry-accepted "definitions" of what copper tubing is, how to test asphalt paving, what steel pipe is, etc., to an almost unimaginable wide variety of materials and products. Because reputable manufacturers make products to meet the applicable ASTM standards, you may be assured that the roof shingles (or whatever) you buy are really what the label says they are.

CRUSHING STRENGTH

Crushing strength is the resistance of the pipe to collapse from the pressure of the earth on the pipe. Minimum strength requirements and test methods are described in the standard. Random samples of production pipe and fittings are tested for conformance to all the requirements of the standards.

ABSORPTION

Absorption is the characteristic of a material to hold water in its pores. These pipes must have a very low absorption rate. Consequently, they will not form bacterial breeding sites. Specifications for clay pipe require that absorption cannot exceed eight percent. Further, the pipe may be glazed with a fired-on glaze coating, for additional resistance to absorption.

HYDROSTATIC PRESSURE TESTING

Product samples must be pressure tested to demonstrate soundness of production runs. The pipe or fitting is subjected to the required pressure exerted on the inner walls for a stated minimum period of time and must not show any leakage after the test is completed.

ACID RESISTANCE

Acid resistance is the ability of material to be in contact with acids without being affected. The standard describes the acid type, temperature, duration of test, and method of measuring any acid attack.

DIMENSIONS

Dimensions of the finished products are shown in the standards. These include diameter; wall thickness; length; hub geometry where applicable; and, for fittings, angles and laying lengths.

TOLERANCES AND DEFECTS

Tolerances and defects permitted are also listed in the standard. Since these products are placed in forms and cured in the manufacturing process, each item will be unique. By knowing the permitted tolerances, the factory or the field inspector can intelligently choose which pipe or fitting can be used and which must be rejected. Permitted tolerances concern dimensions, out-of-straight-alignment limits, and maximum size and number of surface blemishes.

STRAIGHTNESS

The pipe must not be more than 1/16" per foot out of straight alignment.

BLISTERS

The finish must not be blistered over 3" in diameter.

CHIPS

Chips must not exceed one-quarter of the wall thickness.

PLASTIC PIPE AND FITTINGS

Plastic piping materials have taken giant strides for acceptance in the last 30 years. The choices of available plastic materials have increased dramatically over the years, increasing productivity and decreasing costs for systems using these products. Material changes over the past century can be seen as a series of transitions from lead to cast-iron or steel, to brass, to copper, and to plastics. Plastics for DWV uses have become a large part of plumbing, especially in residential systems.

Plastics are used in water supply, water distribution, drain-waste-vent, storm drainage, underground gas distribution, and, in some cases, sewage trunk and lateral-line systems. Since the general topic of this lesson is on waste and vent applications, the balance of this presentation will be on that usage — the discussion of pressure-piping is in Lesson 17.

THERMOPLASTIC

Thermoplastic pipe is a major subgroup of plastics. This material softens at higher temperatures and becomes hard at lower temperatures. Thermoplastic pipe dates back to 1923 when PVC was first produced in Europe. It was found to be smooth and chemically resistant to many common materials that are typically present in building sewer systems. The plumbing industry in the USA started working with plastics in the early 1950's. Varieties include the following:

1. PVC — polyvinyl chloride

2. ABS — acrylonitrile butadienestyrene

3. CPVC — chlorinated polyvinyl chloride

4. PE — polyethylene

5. PB — polybutylene

6. PEX — cross-linked polyethylene

7. PEX-AL-PEX — polyethylene-aluminum-polyethylene

The first three of the above list are commonly used in drainage work, as well as in pressure applications. The latter group of four types are usually found in pressure applications. For a complete view of these products, all seven will be described briefly, but the pressure types were described in Lesson 17.

PVC, ABS, and CPVC should not be used for compressed gases because the failure mode in some circumstances for these three materials may be to shatter suddenly. Such a failure can propel fragments at some distance. For the same reason, DWV applications of these piping materials should not be tested with compressed air, even at pressures as low as 5 psi.

PVC (POLYVINYL CHLORIDE)

PVC is a tough material that is highly resistant to many chemicals that are typical in building waste streams. Pipe is manufactured by extrusion processes.

PVC is available in most IPS (Iron Pipe Size) sizes. This sizing method is described in Lesson 17. IPS was adopted by the plastics industry to conform with the sizes of steel pipe so that in many installations steel pipe, and especially steel pipe fittings — are used where advantageous in the pipe run.

There are many applications for PVC material that do not need the same dimensions as steel or cast-iron pipe, so in those cases, PVC products made to tubing dimensions are preferred. Thus, a great range of product is available so that many, many applications can be served with PVC pipe and tubing.

PVC is also suitable for highly corrosive industrial applications involving acids, alkalies, alcohols, salts, and other chemicals. It is also used in gas transmission and salt-water disposal systems.

As a general statement, all plastics must be protected from ultra-violet light. This is the most active component of sunlight. PVC is no exception, and pipe that is to be installed outdoors must have some alloying chemical in the mix when it is manufactured to achieve this protection.

Other advantages of PVC include these: it will not spark when struck; it is safe around explosives and flammable vapors; it has not been proven to build up water contaminants.

PVC comes in ivory, gray, white, and yellow (standard colors), but white is the standard color for plumbing use.

The favorite joining method for pressure applications is to solvent weld the fittings and pipe together (see Figure 18-O).

PVC is also available in SDR sizes (Standard Dimension Ratio), which is a ratio between the outside diameter and the wall thickness.

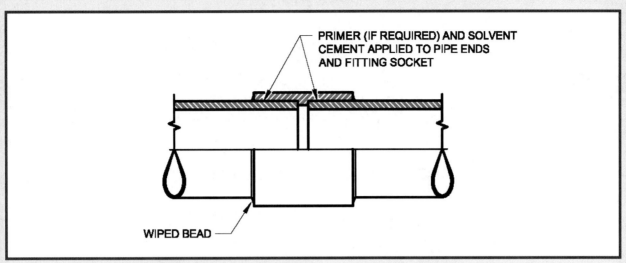

Figure 18-O – A Solvent Cement Joint in Plastic DWV or Water Piping

PVC DRAINAGE FITTINGS

Drainage fittings available are similar to those of copper and include the following:

- ells — 22½°, 45°, 60°, and 90° patterns
- street configurations for the 45° and 90° pattern
- tees, wyes, tee-wyes — double and single configurations
- adapters — male and female
- reducers
- slip and standard couplings
- P-traps, solvent welded and screwed
- increasers
- cast-iron hub adapters
- closet flanges
- side outlet closet ells

Joints may be made with solvent cement methods (see Figure 18-O), with elastomeric gaskets (see Figure 18-P), or with compression bands and elastomeric gaskets (see Figure 18-Q).

ABS (ACRYLONITRILE BUTADIENE STYRENE)

Applications for ABS are similar to the drainage uses for PVC. It is suitable for inorganic acids, bases, salts, hydrocarbons, glycols, and some alcohols. It is also good for transportation of salt water, crude oil, and gas and is used extensively for sewage systems.

Pipes and fittings are injection or extrusion molded. It is available in IPS sizes, similar to PVC, and in various wall thicknesses, depending on the job requirements.

ABS has good chemical resistance, is tough and has excellent tensile strength.

ABS by nature is colored ivory. Carbon black is added to increase resistance to ultra-violet deterioration.

Fittings available are similar to those of PVC.

Pipe and fittings are joined by cementing (see Figure 18-O), or by compression bands and gasket (see Figure 18-Q).

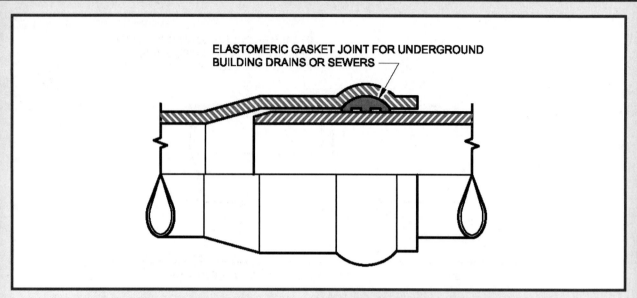

Figure 18-P – An Elastomeric Gasket Joint for Underground Plastic DWV Piping

THERMOSETTING

Thermosetting is the second type of plastic pipe available. This material changes chemically when initially formulated and cannot be reshaped by heating.

Thermoset plastics are used occasionally in chemical plants, etc., where reactive materials and the heat generated from the reactions only help to stabilize the pipe. Thermoset pipe tends to harden with more heat while thermoplastic pipe softens at higher temperatures. Thermoset pipe is reinforced with glass fibers.

Plastic materials are available in many types and configurations. In many cases, making the right selection is truly difficult. Study the characteristics of the plastics you work with to become familiar with them. Typical joints are found in Figures 18-O through 18-T.

Thermoset pipe is made of two major materials, epoxy or polyester.

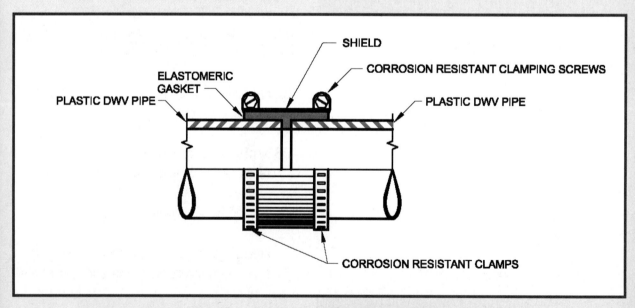

Figure 18-Q – A Shielded Coupling on Plastic DWV Piping

EPOXY

Epoxy pipe is generally made in sizes 2" to 12" I.D. and is used for waste process lines, chemical plants, and salt water transport. Unlike thermoplastic pipe, it will not build up paraffin; therefore, it is used in many food and beverage plants. Fittings are standard threaded IPS sizes or socket solvent-weld sizes.

POLYESTER

Polyester pipe is made in diameters up to 60" I.D. It is suitable for temperatures up to 250° F and is used extensively for acid, alcohol, bleaches, alkalies, and organic transports.

MANUFACTURING METHODS

Thermoset pipe is manufactured by one of three methods:

Centrifugal cast

The plastic material is poured into a spinning metal mold. The spinning forces the material to the sides for uniform wall-thickness pipe.

Hand lay-up process

Multiple layers of glass reinforcing material and plastic resin are applied over a core (or inside mold).

Filament winding process

A plastic-coated fiberglass filament is interwoven around a spindle. This is then coated again with plastic, heated, and treated.

STANDARDS

Throughout this lesson (and in many others), reference is made to standards. Standards are developed by industry groups made up of individuals representing various interests. Thus, the standards reflect a balance of the requirements of manufacturers, contractors, engineers, inspectors, labor, government, and the public. Standards have been developed for specific products, for installation methods, for systems, and for test methods. Since most of the persons working to develop such standards are volunteers, these standards are frequently called voluntary standards.

STANDARD OVERSEEING ORGANIZATIONS

Many organizations sponsor or oversee these standards-preparing activities. Among the principal groups that are active in the plumbing and heating industry standards-writing are these:

AGA — American Gas Association
Suite 450
400 N. Capitol St., NW
Washington, D.C. 20001
202-824-7000
http://www.aga.org

AHAM — Association of Home Appliance Manufacturers
Suite 402
1111 19th Street., NW
Washington, D. C. 2000
202-982-5955
http://www.aham.org

ANSI — American National Standards Institute
1819 L Street, NW
Suite 600
Washington, D.C. 20036
202-293-8020
http://www.ansi.org

ASME — American Society of Mechanical Engineers
Three Park Avenue
New York, New York 10016-5990
212-591-7000
http://www.asme.org

ASPE	—	American Society of Plumbing Engineers 8614 W. Catalpa Ave., Suite 1007 Chicago, IL 60656-1116	CSA	—	Canadian Standards Association 5060 Spectrum Way Mississauga, Ontario L4W 5N6 Canada 416-747-4000 http://www.csa.ca
ASSE	—	American Society of Sanitary Engineering International Office 901 Canterbury, Suite A Westlake, Ohio 44145 440-835-3040 http://www.asse-plumbing.org	FS	—	Federal Specification General Services Administration Specification Section Room 6039 7th & D Streets Washington, D.C. 20407 http://www.gsa.gov
ASTM	—	American Society for Testing and Materials 100 Barr Harbor Drive West Conshohocken, Pennsylvania 19428-2959 610-832-9585 http://www.astm.org	GAMA	—	Gas Appliance Manufacturers Association 2107 Wilson Boulevard, Suite 600 Arlington, Virginia 22201 703-525-7060 http://www.gamanet.org
AWS	—	American Welding Society 550 N. W. LeJeune Road Miami, Florida 33126 305-443-9353 http://www.aws.org	IAPMO	—	International Association of Plumbing and Mechanical Officials 5001 E. Philadelphia Street Ontario, California 91761 909-472-4100 http://iapmo.org
AWWA	—	American Water Works Association 6666 W. Quincey Avenue Denver, Colorado 80235 303-794-7711 http://www.awwa.org	NFPA	—	National Fire Protection Association 1 Batterymarch Park Quincy, Massachusetts 02169-7471 617-770-3000 http://www.nfpa.org
CISPI	—	Cast Iron Soil Pipe Institute Suite 419 5959 Shallowford Road Chattanooga, Tennessee 37421 423-892-0137 http://www.cispi.org	NSF	—	National Sanitation Foundation NSF International P. O. Box 130140 789 N. Dixboro Road Ann Arbor, Michigan 48113-0140 734-769-8010 http://www.nsf.org

PDI — Plumbing & Drainage Institute
Suite 300
800 Turnpike Street
North Andover,
Massachusetts 01845
978-557-0720
http://www.pdionline.org

UL — Underwriters' Laboratories, Inc.
333 Pfingsten Road
Northbrook, Illinois 60062-2096
847-272-8800
http://www.ul.com

Many other organizations contribute time and effort to the standards preparation process. Consensus standards have been developed that describe and define our products and methods. As a consequence, everyone involved in the manufacture, sale, installation, and use of plumbing materials can have access to the knowledge of the details, limits, and parameters of our products and methods.

To aid in recognizing that products meet standards requirements, many organizations authorize the use of their symbols on products that conform to standards requirements. Such symbols include the following:

1. ASME (see above)

2. ASSE (see above)

3. ASTM (see above)

4. CSA (see above)

5. NSF (see above)

6. UL (see above)

7. UPC (Uniform Plumbing Code produced by IAPMO)

Such symbols on our products assure us that they are of high quality and suitable for plumbing applications. Look for these markings and related standards numbers when purchasing pipe, fittings, and other plumbing products.

LESSON 19

JOINING METHODS AND MATERIALS FOR DWV PIPING

This lesson presents joining methods for cast-iron soil pipe fittings; cast-iron drainage fittings; and glass, aluminum, clay, and concrete pipe and fittings.

CAST-IRON SOIL PIPE — HUBBED

Cast-iron soil pipe in the hubbed style is assembled by either caulked joints or by compression gasket joints. First the pipe is measured and cut to required length. The true running length of the pipe is measured from the base of the hub, not the end of the hub. The pipe may be cut with a hammer and cold chisel, a snap cutter, a ratchet cutter, or a power saw — all as described in Lesson 4.

Saw cuts are made using special abrasive wheel saws. Safety goggles are mandatory for these operations. Follow manufacturer's instructions when using these powerful machines.

To cut with a hammer and chisel, first mark completely around the pipe, using a grease pencil. Score this mark by using light hammer blows on a chisel until the entire circumference is **dented** by the chisel. Continue the scoring with a stronger force until the pipe breaks. Use a block of 2" x 4" lumber as a pipe bedding for this operation. Always wear safety equipment when working with cast iron. Certain types of spun pipe cannot be cut this way, but must be sawed.

To cut with a snap cutter or ratchet chain cutter, position the chain around the pipe so that the maximum number of wheels are in contact with the pipe. Before cutting the pipe, score it by applying light pressure with the cutter and rotating the cutting tool a few degrees, then make the final cut by using the leverage of the cutter mechanism to force the cutter wheels into the pipe wall.

After cutting the pipe, peen (with a hammer) or file the inside and outside diameters to remove all burrs.

CAULKED (See Figure 19-A)

If a caulked joint is to be made, the spigot (male) end of the pipe is placed in the hub or bell (female) socket, aligned, and two or three rounds of oakum are pushed to the bottom of the hub with a yarning iron. (Oakum is a rope-like material which swells and seals the joint when wetted.) The oakum is then tightened in the hub with a packing iron and a hammer. Oakum is again added, pushed in with the yarning iron, and packed with the packing iron until one inch of hub depth remains. The joint is then ready for the molten lead. If a horizontal joint must be poured, a heat-resistant runner is placed adjacent to the hub, with the gate packed for a good pour.

Prior to packing the joint, you should activate the furnace to heat the lead in the pot, as it takes time for the lead to become molten. Lead melts at 622° F. It will become goldish-silver when it is ready to be poured. To check the lead, pre-heat the ladle by rotating it in the lead.

Never place a wet ladle or wet lead into the molten lead.

Scoop a small amount of lead into the ladle and swirl it around. If the lead is cold, it will stick to the ladle and must be heated further.

Turn down the supply of heat on the furnace once the lead is hot. This will prevent the lead from burning and caking. The caked material should be removed before gathering a ladle of lead.

Scoop up a full ladle of lead and pour it quickly and smoothly in one continuous stream into the joint until it tops out. Avoid spilling this material, as it will splatter and burn. **BEFORE POURING BE SURE THAT THE OAKUM AND JOINT ARE DRY.** If water is present, the water will explode because the molten lead is so much hotter than the boiling point of water. The force exerted by the steam formed will scatter molten lead in all directions! Keep your head and

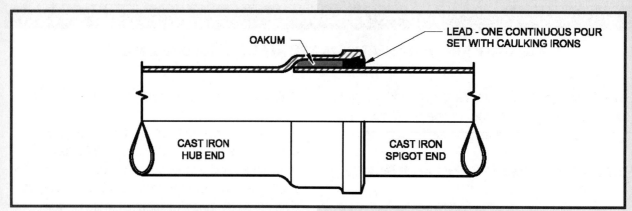

Figure 19-A – A Lead Caulked Joint in Cast Iron Soil Pipe

face away from the joint as you pour the lead. This precaution will reduce the risk of serious injury if the lead does explode.

ALWAYS WEAR SAFETY GOGGLES WHEN WORKING WITH CAST-IRON SOIL PIPE.

After the lead has cooled, set the lead into the hub with inner and outer caulking irons and light hammer blows. There are caulking irons for the inside diameter and the outside diameter of the lead ring. Caulk the inner ring face, then the outer ring face. Chisel off any excess lead formations or lead gates.

This joint holds because the oakum swells once water is applied, causing an extremely tight joint.

Plastic pipe can be joined to a cast-iron hub using the same method. When plastic is assembled into this type of joint, allow the joint extra time to cool before caulking.

COMPRESSION (See Figure 19-B)

If a compression-gasket joint is to be made, a gasket is first placed inside the hub. This gasket must be sized properly for the diameter and weight of pipe being used. Lubricant is applied to the inside of the gasket and on the spigot of the pipe. The pipe end or fitting is pushed into the hub. The pipe or fitting can be driven in with a lead maul, or with special pulling tools.

CAST-IRON SOIL PIPE — HUBLESS

Hubless joints were developed to speed the assembly of cast-iron systems, and also to reduce the space required for this class of piping. An advan-

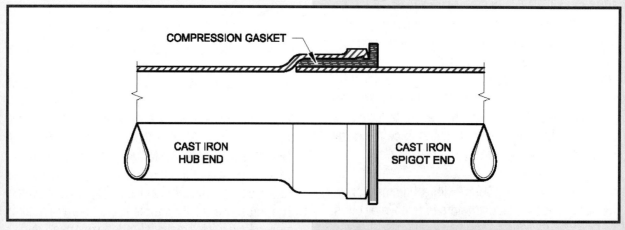

Figure 19-B – A Cast Iron Hubbed Joint With a Compression Gasket

Plumbing Apprentice Student Workbook Year One
Fourth Edition

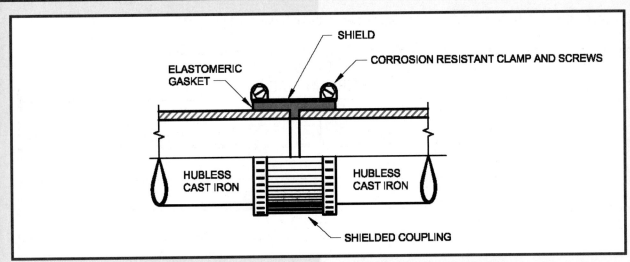

Figure 19-C – A Shielded Coupling on Hubless Cast Iron Soil Pipe

tage of this system is that any joint can be readily disassembled for repair, cleaning, or changes. Figure 19-C shows a typical joint.

Hubless pipe is cut like any other soil pipe. After the cut ends are peened or filed, the joints are made up with a neoprene gasket and stainless steel sleeve and clamps. The clamps are tightened with a torque wrench that will tighten to 60 inch-pounds (some couplings require 80 or 125 inch-pounds), and then the wrench clutch will slip. Thus, the mechanic is assured that the fitting is tight enough, but not over-tightened. The clamp actually relaxes to a lesser value but will still be water and gas tight.

Other variations of the hubless coupling are presently being marketed. Variations such as cast-iron supporting sleeves, heavier duty stainless sleeves, interlocking bands and sleeves, or sleeveless couplings are available. Consult manufacturer's recommendations for proper installation, as some products use special 80 or 125 in-lb torque wrenches.

CAST-IRON DRAINAGE

Cast-iron drainage fittings are used with galvanized steel pipe in the Durham drainage system. These fittings are either threaded, or they are a combination of thread and hub.

A 2" male tucker fitting consists of a 2" hub x 2" male iron pipe thread. The tucker fitting allows the adaptation of a caulked joint to a threaded joint. The caulked joint is made by one of the methods previously discussed.

Threaded fittings made of cast iron are machined with male or female iron pipe threads and are used in waste lines. To attach a threaded pipe to a threaded fitting, thread-sealant is applied to the male pipe end and the fitting is turned onto the pipe with a clockwise motion. After two to three turns, the engagement should be too tight to continue turning by hand. The fitting is then tightened by wrenches.

It is important to use the proper size wrench for the pipe size with which you are working. An undersized wrench will not be able to get the joints tight enough, and an oversized wrench can distort (that is, flatten) the pipe or the fitting. If either component is out-of-round by the slightest amount, the joint is likely to leak.

Appropriate wrench sizes for pipe sizes are shown in the following table.

TABLE 19-A WRENCH SIZE	
WRENCH SIZE (INCHES)	PIPE SIZE (INCHES)
6	1/8
8	1/8, 1/4
10, 12	3/8, 1/2
14	3/4, 1
18	1 1/4, 1 1/2
24	2, 3
36	3+

NOTE: These size recommendations are for *assembling* new work. Larger wrenches may be required to disassemble existing work. It should be understood that such disassembled material may be damaged and *should not be reused*.

Cast-iron fittings will crack suddenly if they are over-tightened. They are also likely to crack when they are reused. It is generally poor practice to install used cast-iron fittings.

GLASS PIPE

Glass pipe joints are similar in design to those for hubless cast-iron. To make a joint, the pipe must be cut to the proper length by scoring and then applying heat from a torch. The heat will cause the glass pipe to break at the score line. The cut end is then smoothed with emery cloth, or sometimes, it is finished in a beading machine which forms a slightly enlarged, rounded end on the pipe.

Early glass pipe joints required this bead, but couplings are now available that form satisfactory joints without the bead. This development means that glass pipe can be field fabricated without requiring the cost of renting and using a beading machine. Whether a pipe is beaded or not beaded, clamps are available to adapt to connections such as these:

- Beaded to beaded
- Plain end to beaded
- Tailpiece to beaded
- Reducer sizes such as 2" x 1½"

The actual joint coupling is a neoprene or other gasket with an external sleeve that is bolted together to specified torque. The gasket material must be specified to tolerate the fluids conveyed.

CLAY PIPE

Clay pipe joints are hub and spigot type. Clay pipe is cut with hammer and chisel or a wheeled cutter. In the past, cut ends of clay pipe had to be joined with concrete (sand aggregate) as the filler material. This method is rarely used now — mostly for emergency repairs or cut-ins.

The common clay pipe joints used today are proprietary (i.e., each manufacturer designs his own joint specifically for his pipe). You cannot be certain that pipe from two manufacturers will join together satisfactorily. The inside and outside elements of the joint are similar in that part of the joint is cast in the hub and part is cast on the spigot end of the pipe. A common feature of these designs is that they go together relatively easily, but they have great strength against pulling apart.

The actual joint material is a flexible plastic that only needs a lubricant to facilitate assembly. After lubricating the spigot end, it is pushed into the bell. The joint is immediately ready for use, and it will tolerate

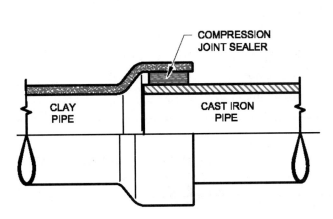

NOTES:
1. The rubber rings must be sized to adapt to the different outside diameters (O.D.) of the different piping materials being joined.
2. There are hundreds of different adapters available for joining different materials, different pipe sizes, and different wall thickness, as well as clay pipe from different manufacturers.

Figure 19-D – A Rubber Ring Transition Joint to Vitrified Clay Pipe

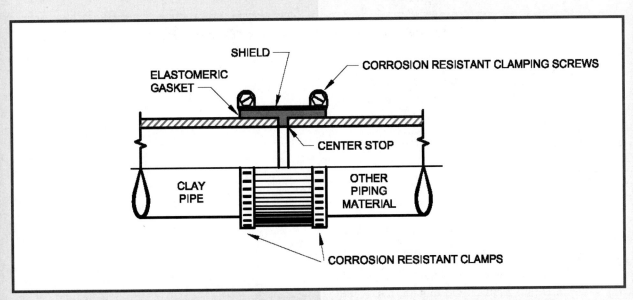

Figure 19-E – An Externally Clamped Transition Coupling Joint to Vitrified Clay Pipe

up to 5° misalignment without leaking. Figure 19-D shows an example. Modern clay pipe sewer lines, with properly assembled joints, are leak-tight to 10' or 20' water head.

These tight sewers assure protection from contamination caused by the pipe contents leaking out of the line, and they will not contribute to diluting and/or overloading the sewer treatment plant by having ground water leak into the pipe.

TRANSITION JOINT

Transition joints are fittings made to connect building drains to sewer lines. They are heavy neoprene or similar gaskets with stainless steel clamps. They are sized to the proper diameters for cast-iron or plastic to join with tile, or plastic sewer pipe, or concrete, or asbestos cement or other pipe. Figure 19-E shows a typical example.

An improved joint is made with an elastomeric "O" ring in the hub-spigot joint. As with the other joints of this type, moderate angular misalignment or settling is tolerated without leaking. Figure 19-F shows an example of this joint type.

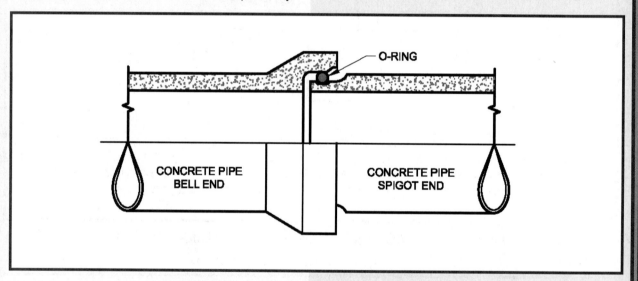

Figure 19-F – A Joint in Bell and Spigot Concrete Pipe

CONCRETE PIPE

Historically, concrete pipe joints were made with concrete filler in the hub-and-spigot configuration. This type joint will not stay tight if there is any settling or shifting of the line. The major model plumbing codes now prohibit this type of joint; check the code in your area.

SUMMARY

Table 19-B presents typical joint types for the piping described in this lesson.

TABLE 19-B
SUMMARY OF CONNECTIONS

MATERIALS	TYPES OF JOINTS	METHODS OF ASSEMBLY	EXTRA MATERIALS NEEDED	TOOLS REQUIRED
Cast-Iron Soil Pipe	Hubbed	Compression Caulked	Gaskets and Lube Lead and Oakum	Puller, maul Irons and pouring ropes, furnace, pot, ladle Torque wrench
	Hubless	Hubless Clamp	Clamps and Gaskets	Wrenches, puller, packing and pouring, furnace
Cast-Iron Drainage Fittings	Screwed and Hubbed	Screwed and Hubbed Joints	Gaskets and Lead and Oakum	
Glass Pipe	Clamp	Clamp	Clamp	Torque wrench
Clay Pipe	Hubbed Hubless	Ring Joint Hubless Coupling	Rubber ring and lube Hubless Coupling	Puller Torque wrench
Concrete	Hubbed Hubless	Ring Joint Hubless Coupling	Rubber ring and lube	Puller Torque wrench

Study the summary table for an overview of joint characteristics and materials and tooling required.

LESSON 20

PRESSURE PIPE AND FITTINGS

STEEL PIPE AND FITTINGS

GENERAL REMARKS

Steel pipe — and all the various types of fittings to join steel pipe — are still the *model* for this business. Nearly all the other materials we use are substitutes for steel pipe and are evaluated on how they compare with steel pipe.

Steel pipe is used on the larger jobs for water and waste systems, fire protection sprinkler systems, heating and cooling systems, snow making, and snow-melting systems. Steel pipe also has many applications for process piping installations.

Steel pipe is the strongest and toughest piping material. It also has the lowest coefficient of expansion with temperature change, is better able to withstand cyclic stress change, and is most resistant to mechanical or thermal shock. It can be machined and welded. In short, mechanically it is the superior product. Substitute materials are satisfactory, in fact, because steel characteristics *far exceed* the physical requirements for typical piping applications.

See Lesson 17 for further descriptions of steel pipe.

WELDING

Welding is a method of assembly whereby two adjoining pieces of metallic material are fused together by heating to the melting point and filler material is added to complete the joint.

Steel pipe and fittings can be welded together through the electric arc or gas welding methods. This joining method, as well as other types of connecting steel pipe components together, will be explored in later lessons.

Configurations similar to threaded fittings are available for welding purposes. Available fittings can be either butt-welded where the pipe and fitting are placed end-to-end and are welded on the adjoining edges, or socket-welded where the pipe and fittings are assembled similar to soldered fittings; that is, the fitting fits over the pipe, and a fillet weld is applied to the face of the fitting socket and the side wall of the pipe.

It must be noted that welding is usually not an appropriate joining method for piping that is carrying fresh water, because the chemicals in the fresh water will degrade the weld material.

FLANGES

Flanges offer a method of connecting large piping through the use of gaskets and bolts. This method of connecting pipe is extensively used for large diameter pipe or in applications which require the ability to remove sections of the piping installation.

A flange is a fitting which looks like a large rim which projects on a pipe end. This rim has a series of holes in it into which bolts are inserted for connection with another flange. Flanges are connected to pipe ends or fitting ends by threaded or welded joints. There are also flanged fittings (tees, ells, etc.) which include flanged ends as part of the manufacturing processes. Once a flange is connected to a pipe or fitting through some sort of mechanical connection, the flanges themselves can be connected together through the use of bolts and gaskets.

Figure 20-A shows a "companion" flange — so called because it is threaded to attach to a thread-

Figure 20-A – Companion Flange

ed pipe. Figure 20-B shows a reducing flange connected to a standard-sized flange. When flanges are used, it is very important to be sure that the same torque (turning pressure) is exerted on each bolt. Uneven torque on the bolts could mean uneven gasket pressure which could result in leakage.

Figure 20-B – Reducing Flange

CLAMP CONNECTIONS

Clamp connections can be divided into two types:

1. Common type applied to grooved pipe

2. The type which clamps to plain-end pipe — this is a variation of the grooved coupling and it uses a compression gasket to seal plain ends of pipe together.

GROOVED COUPLING

This method of joining steel pipe uses a coupling that engages grooves in the pipe ends. This coupling also contains a gasket which looks like a rubber doughnut. The metal clamps contain the gasket and connect the grooves in the piping. See Figures 20-C and 20-D.

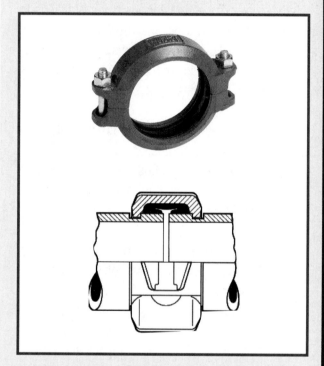

Figure 20-C – Victaulic Grooved Coupling

Figure 20-D – Victaulic Grooved End Fittings

As pipe fittings are made specifically for threaded, welded, and flanged method of joining, there are also fittings made for grooved connections. Grooved fittings are available in patterns like any other fitting type — but unlike the threaded, beaded, or welded fitting — this fitting has a groove machined on its *outside* diameter. The mating pipe is grooved so that the clamp will connect the two grooved ends securely. It is important to note that the grooved-coupling connection system relies on internal pressure to maintain joint tightness.

The grooved coupling method is quick, strong, and easy to assemble and disassemble because, in effect, there is a union at every connection. This assembly method is also excellent when system changes must be made.

The groove is about $3/8$" wide, shallower than pipe thread, and about $5/8$" in from the pipe end. The groove configuration can be rolled into thin-walled pipes where it could not be machined out, but in Schedule 40 pipe or heavier, the material removed to form the groove is insignificant.

The pipe ends (or pipe and fitting) are aligned and the gasket put in place. Then the split clamp is assembled with bolts so that it engages the grooves. The joint is then ready for use.

Although the split clamp engages the grooves, it does not hold them tight. The result is a joint that can tolerate as much as 5° misalignment, and that can move back and forth somewhat to tolerate thermal changes. The standard coupling, as made by several manufacturers, has a pressure rating of 500 psi.

Grooved pipe joints have two limitations: first, the gasket seals by internal line pressure, thus making it unsatisfactory on a line operating at atmospheric pressure (or less); second, (as this is written) gasket temperatures cannot exceed about 200° F, thus preventing these joints from being used on steam piping.

PLAIN-END COUPLING

Plain-end pipe joints are similar to a compression joint. However, they are not limited to small sizes as is the case for copper compression joints. The joint uses a gasket that surrounds the pipe, a tapered outer shell, and a movable end that is drawn up to force the gasket into the taper. This gasket seals tightly to the pipe outside diameter, making a leak-tight joint. In sizes up to 2", the movable end is a threaded nut. In sizes larger than 2", the two movable ends are pulled simultaneously by draw bolts.

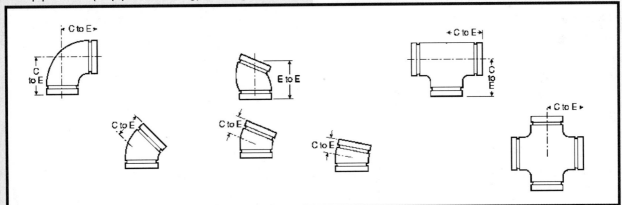

Figure 20-E – Typical Victaulic Fittings

These are extremely versatile fittings and many, many gasket types are available for a wide variety of piped fluids.

The plain-end joint has one serious disadvantage: the pipe is held in the joint solely by gasket friction.

Lines operating above moderate pressures can blow apart suddenly causing injury or property damage. Therefore, in any pressure system, these joints must be supported or held together in some way.

A note of caution when using all forms of steel pipe fittings: inspect all fittings for sand holes. A sand hole is a hole in a fitting which is caused by a casting defect such as gas in the mold, a poor molding job, bad sand, cold metal, or other problems.

All fittings and pipe should be pre-inspected to minimize lost time and possible property damage from a leak.

COPPER TUBING, FITTINGS, AND FITTING DESCRIPTION

COPPER TUBE

Copper is corrosion-resisting to most chemicals found in water or waste systems (except for urine). This corrosion resistance, coupled with the strength of copper and its ductility, make it a suitable material for water and many waste systems. The presentation in this lesson will concern copper as a pressure piping material.

Details of manufacture, standards, fittings, and normal supply-house products in stock are described in Lesson 17.

Illustrations of fittings are shown in Figures 20-F and 20-G. Illustrations of copper pressure fittings are shown in Figure 20-I.

SOLDERING OR BRAZING

Historically, fittings for pressure applications typically used soldering and brazing joining methods. Such fittings are available in ells and street ells (i.e., male and female ends on the same fitting) with 90° and 45° turns, tees, couplings, crosses, caps, adapters (male and female), unions, and reducers.

Many fitting types are also available in reducing patterns.

With regulations appearing that prohibit certain soldering materials, and with the pressure to economize on solder material and the time to produce a soldered joint, we may expect to see new developments in joint methods that will eliminate the need to feed solder into the joint from a roll in your hand. Such techniques have been used in Europe and have recently been introduced in the North American markets.

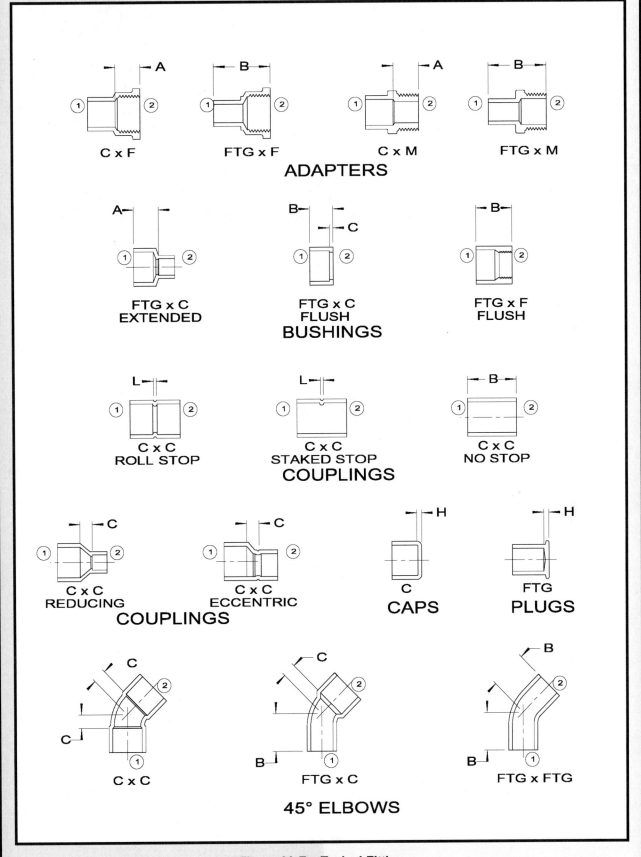

Figure 20-F – Typical Fittings
NIBCO INC

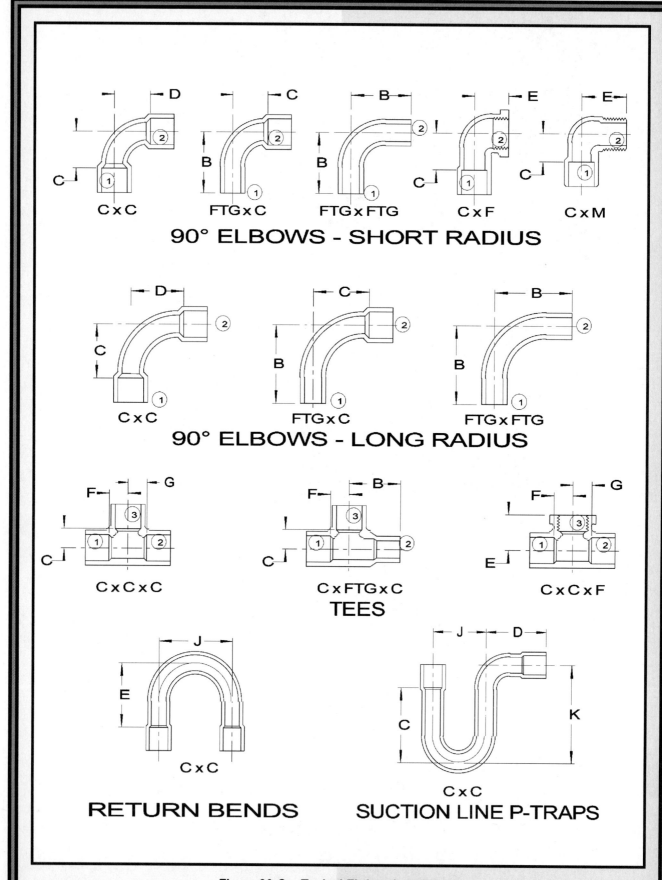

Figure 20-G – Typical Fittings(continued)
NIBCO INC

PRESS CONNECTION JOINING

This method is discussed in Lesson 17.

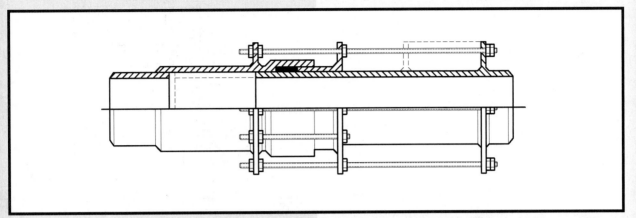

Figure 20-H – A Mechanical Expansion Joint in Pressure Piping

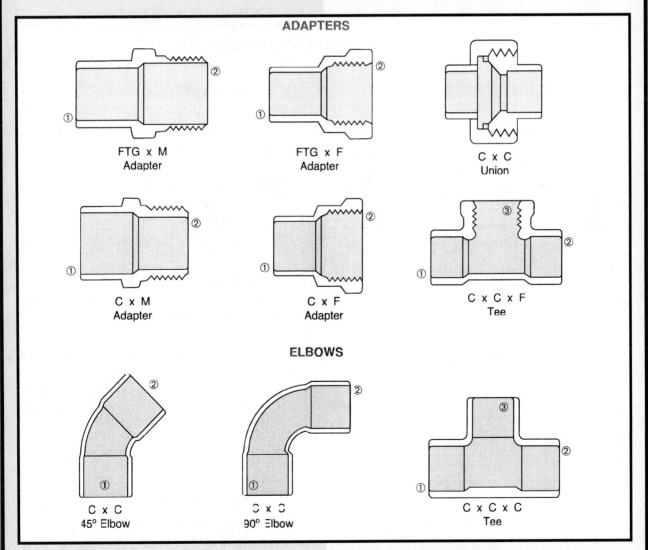

Figure 20-I – Cast and Wrought Copper Fittings NIBCO INC

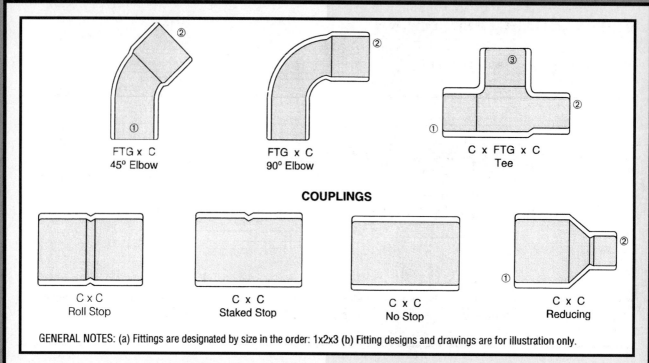

Figure 20-I(2) – Cast and Wrought Copper Fittings NIBCO INC

FITTING IDENTIFICATION

Fittings must be *read* in a standard way so that we can discuss with each other the details of an installation, order materials from supply houses or the warehouse, or understand code provisions. Develop the habit of ordering fittings with full and accurate description. Specify quantity, size(s), configuration, pressure or drainage, joint type, material and finish (if applicable).

EXAMPLES

To order one hundred solder fittings to turn a 90° angle for 1/2" copper tube, write:

100 - ½" sweat copper ells (or 90° ells)

For a fitting to connect tubing to a male pipe thread, the following would be ordered:

1 - ½" sweat x female adapter

A 3" drainage pipe line that requires a tee, an ell, and two 45's (one a street fitting which has a male and a female connection) would have the following material list:

1 - 3" sweat copper 45 DWV
1 - 3" sweat copper street 45 DWV
1 - 3" sweat copper tee DWV
1 - 3" sweat copper ell DWV

Note that *ell* alone is understood to be a 90° ell, and if pressure or DWV is not indicated, pressure would normally be understood.

For tees or crosses, read the straight path first, starting with the larger opening, then read the side branches. Thus a tee that has ¾" and 1" openings on the straight path, and a ½" side opening, would be read:

1" x ¾" x ½" tee

If a tee is 1" x 1" x ½," reading it as a 1" x ½" tee is understood to mean the same thing. Be sure to understand that a 1" x ½" x 1" tee is **not** the same thing and must be called out in full.

Practice identifying fittings and ask your fellow workers about any you are unsure of. The reading of fittings is not very complicated, but you will not be a professional if you don't learn how to do it.

SUPPLEMENT

At the end of this lesson there is the following information from NIBCO:

Copper Tube Fittings
What Makes a Plumbing System Fail?
The Fine Art of Soldering
The Fine Art of Brazing

Note that much of what has been described above is treated in these papers. A careful review of the material will enhance your understanding of the applications of copper tubing and give you a perspective of how to make copper installations the very best that they can be.

TRANSITION FITTINGS

The following illustration show the method used to join different piping materials by using the proper fittings and adapters. Check with your supervisor for other examples.

PLASTIC PIPE AND FITTINGS

PVC DRAINAGE FITTINGS

Drainage fittings are described in Lesson 18.

PVC PRESSURE FITTINGS

Pressure fittings are described in Lesson 17.

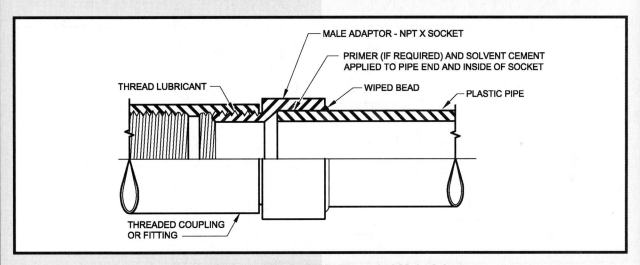

Figure 20-J – A Plastic DWV Threaded Male Adapter

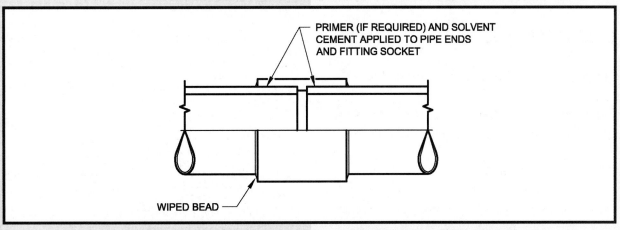

Figure 20-K – A Solvent Cement Joint in Plastic DWV or Water Piping

ABS (ACRYLONITRILE BUTADIENE STYRENE)

Pipe and fittings are cemented (see Figure 20-K and Figure 20-L), or connected by compression bands and gasket (see Figure 20-M).

PB (POLYETHYLENE)

Polybutylene tubing was once popular for water distribution in buildings and gas distribution outside buildings underground. Joints can be made with

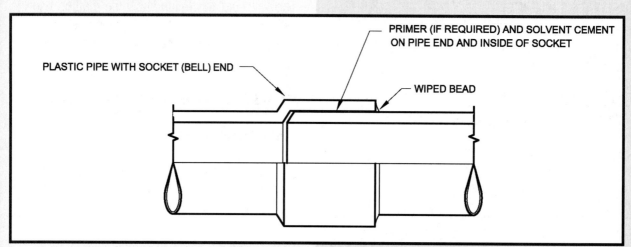

Figure 20-L – A Solvent Cement Joint in Socket (Bell) End Plastic Pressure Pipe

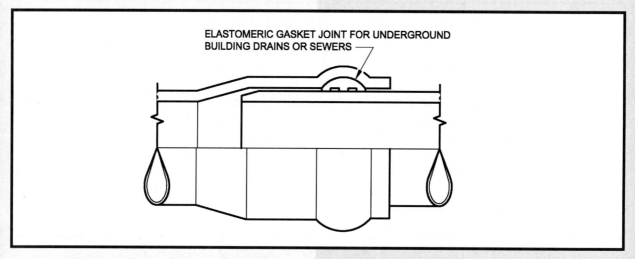

Figure 20-M – An Elastomeric Gasket Joint for Underground Plastic DWV Piping

CPVC (CHLORINATED POLYVINYL CHLORIDE)

These products are described in Lesson 17.

Some pipe manufacturers provide pipe with a socket for joining formed on one end (see Figure 20-L).

PE (POLYETHYLENE)

Polyethylene methods are described in Lesson 17.

insert fittings (Figure 20-N) or with heat fusion (Figure 20-O).

PEX

See also Lesson 17.

Cross-linked polyethylene is a relatively recent offering to the U.S. plumbing water distribution market, although it has been used for a number of years in Europe. Its pressure and temperature

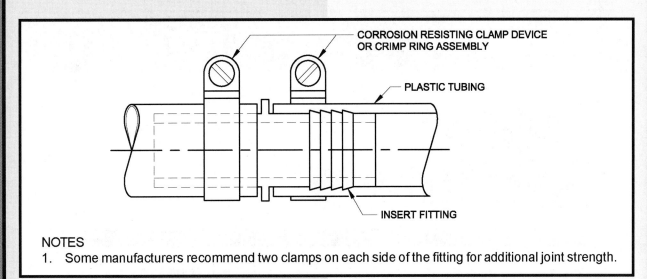

Figure 20-N – An Insert Fitting Joint in Plastic Tubing

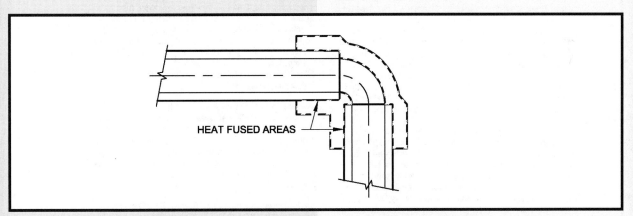

Figure 20-O – A Heat Fused Joint in Plastic Water Piping

ratings are higher than other coiled plastic materials. Joints are special insert type. Manifold systems are available with this material.

PEX-AL-PEX

See also Lesson 17.

Polyethylene-aluminum-polyethylene tubing for water distribution is comprised of a sandwiched tubing of cross-linked polyethylene extruded on both sides of a thin aluminum tube. The semi-rigid nature of this tubing allows it to retain its shape when bent. It can be bent manually, but the use of special tools is preferred. Brass insert fittings are used to assemble tubing lengths together.

THERMOSETTING

See also Lessons 17 and 18.

Thermosetting is the second type of plastic pipe available. This material changes chemically when initially formulated and cannot be reshaped by heating.

Engineering Data

Copper Tube Fittings

TYPES OF JOINTS

Flared Joint — The principle of the flared type joint was first developed for copper tube plumbing in 1928 by NIBCO. The flared type joint is wholly a mechanical means of joining copper tubes. The tube nut is placed over the end of the copper tube to be joined; the tube end then is flared out at an approximate 45 degree angel by a flaring tool. The flared end is then drawn up by the tube nut so the inside surface is tightly secured against the ball seat of the fitting. This joint can be readily dismantled at any time and is, in effect, a type of union connection. Its use is generally restricted to soft (annealed) copper tubes since hard drawn tubes would be subject to splitting when flared (if the ends were not previously annealed). The flared ends of NIBCO Flared Fittings are produced to the requirements of ASME B16.26, "Cast Copper Alloy Fittings for Flared Copper Tube."

Solder Joint — NIBCO pioneered the development of the solder type joint and its application to the field of copper tube piping. Today the solder type joint is widely adopted, as evidenced by the majority of cities and states that have written codes to include copper tube and solder joints as desirable for general plumbing, water lines, vent, stack, waste and drain lines, as well as other uses in industry. Testing has shown that often the solder joint has greater strength than the tubes being joined, depending upon the soldering alloy selected. While the method of preparing a solder joint is an exacting art to insure a full strength joint, it can be readily mastered by skilled tradesmen. It is for this reason — to insure the public of the protection afforded by properly prepared joints — that NIBCO products are marketed through the reputable sources of supply to the piping trades. Important procedures for preparing a solder joint are graphically illustrated in this catalog on page 95.

Brazed Joint — This type of joint has long been used wherever and whenever critical situations have been encountered in copper piping. The joint itself is completed much in the same manner as the solder joint; however, considerably more heat and several refinements of technique require separate procedures that are described further in this catalog on pages 96-97.

Threaded Ends — To adapt copper tube to equipment having National Standard Pipe Taper (NPT) threads or to add copper tube to existing iron pipe installations or other threaded connections, NIBCO provides fittings having both external and internal NPT threads. These threaded ends are produced to the requirements of ASME B1.20.1, "Pipe Threads, General Purpose (Inch)."

Flanges — To adapt copper tube to equipment having flanged connections, or to add copper tube to flanged pipe installations or other purposes, NIBCO provides flanges. The flanges are produced in two standard types widely used in this field where copper tube can serve — Class 150, comply with ASME B16.24, "Cast Copper Alloy Pipe Flanges and Flanged Fittings Class 150, 300, 400, 600, 900, 1500 and 2500"; and Class 125, which conform to MSS SP 106, "Cast Copper Alloy Flanges and Flanged Fittings Class 125, 150 and 300."

Barbed Insert Fittings for Polybutylene (PB) — NIBCO offers a complete line of copper barbed insert fittings for joining PB tube. The insert fittings are produced to the requirements of ASTM F 1380. Along with the insert fittings are copper crimp rings, which, when properly installed, provide a leak-tight mechanical joint. Transition fittings are available for adapting to new or existing threaded or solder joint ends.

Barbed Insert Fittings for PEX — NIBCO offers a complete line of copper barbed insert fittings for joining PEX tube. The insert fittings are produced to the requirements of ASTM F 1807-98. Along with the insert fittings are copper crimp rings, which, when properly installed, provide a leak-tight mechanical joint. Transition fittings are available for adapting to new or existing threaded or solder joint ends.

Fitting Terms and Abbreviations

C	Female solder cup
Ftg	Male solder end
F	Female NPT thread
M	Male NPT thread
Hose	Standard hose thread
Hub	Female end for soil pipe
Spigot	Male end for soil pipe
No Hub	Used with mechanical coupling
O.D. Tube	Actual tube outside diameter
S	Straight thread
SJ	Slip joint

NIBCO NIBCO INC. World Headquarters ■ 1516 Middlebury St. ■ Elkhart, IN 46516-4714 U.S.A. ■ www.nibco.com

Figure 20-P – NIBCO Copper Tube Fittings

Engineering Data

WHAT MAKES A PLUMBING SYSTEM FAIL?

Failure in a copper plumbing system is rare, but may occur due to a variety of reasons. The most common causes of failure are:

1. **Excessive fluid velocity** causes erosion-corrosion or impingement (to strike or hit against) attack in the tube and/or fitting. For this reason, the copper plumbing industry has establish design velocity limits for copper plumbing systems to the following:

Hot Water > 140°F (60°C)	2 to 3 feet per second (0.6 to 0.9 meters per second)
Hot Water ≤ 140°F (60°C)	4 to 5 feet per second (1.2 to 1.5 meters per second)
Cold Water	5 to 8 feet per second (1.5 to 2.4 meters per second)

2. **Workmanship** system life by creating localized high velocities and/or turbulence. The presence of a dent, tube ends which are not reamed or deburred before soldering, and sudden changes in direction can all cause localized high velocity conditions.

3. **Flux Corrosion** is typified by pin hole leaks, generally in the bottom of a horizontal line. Fluxes are mildly corrosive liquid or petroleum-based pastes containing chlorides of zinc and ammonia. Unless the flux is flushed from the system, it will lay in the bottom of the tube and remain active. ASTM B813, "Liquid and Paste Fluxes for Soldering Applications of Copper and Copper-Alloy Tube," limits the corrosivity of soldering fluxes and ensures that these fluxes are flushable in cold water, which facilitates easy removal of flux residue following installation.

4. **Galvanic Corrosion** may be defined as the destruction of a material by electrochemical interaction between the environment and the material. Generally, it is slow but persistent in character and requires the presence of dissimilar metals. Galvanic corrosion requires the flow of an electric current between certain areas of dissimilar metal surfaces. To complete the electric circuit, there must be two electrodes, an anode and a cathode, and they must be connected by an electrolyte media (water) through which the current can pass. The amount of metal which dissolves at the anode is proportional to the number of electrons flowing, which in turn is dependent upon the potential and resistance of the two metals. The use of dissimilar metals in a plumbing system may or may not create a problem. For instance, copper and steel are perhaps the most common dissimilar metals found together in a plumbing system. In closed systems, such as a chilled or heating water piping, the use of dissimilar metals may not create a serious problem; this is because there is virtually no oxygen in the water and corrosion relations tend to be stifled. Where dissimilar metals must be used, some codes require that they should be separated by dielectric union or a similar type of fitting. The effectiveness depends upon; distance between the metals on the electromotive-force series (EMF) chart, ratio of cathode to anode area, degree of aeration, amount of agitation, temperature, presence of dissolved salts, and other factors.

ABBREVIATED EMF SERIES
(Electromotive-Force Series; Common Piping Materials in Sea Water)

CATHODE (+) Passive
- GOLD – Fixtures, Faucets, Plating
- PLATINUM
- SILVER – Brazing alloys, Silver-bearing solders
- TITANIUM – Condenser tubes
- MONEL (67% Ni - 33% Cu) – Specialty piping & equipment
- CUPRO-NICKEL – Condensers, Marine, Nuclear
- COPPER – Pressure, DWV, Gases, Air, Refrigeration, etc.
- BRASS (85/15 - Red) – Cast fittings, Valves
- BRASS (70/30 - Yellow) – Gas-cocks, Fittings, Connectors
- LEAD – Solder, Pipe, Sheet, Coating, Lining
- TIN – Solders, Coating, Lining
- CAST IRON – Pressure, DWV
- WROUGHT IRON – Pressure
- MILD STEEL – Fire Protection
- ALUMINUM – Refrigeration, Irrigation, some Solar
- GALVANIZED STEEL – Pressure, DWV
- ZINC – Coatings, Linings, some Fittings
- MAGNESIUM – Water Heater Anodes, Cathodic protection for pipelines

ANODE (-) Active; Sacrificial Material

Galvanic corrosion may be defined as "the destruction of a material by electrochemical interaction between the environment and the material." Generally it is slow but persistent in character. The basic cause of corrosion is the instability of metals in their refined forms. The metals tend to revert back to their natural states through the processes of corrosion through transformation from the metallic to the ionic state.

5. **Dezincification** is a type of corrosion in which brass dissolves as an alloy and the copper constituent redeposits from solution onto the surface of the brass as a metal, but in the porous form. The zinc constituent may be carried away from the brass as a soluble salt, or may be deposited in place as an insoluble compound. Dezincification is normally associated with brass valves where the zinc content exceeds 15%. Generally, areas of high stress, such as valve stems and gate valve bodies, are primary targets of attack.

6. On rare occasion problems of corrosion by aggressive water, possibly aggravated by poor design or workmanship, do exist. **Aggressive, hard well waters** that cause pitting can be identified by chemical analysis and treated to bring their composition within acceptable limits. Typically these hard waters are found to have high total dissolved solids (t.d.s.) including sulfates and chlorides, a pH in the range of 7.2 to 7.8, a high content of carbon dioxide (CO_2) gas (over 10 parts per million, ppm), and the presence of dissolved oxygen (D.O.) gas. Soft acidic waters can cause the annoying problem of green staining of fixtures or "green water". Raising the pH of such waters to a value of about 7.2 or more usually solves the problem, but a qualified water treatment specialist should be consulted.

7. **Aggressive soil conditions** can be a cause for external corrosion of copper piping systems. Non-uniform soil characteristics, such as different soil aeration, resistivity, or moisture properties, between adjacent sections of tube can create galvanic corrosion cells. Soils contaminated with high concentrations of road salts or fertilizers containing ammonia, chlorides, and nitrogen are known to combine with water to form acids. Any metal pipe laid in ash or cinders is subject to attack by the acid generated when sulfur compounds combine with water to form sulfuric acid.

U.S. customary units in this document are the standard; the metric units are provided for reference only. The values stated in each system are not exact equivalents.

NIBCO INC. World Headquarters ■ 1516 Middlebury St. ■ Elkhart, IN 46516-4714 U.S.A. ■ www.nibco.com **NIBCO**

Figure 20-Q – NIBCO What Makes a Plumbing System Fail?

Installation Instructions

The Fine Art of Soldering

When adjoining surfaces of copper and copper alloys meet under proper conditions of cleanliness and temperature, solder will make a perfect adhesion. The strength of joint is equal to or even greater than the strength of tube alone. Surface tension seals the joint. Capillary attraction draws solder into, around, and all about the joint. It's easy to learn to make a perfect solder joint when you use NIBCO Fittings.

WITH 95-5 SOLDER AND INTERMEDIATELY CORROSIVE FLUX

1. Cut tube end square, ream, burr and size.

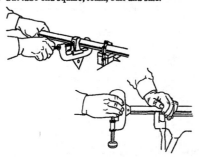

2. Use sand cloth or steel wire brush to clean tube and cup to a bright metal finish.

3. Apply solder flux to outside of tube and inside of cup of fitting carefully so that surfaces to be joined are completely covered. **Use flux sparingly.**

4. Apply flame to the fitting to heat tube and solder cup of fitting until solder melts when placed at joint of tube and fitting.

5. Remove flame and feed solder into the joint at one or two points until a ring of solder appears at the end of the fitting. The correct amount of solder is approximately equal to 1½ the diameter of the fitting... ¾" (20mm) solder for ½" fitting, etc.

6. Remove excess solder with a small brush or wiping cloth while the solder is plastic.

NIBCO INC. World Headquarters ■ 1516 Middlebury St. ■ Elkhart, IN 46516-4714 U.S.A. ■ *www.nibco.com*

Figure 20-R – NIBCO The Fine Art of Soldering

Installation Instructions

The Fine Art of Brazing

Best results will be obtained by a skilled operator employing the step-by-step brazing technique that follows:

1. The tube should be cut to desired length with a square cut, preferably in a square-end sawing vise. The cutting wheel of the type specifically designed for cutting copper tube will also do a satisfactory job. The tube should be the exact length needed, so that the tube will enter the cup of the fitting all the way to the shoulder of the cup. Remove all slivers and burrs left from cutting the tube, by reaming and filing, both inside and outside.

2. To make a proper brazing joint, the clearance between the solder cup and the tube should be approximately 0.001" to 0.010" (0.0254mm to 0.254mm). Maintaining a good fit on parts to be brazed insures:

   Ease of Application — Excessively wide tolerances tend to break capillary force; and, as a result the alloy will either fail to flow throughout the joint or may flush out of the joint.

   Corrosion Resistance — There is also a direct relation between the corrosion resistance of a joint and the clearance between members.

   Economy — If brazing alloys are to be used economically, they, of necessity, must be applied in the joint proper and in minimum quantities, using merely enough alloy to fill the area between the members.

3. The surfaces to be joined must be clean and free from oil, grease and heavy oxides. The end of the tube need be cleaned only for a distance slightly more than it is to enter the cup. Special wire brushes designed to clean tube ends may be used, but they should be carefully used so that an excessive amount of metal will not be removed from the tube. Fine sand cloth or emery cloth may also be used with the same precautions. The cleaning should not be done with steel wool, because of the likelihood of leaving small slivers of the steel or oil in the joint.

4. The cup of the fitting should be cleaned by methods similar to those used for the tube end, and care should be observed in removing residues of the cleaning medium. Attempting to braze a contaminated or an improperly cleaned surface will result in an unsatisfactory joint. Brazing alloys will not flow over or bond to oxides; and oily or greasy surfaces tend to repel fluxes, leaving bare spots which will oxidize, resulting in voids and inclusions.

5. Flux should be applied to the tube and solder cup sparingly and in a fairly thin consistency. Avoid flux on areas not cleaned. Particularly avoid getting excess flux into the inside of the tube itself. Flux has three principal functions to perform:

   A. It prevents the oxidation of the metal surfaces during the heating operation by excluding oxygen.

   B. It absorbs and dissolves residual oxides that are on the surface and those oxides which may form during the heating operation.

   C. It assists in the flow of the alloy by presenting a clean nascent surface for the melted alloy to flow over. In addition, it is an excellent temperature indicator, especially if an indicating flux is used.

6. Immediately after fluxing, the parts to be brazed should be assembled. If fluxed parts are allowed to stand, the water in the flux will evaporate, and dried flux is liable to flake off, exposing the metal surfaces to oxidation from the heat. Assemble the joint by inserting the tube into the cup, hard against the stop. The assembly should be firmly supported so that it will remain in alignment during the brazing operation.

NIBCO NIBCO INC. World Headquarters ■ 1516 Middlebury St. ■ Elkhart, IN 46516-4714 U.S.A. ■ *www.nibco.com*

Figure 20-S – NIBCO The Fine Art of Brazing

Installation Instructions

7. Brazing is started by applying heat to the parts to be joined. The preferred method is by the oxyacetylene flame. Propane and other gases are sometimes used on smaller sizes. A slightly reducing flame should be used, with a slight feather on the inner blue cone; the outer portion of the flame, pale green. Heat the tube first, beginning at about one inch from the edge of the fitting. Sweep the flames around the tube in short strokes up and down at right angles to the run of the tube. It is very important that the flame be in continuous motion and should not be allowed to remain on any one point to avoid burning through the tube. Generally, the flux may be used as a guide as to how long to heat the tube, continuing heating after the flux starts to bubble or work, and until the flux becomes quiet and transparent, like clear water. The flux will pass through four stages:

   A. At 212°F (100°C) the water boils off.
   B. At 600°F (315.6°C) the flux becomes white and slightly puffy and starts to work.
   C. At 800°F (426.7°C) it lays against the surface and has a milky appearance.
   D. At 1100°F (593.3°C) it is completely clear and active and has the appearance of water.

8. Now switch the flame to the fitting at the base of the cup. Heat uniformly, sweeping the flame from the fitting to the tube until the flux on the fitting becomes quiet. Avoid excessive heating of cast fittings.

9. When the flux appears liquid and transparent on both the tube and the fitting, start sweeping the flame back and forth along the axis of the joint to maintain heat on the parts to be joined, especially toward the base of the cup of the fitting. The flame must be kept moving to avoid burning the tube or the fitting.

10. Apply the brazing wire or rod at a point where the tube enters the socket of the fitting. The temperature of the joint should be hot enough to melt the brazing alloy. Keep the flame away from the rod or wire as it is fed into the joint. Keep both the fitting and the tube heated by moving the flame back and forth from one to the other as the alloy is drawn into the joint. When the proper temperature is reached, the alloy will flow readily into the space between the tube outer wall and the fitting socket, drawn in by the natural force of capillary attraction. When the joint is filled, a continuous fillet of brazing alloy will be visible completely around the joint. Stop feeding as soon as the joint is filled, using table on page 54 as a guide for the alloy consumption.

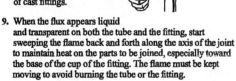

NOTE: For tubing one inch and larger, it is difficult to bring the whole joint up to heat at one time. It frequently will be found desirable to use a double-tip torch to maintain the proper temperature over the larger area. A mild pre-heating of the whole fitting is recommended. The heating then can proceed as in steps 7, 8, 9, and 10. If difficulty is encountered in getting the whole joint up to heat at one time, then when the joint is nearly up to the desired temperature the alloy is concentrated in a limited area. At the brazing temperature the alloy is fed into the joint and the torch is then moved to an adjacent area and the operation carried on progressively all around the joint.

HORIZONTAL JOINTS — When making horizontal joints, it is preferable to start applying the brazing alloy at the 5 o'clock position, then move around to the 7 o'clock position and then move up the sides to the top of the joint, making sure that the operations overlap.

VERTICAL JOINTS — On vertical joints, it is immaterial where the start is made. If the opening of the cup is pointed down, care should be taken to avoid overheating the tube, as this may cause the alloy to run down the tube. If this condition is encountered, take the heat away and allow the alloy to set. Then reheat the solder cup of the fitting to draw up the alloy.

After the brazing alloy has set, remove residual flux from the joint area as it is corrosive and presents an unclean appearance and condition. Hot water or steam and a soft cloth should be used. Wrot fittings may be chilled; however it is advisable to allow cast fittings to cool naturally to some extent before applying a swab. All flux must be removed before inspection and pressure testing.

TROUBLE SPOTS

If the alloy fails to flow or has a tendency to ball up, it indicates oxidation on the metal surfaces, or insufficient heat on the parts to be joined. If work starts to oxidize during heating, it indicates too little flux, or a flux of too thin consistency. If the brazing alloy refuses to enter the joint and tends to flow over the outside of either member of the joint, it indicates this member is overheated, or the other is underheated, or both. In both cases, operations should be stopped and the joints disassembled, recleaned and fluxed.

NIBCO INC. World Headquarters ■ 1516 Middlebury St. ■ Elkhart, IN 46516-4714 U.S.A. ■ *www.nibco.com* **NIBCO**

Figure 20-S – NIBCO The Fine Art of Brazing (continued)

LESSON 21

REVIEW OF LESSONS 8-11

Mathematics is the study of measuring and calculating with numbers. To solve mathematical problems, you must proceed logically from the information you know. Follow these practices in arriving your answers:

1. Read the problem carefully and determine what is asked for.

2. Write down the given information and what must be determined.

3. Decide which function(s) must be used to solve the problem.

4. Estimate the answer.

5. Perform the operation completely.

6. Check your answer against the estimate.

7. Label your answer.

8. Check your work.

READING AND WRITING NUMBERS

ARABIC SYSTEM

The Arabic system is the most widely used counting system in the world. The digits

0, 1, 2, 3, 4, 5, 6, 7, 8, 9

are units of the Arabic system. A combination of these units and the value of position denote any numerical amount, as shown in Table 21-A.

LABELING

In any problem dealing with physical things, your answer is incomplete if you do not indicate the units after your calculations. Be sure you provide the physical unit as a label for all your answers.

WHOLE NUMBER

A whole number is a complete number with no fractional part.

TABLE 21-A
POSITION NAMES

Etc	Millions	Hundred Thousands	Ten Thousands	Thousands	Hundreds	Tens	Units	Decimals
	0 to 9,000,000	0 to 900,000	0 to 90,000	0 to 9,000	0 to 900	0 to 90	0 to 9	.

Thus, the number

4,242,102

is read this way: four million, two hundred forty two thousand, one hundred two

Write out the digits in the proper value positions for this number:

fifteen million, two hundred forty eight thousand, eight hundred fifty three

_____

191 Plumbing Apprentice Student Workbook Year One
Fourth Edition

MIXED NUMBER

A mixed number is a whole number and part of a whole number (a fraction).

DECIMAL

A decimal is a fraction where the denominator is a multiple of ten. Decimals are written in the form 5.75 where "5" is the whole number and "75" is the decimal number.

FRACTIONS

A fraction is a part of a whole number. Fractions express part over the whole in the form:

$$\frac{a}{b} = \frac{numerator}{denominator} = \frac{part}{whole}$$

It is important in our work to become used to working with fractions so that we can dimension pipe and other material accurately.

FRACTIONS — SIMPLEST TERMS

Generally, fractions should be expressed in lowest or simplest terms. A fraction is not changed if the numerator and denominator are multiplied (or divided) by the same number. These examples demonstrate the idea.

$$\frac{2}{4} = ? \qquad \frac{\frac{2}{2}}{\frac{4}{2}} = \frac{1}{2} \qquad \frac{4}{8} = ? \qquad \frac{30}{50} = \frac{30(\frac{1}{10})}{50(\frac{1}{10})} = \frac{3}{5}$$

$$\frac{2}{4} = \frac{3(1)}{3(5)} = \frac{3}{15} \qquad \frac{4}{8} = \frac{\frac{4}{4}}{\frac{8}{4}} = \frac{1}{2}$$

IMPROPER FRACTIONS

An improper fraction is one in which the numerator is larger than the denominator.

22/16 is an example of an improper fraction

It is sometimes best to express an improper fraction as a mixed number, i.e., a whole number plus a fraction. This is done by dividing the denominator into the numerator, then writing a whole number outside the fraction and fractional result over the denominator. The fraction must then be reduced to its lowest terms. For example,

22/16 = 16/16 + 6/16 = 1 6/16 = 1 3/8

CALCULATIONS WITH FRACTIONS

ADDITION OF FRACTIONS

Addition of fractions can be performed only when all of the fractions to be added are in terms of a common denominator. A common denominator is a number into which all of the denominators of the fractions to be added can divide evenly. For example: a common denominator of the following fractions is the number 40.

$$\frac{3}{8} + \frac{2}{5} + \frac{3}{4} = ?$$

An easy way to find a common denominator is to multiply all the denominators together.

$$8 \times 5 \times 4 = 160$$

In this case, if we stop with the first two denominators we have the lowest common denominator (because 40 is also divisible by 4).

$$8 \times 5 = 40$$

It is not always possible to find the common denominator by multiplying only some of the denominators together. (Once in a while, we just get lucky.)

If we convert the fractions to have the common denominator of 40, we have:

$$3/8 = 15/40 \qquad 2/5 = 16/40 \qquad 3/4 = 30/40$$

Adding the fractions, we get

$$61/40$$

for we only add the numerators and keep the denominator constant. If we change this improper fraction to a whole number plus a fraction, we get

$$\frac{40}{40} + \frac{21}{40}$$

or

$$1\frac{21}{40}$$

Remember that addition of fractions can only take place when all the fractions are expressed with a common denominator.

SUBTRACTION OF FRACTIONS

Subtraction of fractions follows similar rules. The fractions must be expressed in terms of a common denominator prior to subtraction. For example, subtract

5/16 from 25/32

Since 16 goes into 32 twice, the lowest common denominator is 32. Rewrite the equation.

This fraction is already reduced to its lowest form.

$$\frac{25}{32} - \frac{10}{32} = \frac{15}{32}$$

MULTIPLICATION OF FRACTIONS

Multiplication of fractions is a simple process. You must do the following operations:

1. Multiply the numerators
2. Multiply the denominators

(Optional) Reduce the fraction

EXAMPLE Multiply

$$\frac{3}{5} \times \frac{5}{9} = \frac{15}{45} = \frac{1}{3}$$

Cancellation can make the multiplication easier. In the last problem, we see:

$$\frac{3}{5} \text{ and } \frac{5}{9}$$

We divide the 5's by 5 and divide both the numerator and denominator by 3.

$$\frac{\cancel{3}^1}{\cancel{5}_1} \times \frac{\cancel{5}^1}{\cancel{9}_3} = \frac{1}{1} \times \frac{1}{3} = \frac{1}{3}$$

DIVISION OF FRACTIONS

Division of fractions is achieved by inverting the divisor and multiplying the fractions.

EXAMPLE

Divide   9/25 by 13/15   9/25 ÷ 13/15   9/25 x 15/13 =

9/5 x 3/13 = 27/65

Two other important rules for multiplication or division of fractions are:

1. Change all mixed numbers to improper fractions prior to working the function.

2. Put all whole numbers in fractional form by putting them over the number (1).

$$2 = \frac{2}{1} \text{ (fractional form)}$$

CALCULATIONS WITH DECIMALS

Decimals are expressions of part of the whole with an *understood* denominator. Understanding of the decimal system is important for it is similar to the metric system.

The decimal system is based on the root number 10 or its multiple as the *understood* denominator. The placement of the decimal point denotes how many zeros are behind the (1) placed in the denominator.

$$\frac{3}{10} = 0.3$$

EXAMPLE

Decimals can be written in mixed number or decimal form. For example:

$$\frac{81}{1000} = \frac{8.1}{100} = \frac{0.81}{10} = 0.081$$

Express $347\frac{91}{100}$ as a decimal $347\frac{91}{100} = 347.91$

Reading decimals is similar to reading whole number columns. The columns represent position values in this manner:

TABLE 21-B POSITION NAMES							
Etc	Hundreds	Ten	Units	Decimal	Tenth	Hundredths	Thousandths	Etc.
	0 to 900	0 to 90	0 to 9	0 to .	0 to .09	0 to .009	0 to .0009	.

EXAMPLE

525.125   is read five-hundred twenty-five and one hundred twenty-five thousandths

or

five-two-five point one-two-five

ADDITION

Decimals can be added quite easily provided that you align the decimal points in a vertical column and then add the values.

EXAMPLE Add 2.125 and 62.275

```
   2.125
+ 62.275
  64.400
```
OR 64.4

SUBTRACTION

Decimals can be subtracted in a similar fashion provided the decimal points are vertically aligned.

EXAMPLE Subtract 62.125 from 125.6201

```
  125.6201
-  62.1250
   63.4951
```

MULTIPLICATION

Decimals can be multiplied easily provided that after the numbers are multiplied, you place the decimal point in the answer the same total number of places from the right that are in the multiplier and the multiplicand.

EXAMPLE Multiply 62.1 by 1.8

```
  62.1    1 place from the right
x  1.8    1 place from the right
 111.78   2 places from the right
```

DIVISION

Decimals can be divided by placing the numbers in the divisor box positions; moving the decimal point to the far right in the divisor; moving the decimal the same number of places in the dividend; locating the decimal point in the quotient; and then dividing.

EXAMPLE Divide 0.217 by 0.0007

$$0.0007\overline{)0.2170}$$

```
           310.
00007.)2170.
       21
       ---
        70
        70
        ---
         0
```

Therefore, 0.217 ÷ 0.0007 = 310

CONVERSIONS

To convert a decimal into a fraction, place the amount given in the numerator and then put the number (1) plus the same number of zeros as there were places to the right of the decimal in the denominator position.

EXAMPLE Convert 0.172 into a fraction:

$$\frac{172}{1 + 000}$$ *represents 3 places to the right of the decimal*

To convert a fraction to a decimal, divide the numerator by the denominator.

EXAMPLE: Express **11/20** as a decimal

$$\frac{11}{20} = 20\overline{)11.00} = 0.55$$

LESSON 22

PERCENTS AND DECIMALS

Percents are a particular way to express a decimal value. Percentages can be used in calculations in decimal or fractional form. The last lesson emphasized the fraction form, and this lesson is devoted to operations with percentages in decimal form.

FORMULAS

A formula is an abbreviated method of expressing a combination of mathematical processes which produce the solution to a physical problem when the formula is applied to the problem.

An equation is the expression of a formula for one set of conditions.

Both the formula and the equation are set up as an expression of equality. Some sample formulas are:

$$C = D$$
$$P = B \times R$$
$$a + b = c + d$$

Equations and formulas are written to express mathematical relationships.

Many formulas are known that we can use to aid us in determining the particulars of various geometric shapes. There are formulas to compute the total value of interest for bank loans, the worth of our company inventory, the mileage of each company vehicle (or of the total fleet), and so on.

EXAMPLE

If it is known that one truck carries twice as much material as another, a formula can be written about the capacity of the two trucks.

$$a = 2b$$

where "a" represents the capacity of the first truck, and "b" represents the capacity of the second truck

This formula says that 1 unit of capacity in the first truck is equal to 2 units of capacity in the second truck.

Notice that the formula is an expression of one thing in relation to another.

In the formula

$$P = B \times R$$

Percentage = Base x Rate

the percentage is equal to the base number times the rate.

This expression of equality can also be expressed in terms of the rate

$$R = \frac{P}{B}$$

or, in terms of the base

$$B = \frac{P}{R}$$

PERCENT EXPRESSIONS TO DECIMAL VALUES

A percent is converted to a decimal by removing the percent sign and moving the decimal two places to the left, as shown in these examples:

$$62\% = 0.62 \qquad 18\% = 0.18 \qquad 125\% = 1.25$$

To solve problems, you may find it easier if you express the percent as a decimal, not as a fraction.

EXAMPLE

A contractor pays $65 for a stainless steel sink. The contractor's sale price is 55% above his cost. What is the contractor's selling price?

Multiplying percents can be easily performed by converting the percent to a decimal, then multiplying the base times the rate.

1. Write the percent

$$55\%$$

2. Set up the percentage formula

$$P = B \times R$$

3. Decide which numbers are the base, rate, and percentage.

Base = $65.00
Rate = 55%
Percentage = ?

4. Change the percent to a decimal.

$$55\% = 0.55$$

5. Plug in the numbers.

$$P = 65 \times 0.55$$

6. Multiply, remembering to place the decimal point in the proper place.

```
     65
  x 0.55
    32.5
    325
   35.75
```

p = $35.75

7. Re-read the question and solve the problem accordingly.

```
  $ 65
  + 35.75
  $100.75
```

8. Label the answer.

Selling price = $100.75

Note that in the problem above, the percentage was added to the base because it asked

"What is the contractor's selling price?"

It is very important to read the problem and determine what is needed before attempting to solve the problem.

Problems with percents that contain fractions are treated in the same fashion.

EXAMPLE

A plumber returns 20 valves to his stockroom after completing a job. This is about $8\tfrac{1}{2}\%$ of the valves that he figured were needed for the job. How many valves were actually used on the job?

First, it is important to determine what is given in the problem, and **what is requested of the answer**.

Given:
Rate = $8\tfrac{1}{2}\%$
Percentage = 20 (valves)

Requested: **Number of valves used on the job?**

In order to find the number of valves used on the job, the **total** number of valves must first be determined, then the number of valves used can be calculated.

1. Write the formula. $P = B \times R$

2. Plug in the numbers. $20 = B \times 8\frac{1}{2}\%$

3. Convert the percent to a decimal amount. $20 = B \times .085$

4. Divide both sides of the equation by .085. $20/.085 = B = 235$ valves

5. Determine what was asked of the problem:

   How many valves were used on the job?

6. Perform the operation and label the answer.

   235 valves
   - 20 valves
   215 valves

215 valves were used on the job

EXAMPLE

If an older water closet uses 3½ gallons per flush and a replacement water closet uses 1.6 gallons per flush, what is the percent rate of water saved with the new model?

1. Write the formula. $P = B \times R$

2. Determine the information given and plug it into the formula. 1.6 gallons per flush versus 3½ gallons per flush, yields savings of

   3.5 - 1.6 = 1.9 gal saved per flush

   Thus, $1.9 = 3.5 \times R$

3. Solve for R by dividing both sides of the formula by 3.5: $1.9/3.5 = R$

4. Divide the fraction. $0.543 = R$

5. Convert the decimal form back to a percent by placing the decimal over 2 places to the right, and adding a percent (%) sign,

   R = 54.3%

6. Label and answer the question. **Water saved = 54.3%**

PROBLEMS

Write the decimal equivalent of the following:

1. 0.3%

2. 26.5%

3. 350%

Write the percent equivalent of the following:

4. 0.16

5. 0.0002

6. 5.7

LESSON 23

SQUARES, SQUARE ROOTS, AND CIRCLES

SQUARES AND SQUARE ROOTS

This section of this lesson is about using the square of a number, and the square root of a number. The first part will discuss squares.

SQUARES

The *square* of a number is the product of the number multiplied by itself. We will find many applications where the square of a number must be calculated. Numerous examples exist in the physical world. For example, the energy that the brakes on your car must absorb (to have the car stop) is determined by the *square* of the speed of the car.

A more difficult calculation is to find that value which, when squared, equals a given number. Such a value is called the *square root* of the given number. As with squares, many problems are encountered where square roots must be calculated.

Squares and square roots are used quite often in plumbing in *on-the-job* calculations and are also used for formula and equation derivations.

EXAMPLE Consider **4 "squared" = 16**

That is, **4 x 4 = 16**

The number "4" is multiplied by itself, which yields the value of 16. Thus, the square of 4 is 16.

Squares are written in terms of the base number and the exponent in the form: A^2

where   A is the base number, and 2 is the exponent

The exponent is the number of times that the base number is multiplied by itself. In the case of squaring, the exponent is 2.

EXAMPLE

If the number 2 is to be squared, it would be written as "2^2". $2^2 = 4$, because 2 multiplied by itself is 4.

The squaring functions, therefore, are written in the form A^2, which means A x A. All numbers can be squared:

$(\frac{1}{2})^2 = \frac{1}{2} \times \frac{1}{2} = \frac{1}{4}$ $(16)^2 = 16 \times 16 = 256$

$(3.2)^2 = 3.2 \times 3.2 = 10.24$ $(1\frac{1}{2})^2 = \frac{3}{2} \times \frac{3}{2} = \frac{9}{4} = 2\frac{1}{4}$

$(1.5)^2 = 1.5 \times 1.5 = 2.25$

SQUARE ROOTS

If an equation reads $A^n = C$, and you know what C is, and also that n is equal to 2, how do you determine what A is equal to? To answer this question, we must know the method of calculating square roots.

Finding the square root of a number is the operation in which you are looking for the figure that, multiplied by itself, yields that number.

In the equation $A^2 = C$,   or $A \times A = C$ if $C = 4$, then $A \times A = 4$

Exploring this equation, the square root of the numbers and letters on both sides of the equal sign may be determined.

$$A^2 = 4$$
$$A = \sqrt{4}$$
$$A = 2$$

It is also true that the square root of 4 is negative 2 (-2). A negatives times a negative is a positive. As we concern ourselves with dimensions, we are interested in the positive square roots only.

The square root of a number is symbolized by the square root radical $\sqrt{}$

When a number is placed under the radical sign, it means that you are looking for the number that, when it is multiplied by itself, will equal the number under the radical.

EXAMPLE

$3 \times 3 = 9$ $\sqrt{9} = 3$

Important rules to remember when working with square roots are:

1.   When a number is written under the radical sign, the square root is meant.

2. If a number or letter written under a radical is squared, that number is removed from the radical, and the radical disappears.

In the example above, $A = \sqrt{9}$ can be written as $A = \sqrt{3^2}$

The square root of 1 is 1.

To determine the value of a number under the radical all squared numbers should be factored and removed from the radical. The remaining factors remain under the radical.

EXAMPLE Solve for whole numbers and radical amounts:

$\sqrt{8} =$

$\sqrt{8} = \sqrt{2^2 \times 2}$

$= 2\sqrt{2}$

$\sqrt{54} =$

$\sqrt{54} = \sqrt{9 \times 6}$

$= \sqrt{3^2 \times 6}$

$= 3\sqrt{6}$

Notice that the squared amounts are factored out, thus simplifying the expression.

It is important to be able to know how to factor the terms under the radical so that the radical can be simplified.

EXAMPLE

Simplify this radical in whole numbers and radical amount.

$\sqrt{768} =$

$\sqrt{768} = \sqrt{256 \times 3}$

$= \sqrt{(2)(2) \times (2)(2) \times (2)(2) \times (2)(2) \times 3}$

$= \sqrt{2^2 \times 2^2 \times 2^2 \times 2^2 \times 3}$

$= 2 \times 2 \times 2 \times 2\sqrt{3}$

$= 16\sqrt{3}$

Square roots can also be solved with a manual method which is cumbersome, but simple if the steps are followed carefully.

EXAMPLE   Find the square root of 2916.

a.   Write the radical amount.

$$\sqrt{2916}$$

b.   Add a decimal point if it does not appear and add zeros as needed.

$$\sqrt{2916.00}$$

c.   Place a decimal point above the radical box, immediately above the decimal in the radical.

$$\sqrt{2916.00}$$

d.   ***Start at the decimal*** and place prime marks after every 2 digits both to the right and left of the decimal point.

$$\sqrt{29'16.00'}$$

e.   Determine the number which, when squared, will be less than, or equal to, the first **primed** amount (29). Write this number above the radical.

$$\overset{5\ \ .}{\sqrt{29'16.00'}}$$

f.   Square this number and write it under the **primed** figures.

$$\overset{5\ \ .}{\sqrt{29'16.00'}}$$
$$25$$

g.   Subtract the square and bring down the next <u>two</u> figures.

$$\overset{5\ \ .}{\sqrt{29'16.00'}}$$
$$\underline{25}$$
$$4\ 16$$

h.   Double the figure over the radical and place it outside the radical frame directly in front of the newly formed number.

$$\overset{5\ \ .}{\sqrt{29'16.00'}}$$
$$\underline{25}$$
$$10\qquad 4\ 16$$

i. Determine how many times 10 plus another ***digit*** will go into 416, such as 101, 102, 103, 104. Place this number above the radical sign, and also next to the 104 at the left of the number 416. Multiply the number 104 by the number 4 and put the result under the 416 in the chain of the calculation.

$$\begin{array}{r} 5. \\ \sqrt{29'16.00'} \\ 25 \end{array}$$
104   4 16

j. In this case, 4 times 104 equals 416, which is placed below the 416 calculated above. As the problem works out evenly, the square root of 2916 is 54.

EXAMPLE

Find the square root of **580.85**

a. Write the radical amount. $\sqrt{580.85}$

b. Add a decimal if needed; add zeros as needed.

c. Place a decimal point above the decimal point in the radical. $\sqrt{580.85}$

d. Start at the decimal point and make prime marks after every two columns, both to the right and left of the point.

$$\sqrt{5'80.85'}$$

e. Determine the number which, when squared, will be less than or equal to the first number (5). Write the number above the radical.

$$\sqrt{5'80.85'} \;\; 2$$

f. Square this number and write it below the first primed number.

$$\begin{array}{r} 2 \\ \sqrt{5'80.85'} \\ \underline{4} \end{array}$$

g. Subtract and bring down the next pair of numbers.

$$\begin{array}{r} 2 \\ \sqrt{5'80.85'} \\ \underline{4} \\ 1\;80 \end{array}$$

h. Double the figure over the radical (2) and place it outside the radical frame directly in front of the newly-formed number.

$$\begin{array}{r} 2\ 4\ . \\ \sqrt{5'\ 80.85'} \\ \underline{4} \\ 4\quad\quad 1\ \ 80 \end{array}$$

i. Determine how many times 4_? will go into 180. Place this number above the second **primed** set and multiply.

$$\begin{array}{r} 2\ 4\ . \\ \sqrt{5'\ 80.85'} \\ \underline{4} \\ 44\quad\ 1\ \ 80 \\ \underline{1\ \ 76} \\ 4\ \ 85 \end{array}$$

j. Subtract this figure and lower the next primed set.

$$\begin{array}{r} 2\ 4\ . \\ \sqrt{5'\ 80.85'} \\ \underline{4} \\ 44\quad\ 1\ \ 80 \\ \underline{1\ \ 76} \\ 48\quad\quad\ \ 4\ \ 85 \end{array}$$

k. Double the figure above the radical (24) and place accordingly.

$$\begin{array}{r} 2\ 4\ .\ 1 \\ \sqrt{5'\ 80.85'} \\ \underline{4} \\ 44\quad\ 1\ \ 80 \\ \underline{1\ \ 76} \\ 481\quad\ 4\ \ 85 \\ \underline{4\ \ 81} \end{array}$$

l. Determine how many times 48? will divide into 485 and place this number above the radical.

m. Add zeros to carry this square root to the accuracy desired.

$$\begin{array}{r} 2\ 4\ .\ 1 \\ \sqrt{5'\ 80.85'\ 00} \\ \underline{4} \\ 44\quad\ 1\ \ 80 \\ \underline{1\ \ 76} \\ 481\quad\ 4\ \ 85 \\ \underline{4\ \ 81} \\ 4\ \ 00 \end{array}$$

n. Answer the problem.

The square root of 580.85 is approximately 24.1

PROBLEMS

Find the square root of the following numbers.

1. 100

2. 1000

3. 6352

4. 10.62

SQUARE ROOTS II

The equations relating squares and square-roots to each other are the following:

If $x = y^2$ then $\sqrt{x} = \sqrt{y^2}$ or $\sqrt{x} = y$

If numbers are substituted for x and y, then the equation becomes, for example,

If $x = 9$, then $9 = y^2$, and $y = 3$

The number 3 is the square root of the number 9 because it is that value, which, when multiplied by itself (3 x 3), will equal the number 9.

EXAMPLES

$\sqrt{25} = \sqrt{5^2} = 5$ $\sqrt{144} = \sqrt{12^2} = 12$

$\sqrt{16} = \sqrt{4^2} = 4$ $\sqrt{169} = \sqrt{13^2} = 13$

Note that the numbers under the radical sign were factored to make it easy to see what could be removed from the radical.

It is important to know that a squared number can be removed from the radical.

The reason is as follows: $\sqrt{a} \times \sqrt{b} = \sqrt{a \times b}$

This expression is always true, and it says that the square root of *a times the square root of b* equals the square root of *(a times b)*. Thus, we can "pull apart" any factors that we please, and treat them independently, without changing the value of the original expression.

Thus, we can factor out any perfect square, place it in its own radical, and extract the square root. The reason the radical sign disappears in the problems above is that all of the terms have whole number square roots.

The square root of (1) is 1 because 1 x 1 = 1.

In most cases, the radical cannot be solved evenly; in such cases there are two methods of solving these problems:

First – Any squares are factored out of the problem and the remaining factors are left under the radical sign. This is the preferred method.

Second – Manually extract the square root by the process as shown in several examples above.

EXAMPLE

Work out the problem. The square root of $\sqrt{18}$ is approximately 4.24.

A square root can also be determined for a fractional expression, as follows:

EXAMPLE

What is the square root of $\frac{16}{25}$?

$$\sqrt{\frac{16}{25}} = \frac{\sqrt{16}}{\sqrt{25}} = \frac{\sqrt{4^2}}{\sqrt{5^2}} = \frac{4}{5}$$

Solve for the value of $\frac{\sqrt{6}}{\sqrt{7}}$

$\sqrt{6}$ equals about 2.45

$\sqrt{7}$ equals about 2.65

$$\sqrt{\frac{6}{7}} = \frac{\sqrt{6}}{\sqrt{7}} = ? \qquad\qquad \sqrt{\frac{6}{7}} = \frac{\sqrt{6}}{\sqrt{7}} = \frac{2.45}{2.65}$$

Therefore, the answer is 0.926

This problem could have been solved by dividing 7 into 6 first, then taking the square root of that figure:

$$\sqrt{\frac{6}{7}} = \sqrt{0.857142}$$

```
              .  9  2  5
           √.85'71'42'
              81
      182    4 71
             3 64
     1845   1 07 42
              92 25
              15 17
```

PRECISION

It is important at this point to address ourselves to the concept of precision, or degree of accuracy, and also the idea of **rounding off** numbers.

Precision refers to how close our calculations approach the true answer. In the last example, 0.925 was approximated as the square root of **6/7**. Yet, if the approximation was carried out to the next digit, this square root would be 0.9258. The number 0.926 would be a closer approximation to the square root of 6/7.

In practical work, there is no sense in carrying out calculations beyond the precision of our measurements or degree of accuracy of our given information. Therefore, most of our answers should use not more than three significant figures, and usually not more than two decimal places.

Generally, work problems out to one more decimal place than required and **round off** the answer, using the following convention:

If the last digit is greater or equal to 5, drop the last digit and increase the last digit retained by 1.

If the last digit is less than 5, drop the last digit and do not change the last digit retained.

EXAMPLES

0.9258 rounds to three places, 0.926

8.9432 rounds to three decimal places, 8.943

80.68 rounds to three significant figures, 80.7

3.575 rounds to three significant figures, 3.58

14.765 rounds to two decimal places, 14.77

It seems that as a problem is calculated to more places, the precision of the answer is improved, but this is not true if the starting information is not sufficiently precise.

Note that the accuracy of the numbers worked in this trade usually goes to the tenths or hundredths only.

EXAMPLE

Approximate this answer to the nearest hundredths place.

If $$A^2 = \frac{7}{8}$$

Calculate **A**

$$\sqrt{A^2} = \sqrt{\frac{7}{8}}$$

$$A = \sqrt{\frac{7}{8}}$$

Solve the division of the fraction first.

$$7 \div 8 = .875$$

Take the square root to solve for A

```
                    . 9  3  5
                 √ .87'50'00'
                   81
            183    6 50
                   5 49
           1865    1 01 00
                     93 25
```

Solve to the nearest hundredth: **0.935 = 0.94**

Because the digit in the thousandths column is 5, "one" is added to the next column on the left.

Of course, if we are realistic, we will realize that we probably will never figure a square root by the above method. Modern calculators generally have a square root function on them which can give us the answer in less time than it would take for us to write the problem out. If your calculator *dies* sometime, Table 23-A, and the notes accompanying it, may be useful.

TABLE 23-A SQUARE ROOT TABLE										
0	.0	.1	.2	.3	.4	.5	.6	.7	.8	.9
0	0	.316	.447	.548	.632	.707	.775	.837	.894	.949
1	1	1.05	1.10	1.14	1.18	1.22	1.26	1.30	1.34	1.38
2	1.414	1.45	1.48	1.52	1.55	1.58	1.61	1.64	1.67	1.70
3	1.732	1.76	1.79	1.82	1.84	1.87	1.90	1.92	1.95	1.97
4	2	2.02	2.05	2.07	2.10	2.12	2.14	2.17	2.19	2.21
5	2.236	2.26	2.28	2.30	2.32	2.35	2.37	2.39	2.41	2.43
6	2.449	2.47	2.49	2.51	2.53	2.55	2.57	2.59	2.61	2.63
7	2.646	2.66	2.68	2.70	2.72	2.74	2.76	2.77	2.79	2.81
8	2.828	2.85	2.86	2.88	2.90	2.92	2.93	2.95	2.97	2.98
9	3	3.02	3.03	3.05	3.07	3.08	3.10	3.11	3.13	3.15
10	3.162									

To use the square root table, read the square root from the table if the number is between 0.1 and 10. If the number is between 10 and 100, move the decimal point one place to the left, read the square root from the table, and multiply by

$$\sqrt{10}, \text{ i.e., } 3.162$$

For any other number, use a factor of $\frac{1}{10^2}$, 10^2, 10^4, *etc.*

and multiply the table values by $\frac{1}{10}$, 10, 10^2, *etc.*

EXAMPLES $\sqrt{3.6} = 1.90$

$$\sqrt{36} = \sqrt{3.6}\sqrt{10} = 1.90(3.162) = 6.0078 - 6.01$$

(*not 6.000 because of rounding in the table*)

$$\sqrt{360} = \sqrt{3.6}\sqrt{100} = 1.90(10) = 19.0$$

PROBLEMS (Second Group)

Find the following square roots using Table 23-A.

1. 1.3

2. 130

3. 6.6

4. 760

CIRCLES

A circle is a plane figure bounded by a curved line whose every point is equally distant from the center. See Figure 23-A.

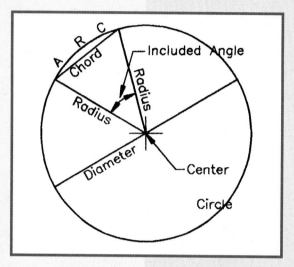

Figure 23-A – Parts of a Circle

Because the circle occurs frequently in nature, and because it is the best shape for some products we use, we need to know the parts of circles, and the mathematical expressions that describe the relationships between parts.

DEFINITIONS

CENTER

The center of the circle is that point where all radii and diameters originate.

RADIUS

The radius is a straight line from the center to any point on the circle.

DIAMETER

The diameter is a straight line drawn through the center to points on opposite sides of the circle.

ARC

An arc is a portion of the circumference of the circle.

CHORD

A chord is the straight line connecting the ends of an arc.

CIRCUMFERENCE

The circumference of the circle is the distance all the way around the circle.

INCLUDED ANGLE

The included angle is that angle formed at the center of the circle by two radii terminating at the ends of a chord or an arc.

CIRCLE

The total included angle in a circle equals 360°.

A complete circle encompasses a total included angle of 360°.

Each degree (°) is subdivided into 60 parts, called minutes.

Minutes are subdivided into 60 parts, called seconds.

Therefore, each degree of a circle is made up of 60 x 60 = 3600 seconds. The entire circle comprises 21,600 minutes or 1,296,000 seconds.

A quarter of a circle contains a 90° (360° divided by 4) included angle, which is called a right angle; the sides of a right angle are perpendicular to each other. A 180° angle (half a circle) is a straight line (the diameter).

It is important to be aware of the angles because fitting angles are based on degree turns. A $1/8$ bend cast-iron fitting is $1/8$ of 360°, or a 45° turn. A 60° turn ($\frac{60}{360}$) is a $\frac{1}{6}$ bend.

CIRCLE FORMULAS

When we discussed formulas before, we learned that if you know the relationship, certain information can be derived if you know other information. The following discussion brings out the relationships that are important with circles:

The following abbreviations will be used in these circle formulas:

C = Circumference

D = Diameter

R = Radius

A = Area

π = 3.1216 — a mathematical constant

The two principal formulas for circles are used to calculate circumference and area. These formulas are:

$$C = \pi D = 2\pi R$$

$$A = \pi R^2$$

CIRCUMFERENCE

The circumference of a circle can be derived in three ways:

1. Direct measurement

2. Derivation from the $C = \pi D$ formula

3. Multiplying the radius by 2, then multiplying by π.

Although we might believe that the easiest way to obtain the circumference of the circle is by direct measurement, it is not always possible due to size, location, etc.

Many times we can obtain the diameter of the circle. When we know the diameter, the circumference can be calculated using the formula

$$C = \pi d$$

EXAMPLE

If the diameter of a circle equals 10 inches, what is the circumference?

1. Set up the formula and solve.

$$C = \pi d$$

$$C = \pi (10")$$

$$C = 31.416"$$

2. The circumference equals 31.416 inches.

If the radius of a circle is known, the diameter can be calculated. The diameter is equal to twice the radius of a circle. If the radius equals 7", the diameter equals 14".

$$D = 2R$$

RADIUS

The radius can be determined in a similar fashion.

1. A radius can be directly measured from the circle diagram.

2. A radius can be determined by taking $1/2$ of the diameter.

3. Indirectly, a radius can be computed by knowing a circumference and then by using a formula substitution.

Since the radius (R) equals $1/2$ of the Diameter (D), then R = $1/2$D, or 2R = D

Substituting 2R for D in the formula or

$$C = \pi D$$
$$C = 2\pi R$$

$$R = \frac{C}{2\pi}$$

EXAMPLE

If the circumference of a circle is 89 inches, what is its radius?

$$R = \frac{C}{2\pi}$$

C = 89 *inches*

π = 3.142

$$R = \frac{89 \text{ in.}}{2 \times 3.142}$$

$$R = \frac{89}{6.284}$$

R = 14.16 *inches*

Therefore, the radius of this circle equals 14.16 inches.

The radius of a circle may also be calculated if the area of the circle is known by using the formula

$$A = \pi R^2$$

$$\frac{A}{\pi} = R^2$$

then if both sides are divided by π,

$$\sqrt{\frac{A}{\pi}} = R$$

EXAMPLES

If the area of a circle is 28.28 square inches, what is the length of the radius?

$$R^2 = \frac{28.28 \text{ in.}^2}{3.142}$$

$$= 9 \text{ in.}^2$$

$$R = 3 \text{ inches}$$

The radius of this circle equals 3 inches.

Note that the area of a circle is always written in terms of a squared amount. Therefore, the square root of inches-squared is inches.

If the radius of a circle is 0.64 inch, what is its circumference?

The circumference of this circle equals 4 inches.

$$C = 2\pi R$$

$$C = 2 \times 3.142 \times 0.64 \text{ in.}$$

$$C = 4.0 \text{ inches}$$

AREA

The area of a circle can be computed by one of three methods.

1. Direct measurement

2. From the formula $\quad A = \pi R^2$

3. From the formula $\quad A = 0.7854 D^2$

This last expression was derived by substituting ½D for the R in the formula.

$$A = \pi R^2$$

Solving the formula, we find

$$A = \pi \left(\frac{1}{2}D\right)^2$$

$$A = 3.1416 \left(\frac{1}{4}D\right)^2$$

$$A = 0.7854 \times D^2$$

EXAMPLE

If the diameter of a circle is 10", what is its area?

$$R = \frac{1}{2}D$$

$$R = \frac{1}{2}(10")$$

$$R = 5 \text{ inches}$$

$$A = 3.142 \times (5 \text{ in.})^2$$

$$A = 3.142 \times 25 \text{ in.}^2$$

$$A = 78.55 \text{ in.}^2$$

What is the radius of this circle? Double check the answer using the formula

$$A = 0.7854 \times D^2$$

$$A = 0.7854 \times 10 \times 10 \text{ in.}$$

$$A = 78.54 \text{ square inches}$$

ARC MEASUREMENTS

Arc length can be determined by:

1. Direct measurement

2. The formula   **arc length = 2πR x Included Angle/360**

EXAMPLE

What is the length of an arc with a 90° included angle and a radius of 3 inches?

$$\text{arc length} = 2\pi R \times \frac{\text{included angle}}{360}$$

$$\text{arc length} = 2 \times 3.142 \times 3" \times \frac{90}{360}$$

$$\text{arc length} = 2 \times 3.142 \times 3" \times 0.25$$

$$\text{arc length} = 4.7 \text{ inches}$$

The number of degrees in an arc can be measured by:

1. Direct measurement

2. From the formula $\quad\quad\quad\quad$ included angle = $\dfrac{\text{arc length} \times 360}{2\pi R}$

EXAMPLE

If an arc is 14.6 inches long, and the diameter is 24 inches, how many degrees are in the arc?

$$\text{included angle} = \dfrac{\text{arc length} \times 360}{2\pi R}$$

$$= \dfrac{\text{arc length} \times 360}{\pi D}$$

$$= \dfrac{14.6 \text{ inches} \times 360}{3.142 \times 24}$$

$$= 0.194 \times 360$$

$$\text{included angle} = 69.7°$$

The number of degrees is equal to 70° (rounded off).

It is important to remember these formulas because they will be used for many calculations later. They are convenient because direct measurement of a diameter, area, etc., is cumbersome and may be impossible in some circumstances.

PROBLEMS (Third Group)

The circle appears in many ways in our work. Complete the following problems to build confidence and ease in working out circle relationships.

1. R = 12", find the diameter

2. R = 6', find the circumference

3. D = 4 miles, find the area

4. Area = 1257 square yards, find the diameter

5. Included angle = 56°, R = 80", find the arc length

6. Arc length = 10 meters, R = 8 meters, find the included angle

LESSON 24

ANGLES, RATIOS, AND TRIANGLES

ANGLES

An angle is a figure formed by extending two lines from a point. A straight line forms an angle of 180°.

DEFINITIONS

Angles, like circles, have components which are defined as follows:

VERTEX

The vertex of an angle is the point where the two lines which form the angle meet.

MEASUREMENT

Angles are measured in degrees (°), minutes ('), and seconds ("). Note the symbols we use when referring to these measurements.

$$\text{One degree} = 60 \text{ minutes}$$

$$\text{One minute} = 60 \text{ seconds}$$

A circle is made up of 360 degrees and contains 4 quadrants, each of which contains a 90° angle.

It is possible to change degrees to minutes and minutes to degrees. To change degrees to minutes, multiply the number of degrees by 60 and express as minutes.

EXAMPLE

An angle measures 65° and 10'. How many total minutes is this?

1. Change the total degrees to minutes

$$65 \times 60 \text{ minutes} = 3900 \text{ minutes}$$

2. Add the total minutes

$$\begin{aligned} & 3900 \text{ }\textit{minutes} \\ + & 10 \text{ }\textit{minutes} \\ \hline & 3910 \text{ }\textit{minutes} \end{aligned}$$

Note that 3910 minutes may also be written as 3910'.

When changing minutes to seconds, multiply the minutes by 60 and change the term to seconds.

EXAMPLE

How many seconds in 16°10'30"?

1. Convert 16 to minutes

$$16 \times 60 \text{ minutes} = 960 \text{ minutes}$$

2. Add the minutes

**960 *minutes*
+ 10 *minutes*
970 *minutes***

3. Change to seconds (60 seconds in a minute)

**970 *minutes*
x 60
58,200 *seconds***

4. Total the seconds

**58,200 *seconds*
+ 30 *seconds*
58,230 *seconds***

It is also possible to change a large number of seconds into degrees, minutes, and seconds.

EXAMPLE

Divide 52,256 seconds into degrees, minutes and seconds.

1. Divide 52,256 seconds by 3600 (60 seconds per minute x 60 minutes per degree = 3600 seconds per degree) to determine the degree amount. Stop the division with the whole number amount and write the remainder.

```
              14°R 1856"
       3600 ) 52256
              3600
             16256
             14400
              1856
```

Plumbing Apprentice Student Workbook Year One
Fourth Edition

There are 14 degrees with a remainder of 1856".

2. Divide the remainder (R) of the first division by 60 until a whole number and remainder are determined.

$$60 \overline{)1856} \begin{array}{r} 30' R\ 56" \\ \underline{180} \\ 56 \end{array}$$

3. Write the answer in terms of degrees, minutes, and seconds.

$$52,256" = 14°\ 30'\ 56"$$

ANGLES

If two straight lines intersect at a point, an angle is formed at that point. The four types of angles are listed in terms of 180° or 90°.

STRAIGHT ANGLE

A straight angle is one that is formed by a straight line through points EOF in Figure 24-A.

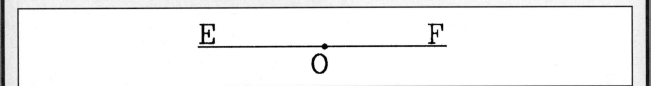

Figure 24-A – Angle EOF – Straight Angle

RIGHT ANGLE

A right angle is formed by the intersection of two lines that are perpendicular. For example, angle GOH in Figure 24-B.

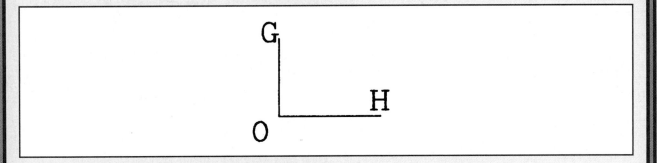

Figure 24-B – Angle GOH – Right Angle

Plumbing Apprentice Student Workbook Year One
Fourth Edition

OBTUSE ANGLE

An obtuse angle is an angle less than 180° but more than 90°, such as angle JOL in Figure 24-C.

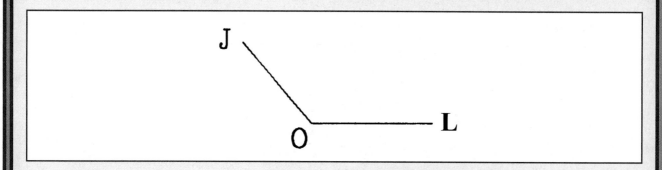

Figure 24-C – Angle JOL – Obtuse Angle

ACUTE ANGLE

An acute angle is one which is less than 90°, for example, angle MON in Figure 24-D.

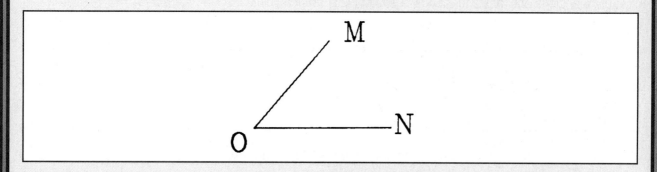

Figure 24-D – Angle MON – Acute Angle

LABELING ANGLES

In order to communicate to others exactly what we mean, it is important that we use standard labels and computations.

When labeling an angle, letters are placed at the vertex (intersection or starting point) and also at the termination of the lines. The angular measurement may also be included.

EXAMPLE

An angle would be labeled as follows:

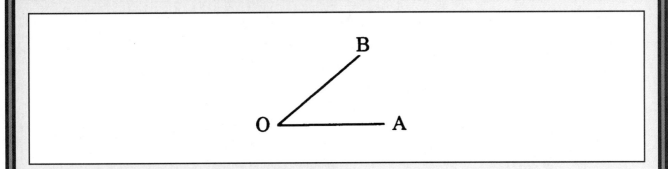

Figure 24-E – Labeling an Angle

We use letters from the alphabet to name each section of the angle. It is common practice to use O as the label for the vertex (the intersection point of the two lines). In this example, we have used A and B to denote the ends of the lines forming the angle. You may use any letters you wish in your own examples; the point is these are aids to help communicate information about the angle.

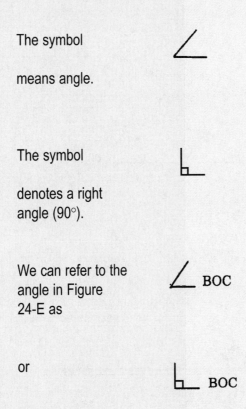

Note that the vertex is always in the middle of the letter identification of the angle.

Angles, when combined, make other geometric figures such as triangles, rectangles, etc. Even a circle is measured in terms of angles. We use angles to figure offsets, refer to fittings, etc. In order to understand the labels that we use for various materials, we must have some background in how these labels came to be.

PROBLEMS

1. If Angle AOB = 82°, it is called a(n) _____ angle.

2. If Angle JOL = 126°, it is called a(n) _____ angle.

3. If Angle EOF = 180°, it is called a(n) _____ angle.

4. If Angle GOH = 90°, it is called a(n) _____ angle.

CAST-IRON PIPING CHANGE OF DIRECTION

Earlier lessons discussed plastic, steel, copper, cast-iron, and other fittings which are used to obtain the needed routings for the flow of water, waste, sewage, or air (in vent systems). Most fittings are made to specifications which require that they be made to conform to appropriate angular relationships. Cast-iron soil pipe fittings have the greatest variety of angle fittings. A fitting used for the purpose of making a bend in a cast-iron soil pipe line is listed in terms of parts of a 360° circle. No doubt, the relative difficulty in making cast-iron joints — and the very limited ability to "fudge" the angle as determined by the fitting — require the wide range of fittings in the marketplace.

Look at the fittings that make a turn in the flow. If a 90° turn is required (that is, you need to turn the flow perpendicular to the original flow), a $1/4$ bend (90°) is needed. The bend name is determined by dividing the required angle by 360°. Figure 24-F shows a quarter bend change of direction in a pipe line.

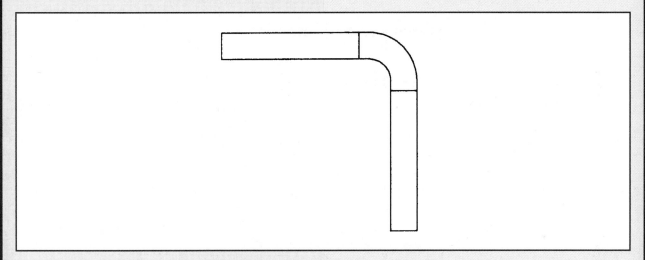

Figure 24-F – Pipe Line 90-Degree Bend

For a change in direction of:

$$60°, \text{ a } \frac{1}{6} \text{ bend is needed } (\frac{60}{360})$$

$$45°, \text{ a } \frac{1}{8} \text{ bend is needed } (\frac{45}{360})$$

$$22\frac{1}{2}°, \text{ a } \frac{1}{16} \text{ bend is needed } (\frac{22\frac{1}{2}}{360})$$

$$72°, \text{ a } \frac{1}{5} \text{ bend is needed } (\frac{72}{360})$$

If a mechanic needs to make a turn, yet wants to make the turn in a gradual sweep rather than at a sharp angle, it is possible to combine fittings to achieve the desired angle.

EXAMPLES

NOTE: Many of the examples given would seldom have any practical application, but they are shown to demonstrate and practice the concept of combining angle turns.

$$\frac{1}{4} \text{ bend} + \frac{1}{16} \text{ bend} = \frac{5}{16} \text{ turn} = 112.5°$$

$$\frac{1}{4} \text{ bend} + \frac{1}{6} \text{ bend} = \frac{5}{12} \text{ turn} = 150°$$

$$\frac{1}{4} \text{ bend} + \frac{1}{5} \text{ bend} = \frac{9}{20} \text{ turn} = 162°$$

$$\frac{1}{4} \text{ bend} + \frac{1}{4} \text{ bend} = \frac{1}{2} \text{ turn} = 180°$$

$$\frac{1}{8} \text{ bend} + \frac{1}{16} \text{ bend} = \frac{5}{16} \text{ turn} = 67.5°$$

$$\frac{1}{8} \text{ bend} + \frac{1}{8} \text{ bend} = \frac{1}{4} \text{ turn} = 90°$$

$$\frac{1}{8} \text{ bend} + \frac{1}{6} \text{ bend} = \frac{7}{24} \text{ turn} = 105°$$

$$\frac{1}{8} \text{ bend} + \frac{1}{5} \text{ bend} = \frac{13}{40} \text{ turn} = 117°$$

$$\frac{1}{6} \text{ bend} + \frac{1}{16} \text{ bend} = \frac{11}{48} \text{ turn} = 82.5°$$

$$\frac{1}{6} \text{ bend} + \frac{1}{6} \text{ bend} = \frac{1}{3} \text{ turn} = 120°$$

$$\frac{1}{6} \text{ bend} + \frac{1}{5} \text{ bend} = \frac{11}{30} \text{ turn} = 132°$$

$$\frac{1}{5} \text{ bend} + \frac{1}{5} \text{ bend} = \frac{2}{5} \text{ turn} = 144°$$

$$\frac{1}{5} \text{ bend} + \frac{1}{16} \text{ bend} = \frac{21}{80} \text{ turn} = 94.5°$$

$$\frac{1}{16} \text{ bend} + \frac{1}{16} \text{ bend} = \frac{1}{8} \text{ turn} = 45°$$

Any multiple combination of bends can be assembled to achieve any desired turn. You should realize that using more than one bend will increase the length of space needed to accomplish the desired change in direction. In most cases, extravagant turns are not used. Forty-five, sixty, and ninety degree turns are most common.

RATIOS

Ratios are useful mechanisms for solving certain problems. They are set up as follows:

$$\frac{a}{b} \text{ as } \frac{c}{d}$$

or

$$\frac{a}{b} = \frac{c}{d}$$

We would read the above ratio "a is to b as c is to d." Ratios are used to solve problems in which you know most of the information and the relationships between the items so that you can find the information you are missing.

EXAMPLES

$$\frac{9}{20} = \frac{x}{360}$$

To solve this problem, cross-multiply and solve for "x".

$$360(9) = 20(x)$$

$$\frac{360(9)}{20} = x$$

$$18(9) = x$$

$$x = 162$$

This problem may also be written

$$360 \cdot 9 = 20 \cdot x$$

$$\frac{360 \cdot 9}{20} = x$$

$$18 \cdot 9 = x$$

$$x = 162$$

EXAMPLE

Three $1/8$ bends are joined to make a turn. What is the total angular deflection?

$$\frac{1}{8} + \frac{1}{8} + \frac{1}{8} = \frac{3}{8}$$

$$\frac{3}{8} = \frac{x}{360}$$

$$\frac{1}{8} = \frac{x}{360} \qquad\qquad x = \frac{360(3)}{8}$$

$$x = 45° \qquad\qquad x = 135° \text{ angle of deflection}$$

ALTERNATE SOLUTION

Since three $1/8$ bends were used, $3 \times 45 = 135°$ angle of deflection

Copper, steel, plastic, and other fittings are available in a variety of angular deflections. These fittings, like cast-iron fittings, are usually called by their degree of deflection. As noted above, however, fittings made of these materials are not available in as extensive a range of angles as fittings made for cast-iron soil pipe.

REVIEW OF ANGLE CONCEPTS

Angles can be:

1. Straight (180°)

2. Obtuse (between 90° and 180°)

3. Right (90°)

4. Acute (less than 90°)

Angles are labeled with letters denoting the lines and the intersection point (vertex).

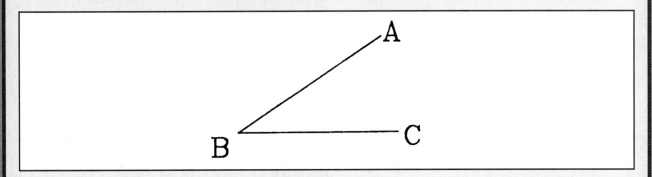

Figure 24-G – Angle

The angle in Figure 24-G is labeled as ABC.

Angles can be added to make new angles. In the same way, as we saw above, fittings can be added to make almost any desired change in direction of flow.

When angles are grouped together, they can form configurations such as triangles, trapezoids, rectangles, and other shapes.

Angles are measured in terms of degrees, minutes, and seconds of a circle.

1. To convert degrees to minutes, multiply by 60.

2. To convert degrees to seconds, multiply by 3600.

3. To convert minutes to seconds, multiply by 60.

RIGHT TRIANGLES

IDENTIFYING PARTS OF RIGHT TRIANGLES

In our work we frequently encounter a triangle with one 90° angle, a right triangle. Many piping problems can be easily solved by using right triangles to obtain the solution.

To describe a right triangle, a common method uses the upper case letters "A" and "B" for the acute angles, and the upper case letter "C" for the right angle. The sides are identified with lower-case versions of the letters opposite the upper case angle, as shown and in Figures 24- H and I.

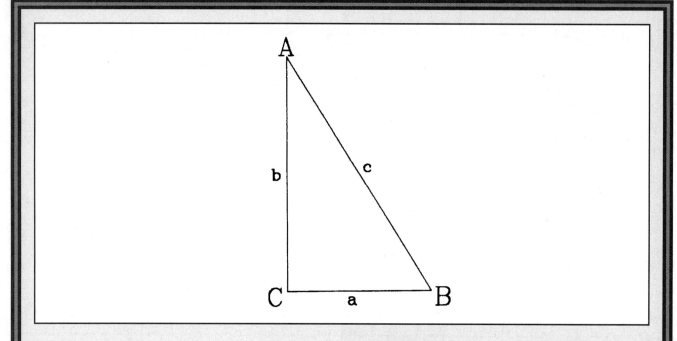

Figure 24-H – Right Triangle

Side a is opposite angle A.

Side b is opposite angle B.

Side c is opposite angle C.

Side b is the *base* of the triangle.

Side a is the *height* or *altitude* of the triangle.

Side c is the *hypotenuse*.

Note in Figure 24-I that if we draw the triangle **on its side**,

Side a is the base.

Side b is the altitude.

We will see that the idea of **base** and **altitude** can be interchanged without difficulty.

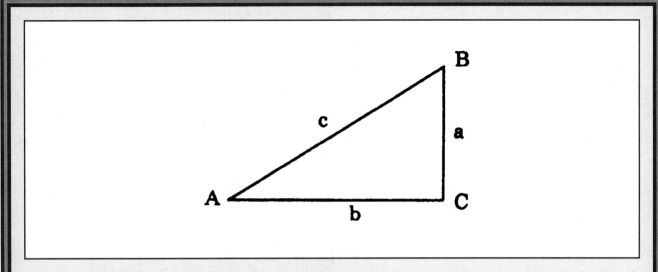

Figure 24-I – Right Triangle "On Its Side"

TYPES OF RIGHT TRIANGLES

ISOSCELES RIGHT TRIANGLES

The base and the height are the same length, as shown in Figure 24-J. Since an isoceles right triangle is a specific triangle (as the two sides are equal), it follows that the two opposite angles are equal. Since the total of all angles in a triangle is 180°, and the right angle is 90°, the other two angles are 90° total, or 45° each. Therefore, the only isoceles right triangle is a 45°-45°-90° triangle.

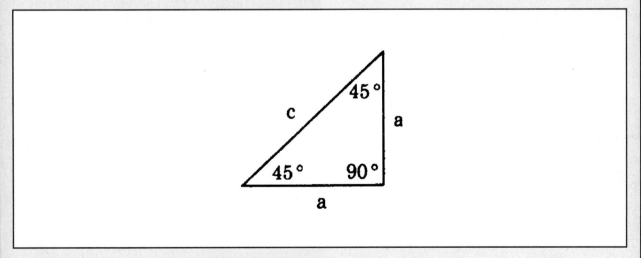

Figure 24-J – Isosceles Right Triangle

SCALENE RIGHT TRIANGLES

The base and the height are not equal. (See Figures 24-H and 24-I.)

CALCULATIONS

SIDES OF A RIGHT TRIANGLE

The Pythagorean Theorem is one of the most widely used formulas in the world of mathematics and that is certainly true in the plumbing industry. This formula states that the sum of the squares of the base and the height of a right triangle is equal to the square of the third side (hypotenuse). The expression is the following:

$$a^2 + b^2 = c^2$$

EXAMPLE

If the base of a right triangle is 12 inches and the height is 5 inches, how long is the hypotenuse?

a = 5 inches   b = 12 inches   c = ?

$$height^2 + base^2 = hypotenuse^2$$

$$a^2 + b^2 = c^2$$

$$5^2 + 12^2 = c^2$$

$$25 + 144 = c^2$$

$$169 = c^2$$

Hypotenuse = 13 inches

EXAMPLE

If a right triangle has a hypotenuse of 5 inches and a base of 4 inches, what is the height?

a = ?   b = 4   c = 5

$$a^2 + b^2 = c^2$$

$$a^2 + 4^2 = 5^2$$

We can rearrange the formula and solve for "a".

$$a^2 + c^2 - b^2 = 5^2 - 4^2$$

$$a^2 = 25 - 16 = 9$$

$$a = 3$$

Height = 3 inches

AREA OF A TRIANGLE

The area of a triangle is the total surface contained within a triangle. The area of a triangle is given by using the equation:

$$A = \frac{1}{2} b \times h$$

EXAMPLE

If a right triangle has a base of 24" and a height of 16", how much area is contained in this triangle?

$$A = \frac{1}{2}bh = \frac{1}{2}(24) \times 16$$

$$A = 12 \times 16 = 192 \text{ sq inches}$$

EXAMPLE

If a right triangle has a hypotenuse of 25 inches and a base of 16 inches, what is the area of the triangle?

$$a = ? \qquad b = 16 \qquad c = 25$$

Find the height (altitude), using $a^2 + b^2 = c^2$.

$$a^2 = c^2 - b^2 = \sqrt{25^2 - b^2}$$

$$a = \sqrt{25^2 - 16^2} = \sqrt{625 - 256}$$

$$a = \sqrt{369}$$

Use your calculator, or the methods discussed previously, to arrive at the square root of 369.

$$a = 19.2$$

Find the area of this triangle: base = 16", altitude or height = 19.2"

$$A = \frac{1}{2} b \times h \qquad\qquad A = \frac{1}{2}(16" \times 19.2")$$

$$\textit{Area} = 153.6 \text{ sq in}$$

EXAMPLE

The area of an isosceles right triangle is 12.5 square inches, what are the lengths of the sides of this triangle? (Remember that an isosceles right triangle is a triangle with the height [altitude] equal to the base.)

$$A = \frac{1}{2}(b \times h) = 12.5 \ in.^2 = \frac{1}{2}b^2$$

$$25 \ in.^2 = b^2 \quad b = \sqrt{25 \ in.^2} = 5 \ in.$$

$$5 \ in. = b = h$$

The lengths of both the base and the height (altitude) are 5 inches.

SHAPES

Triangles can be grouped to form new shapes. Two equal right triangles, when placed together with a common hypotenuse will form a rectangle. If two equal isosceles right triangles are placed together to form a figure, the figure is a square. This is true because the sides of the figure are all the same length and the angles are right angles.

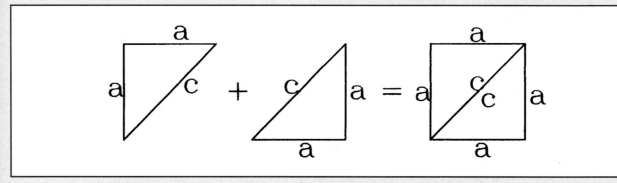

Figure 24-K – Square Formed by Two Isosceles Triangles

If, however, two scalene right triangles of the same size are placed together with a common hypotenuse, the figure formed is a rectangle.

Since the area for a right triangle is equal to one-half the base times the height (altitude),

$$a = \frac{1}{2}(b \times h)$$

The area of two of these triangles is

Area of two equal triangles $= 2 \left(\frac{1}{2}\right)(b \times h)$

$$A = b \times h$$

The formula for the area of a square of side "s" is

$$A = s^2$$

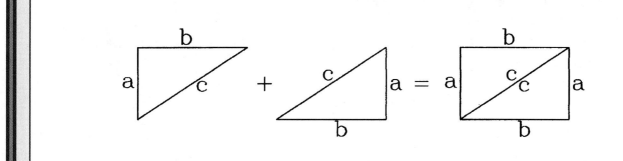

Figure 24-L – Rectangle Formed by Two Scalene Triangles

The formula for the area of a rectangle is **A = l x h**

That is, the area is equal to the length times the height.

PROBLEMS

Calculate the unknown values in the following problems:

Right Triangles				
a = 3	a = 20	a = 100	**a = ?**	a = 15
b = 4	b = 15	**b = ?**	b = 173.2	b = 30
c = ?	**c = ?**	c = 141.4	c = 200	**Area = ?**

Rectangles	
a = 18	a = 12
b = 24	**b = ?**
diagonal = ?	Area = 144
Area = ?	

45° RIGHT TRIANGLES - REVIEW

The Pythagorean Theorem states that the sum of the squares of the two sides of a right triangle is equal to the square of the hypotenuse.

hypotenuse² = base² + height²   or   c² = b² + a²

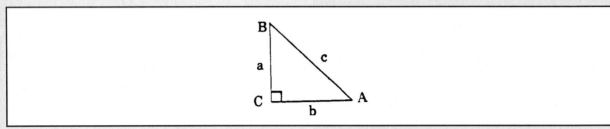

Figure 24-M – Isosceles Right Triangle (a = b)

The area of a right triangle is equal to one-half the base times the height.

$$A = \tfrac{1}{2}(b \times a)$$

A triangle with base equal to height is called an isosceles triangle. Assume that the length of the base and the height of an isosceles right triangle are each 1". See Figure 24-H.

Use the Pythagorean Theorem to find the hypotenuse: a = 1" b = 1" c = ?

Set up the equation and substitute values for a and b

$$c^2 = a^2 + b^2 = 1^2 + 1^2$$

$$c^2 = 1 + 1 \quad \text{(Remember, one squared is one.)}$$

$$c^2 = 2 \qquad \text{Find } c: \qquad c = \sqrt{2}$$

Therefore, the hypotenuse of this triangle is equal to 1.414".

EXAMPLE

What is the hypotenuse of an isosceles right triangle whose sides measures 27 inches?

Set up the equation: $c^2 = a^2 + b^2 = 27^2 + 27^2 = 1458$

Use your calculator or approximate to get the square root of **1458**.

$$c = 38.18 \text{ inches}$$

To check our theory that the hypotenuse of any isosceles right triangle is 1.414 times the side, multiply 27 by 1.414:

$$\begin{array}{r} 1.414 \\ \times\ 27" \\ \hline 9898 \\ 2828\ \ \\ \hline 38.178" \end{array}$$

When the length of the hypotenuse of an isosceles right triangle is required for a problem, instead of using the formula $c^2 = a^2 + b^2$, multiply the length of the side by 1.414.

1.414 is the factor for all 45° measurements.

EXAMPLE

The area of an isosceles right triangle is 41.405 square inches. What is the length of the hypotenuse?

$$A = \tfrac{1}{2}(b \times a)$$

$$41.05\ in^2 = \tfrac{1}{2}\ b \times a = \tfrac{1}{2}\ b^2\ \text{since}\ a = b$$

Therefore, **82.81 in² = b²**

Use your calculator to get the square root of **82.81**.   b = 9.1"

The length of the <u>sides</u> is 9.1 inches.

Multiply **1.414(9.1") = 12.867" = 12.87"** after rounding

SUMMARY

An isosceles right triangle is one in which the two sides (the base and the height) are equal.

The area of this triangle is equal to the base times the height, divided by two.

The length of the hypotenuse for this triangle can be determined by one of two methods:

1. Use the formula: $c^2 = a^2 + b^2$ where   **c = length of the hypotenuse**
   a = length of the base
   b = length of the height

2. Multiply the length of the base (or the height, as they are equal) by 1.414, the 45° isosceles right triangle hypotenuse factor.

PROBLEMS

Determine the unknown items.

45° Right Triangles		
a = 14.1 miles	a = 100 yards	**a = ?**
c = ?	**c = ?**	c = 200'
		Area = ?

45° ANGLE OFFSETS

This part of this lesson will develop the application of the 45° right triangle to the calculations required for segments of a line of piping where the alignment of the piping must be changed to avoid some obstruction in the building.

OFFSET VISUALIZATION

When a pipe route must be altered because of obstructions or for any other reason, the use of 45° fittings will ensure minimum pressure loss. Such fittings will not produce as much pressure drop as will 90° ells, and using these fittings will usually result in less total pipe being installed.

If you visualize what such a situation looks like (see Figures 24-N and 24-O), you see that the piping forms a 45° right triangle. We know we can calculate the lengths of the sides of such a triangle from the information given in this lesson. By using the formulas for triangles, we can figure what length of pipe we will need to produce the offset. The easiest way to do this is by using the 1.414 hypotenuse factor for the 45° triangle.

EXAMPLE

A vertical pipe line is being installed; it must move 12" to the left as it passes from floor to ceiling. Visualize the problem as an isosceles right triangle.

Since the length of one of the sides is known (12" center-to-center), the length of the offset (hypotenuse) can be determined.

```
   1.414
  x  12
   2828
   1414
  16.968
```

The offset length equals approximately 17".

This is the center-to-center length of the diagonal of the offset. The length of pipe is determined by subtracting the part of the offset that is provided by the fittings. After the *offset diagonal length* is calculated, the *pipe length* must be determined.

The pipe must be cut short enough so that the *pipe plus fittings* equals the diagonal along the 45° line.

OFFSET CALCULATIONS

To determine fitting allowance for an offset calculation, examine the fitting and measure the amount of running length not occupied by the pipe after assembly. Tables 24-A and 24-B show the fitting allowance for many standard fittings. Consult manufacturer's literature or measure the fitting itself for any that are not listed in these tables. Always remember that fitting allowances are affected by the threads and the individual doing the installation. You will learn soon if your calculations are correct for the way you install pipe.

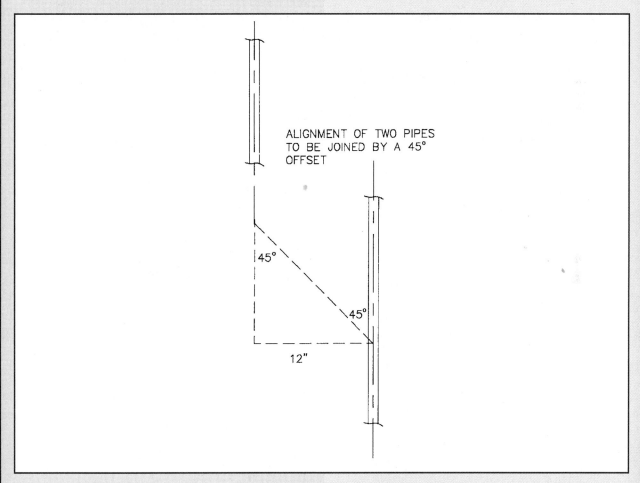

Figure 24-N – 45° Pipe Offset

Fitting allowance is the dimension from the center of the fitting to the end of the pipe when it is placed in the fitting in the final assembled position.

Figure 24-P shows the concepts of fitting dimension, pipe thread make-up and fitting allowance for a $1/2$" threaded 45° ell.

COPPER FITTINGS

To find the fitting allowance for copper fittings, measure from the end of the fitting to the center of the fitting, and then subtract the depth of the female socket well. This is the running length or laying length of the fitting.

The running length of 45° ells is less than for 90° ells. For any given material, 90° ells, tees, and crosses have the same laying length. Various fitting materials are made to different standards, so you will have to measure these dimensions for the material you are using, or obtain the values from the tables at the end of this lesson.

EXAMPLE

If the piping shown in Figures 24-N and 24-O is made of $1/2$" galvanized, what is the laying length of the $1/2$" galvanized 45° ell? From Table 24-B, the fitting allowance for a $1/2$" threaded 45 is $3/8$". What is the length of pipe to cut to make this offset?

Length of pipe = 17" - 2(?) = 17 - 2(3/8)

= 17 - (6/8) = 17 - (3/4) = 16¼"

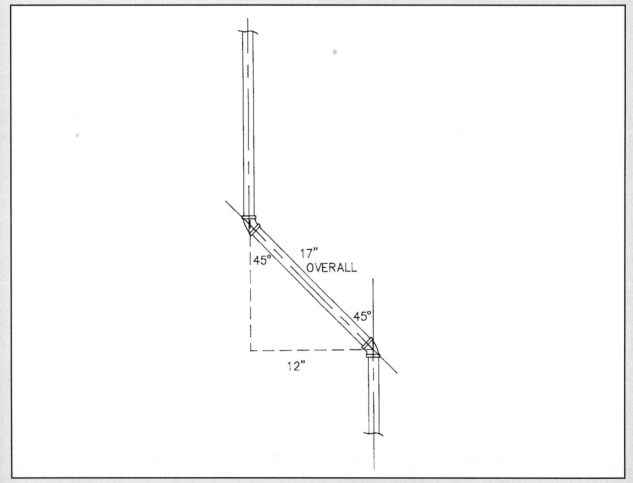

Figure 24-O – 45° Pipe Offset With Fittings

EXAMPLE

Two 1" copper pipes 50" apart are to be connected with a 45° offset. What is the length of the pipe required?

The length of the base of the triangle is 50 inches. Therefore, the diagonal offset length is 70.7" center-to-center.

$$\begin{array}{r} 1.414 \\ \times\ 50 \\ \hline 70.700 \end{array}$$

If two-1" 45°'s are used, the total running length for both fittings of $^5/_8$" must be subtracted from 70.7 inches.

Therefore, the end-to-end measurement of the pipe is or approximately 70 $^1/_8$".

$$\begin{array}{r} 70.7 \\ -0.62 \\ \hline 70.1 \end{array}$$

TABLE 24-A
FITTING ALLOWANCE IN INCHES
DRAINAGE FITTINGS

FITTING (Type)	ANGLE (Deg)	SIZE (In.)	COPPER	PLASTIC	CAST IRON HUB & SPIGOT	CAST IRON HUBLESS	CAST IRON THREADED
¼ Bend, San. Tee	90	1¼	1³⁄₁₆	1⁹⁄₁₆	—	—	1¹⁄₁₆
		1½	1⁷⁄₁₆	1¾	—	4¼	1¼
		2	1¹⁵⁄₁₆	2⁵⁄₁₆	3¼	4½	1⁹⁄₁₆
		3	2⅞	3¹⁄₁₆	4	5	2¹⁄₁₆
		4	3¹³⁄₁₆	3⅞	4½	5½	2¹¹⁄₁₆
		5	—	—	5	6½	3⁵⁄₁₆
		6	—	5⅝	5½	7	3¹⁵⁄₁₆
⅛ Bend	45	1¼	½	1	—	—	⅝
		1½	⁹⁄₁₆	1⅛	—	2⅝	¾
		2	¹³⁄₁₆	1½	1½	2¾	1
		3	1³⁄₁₆	1¾	1¹⁵⁄₁₆	3	1⅛
		4	1⁹⁄₁₆	2³⁄₁₆	2³⁄₁₆	3⅛	1½
		5	—	—	2⅜	2⅞	1⅞
		6	—	3⅜	2⁹⁄₁₆	4¹⁄₁₆	2¼
Short Sweep	90	1¼	—	2¼	—	—	1⁹⁄₁₆
		1½	—	2¾	—	—	1¹³⁄₁₆
		2	—	3¼	5¼	6½	2⅜
		3	—	4¹⁄₁₆	6	7	3¼
		4	—	4¹⁵⁄₁₆	6½	7½	4¹⁄₁₆
		5	—	—	7	8½	4¹⁵⁄₁₆
		6	—	—	7½	9	5¹⁵⁄₁₆
Long Sweep	90	1¼	—	—	—	—	2⁵⁄₁₆
		1½	—	—	—	9¼	2¹³⁄₁₆
		2	—	—	8¼	9½	3⁵⁄₁₆
		3	—	—	9	10	4¼
		4	—	—	9½	10½	5⅛
		5	—	—	10	11½	—
		6	—	—	10½	12	—
Couplings, Straight		1¼	⅛	⅛	—	—	⁹⁄₁₆
		1½	⅛	⅛	—	—	¾
		2	⅛	⅛	—	—	1⅛
		3	⅛	³⁄₁₆	—	—	1⅛
		4	⅛	¼	—	—	1½

Consult Manufacturers' Catalogs for dimensions of other fittings.
All values rounded to the nearest ¹⁄₁₆".

TABLE 24-B
FITTING ALLOWANCES IN INCHES — PRESSURE PIPING

FITTING (Type)	ANGLE (Deg)	SIZE (In.)	THREADED (Class 150)	COPPER (Sweat)	PVC (Sch 40)	FLANGED (Class 150)
Ells, tees, crosses	90	⅛	½	—	—	—
		¼	7/16	¼	—	—
		⅜	9/16	5/16	—	—
		½	9/16	7/16	½	—
		¾	¾	9/16	9/16	—
		1	⅞	¾	11/16	3½
		1¼	1 1/16	⅞	⅞	3¾
		1½	1¼	1	1	4
		2	1 9/16	1¼	1¼	4½
		2½	1¾	1½	1 9/16	5
		3	2 1/16	1¾	1 13/16	5½
		4	2 11/16	2¼	2 5/16	6½
45's	45	⅛	—	—	—	—
		¼	⅜	3/16	—	—
		⅜	⅜	3/16	—	—
		½	⅜	3/16	¼	—
		¾	7/16	¼	5/16	—
		1	7/16	5/16	5/16	1¾
		1¼	⅝	7/16	⅜	2
		1½	¾	½	7/16	2¼
		2	1	9/16	⅝	2½
		2½	1	⅝	11/16	3
		3	1⅛	¾	¾	3
		4	1½	15/16	1	4
Unions Class 150		⅛	⅞	—	—	—
		½	11/16	—	—	—
		⅜	⅞	—	—	—
		½	⅝	—	—	—
		¾	13/16	—	—	—
		1	¾	—	—	—
		1¼	⅞	—	—	—
		1½	1 1/16	—	—	—
		2	1⅜	—	—	—
		2½	1⅜	—	—	—
		3	1 7/16	—	—	—
		4	1 11/16	—	—	—
Couplings, Straight		⅛	9/16	—	—	—
		¼	5/16	1/16	—	—
		⅜	5/16	1/16	—	—
		½	¼	1/16	3/32	—
		¾	7/16	1/16	3/32	—
		1	5/16	1/16	3/32	—
		1¼	9/16	1/16	3/32	—
		1½	¾	1/16	3/32	—
		2	1⅛	1/16	3/32	—
		2½	1	3/16	3/16	—
		3	1⅛	3/16	3/16	—
		4	1½	3/16	3/16	—

Consult Manufacturer's Catalogs for dimensions of other fittings.
All values rounded to the nearest 1/16 inch (except PVC couplings).

PROBLEMS

Fill in the open columns:

CENTER-TO-CENTER MEASUREMENTS
45° OFFSET (IN INCHES)

MATERIAL	SIZE	OFFSET	ALONG DIAGONAL	LAYING LENGTH	PIECE TO CUT
STEEL	2	24		1	
COPPER	1½	30		½	
COPPER	¾	20		¼	
PVC-DWV	2	20		1½	

REVIEW

A 45° offset can be calculated for any problem for an offset piping application if the pipe size is known and the center-to-end dimension of the fitting is known.

To calculate the offset measurement for a 45° angle, multiply the known center-to-center dimension of the offset by 1.414. This will determine the center-to-center measurement along the diagonal.

To calculate the end-to-end measurement of the pipe, subtract the running length of both the fittings. The running length is the total measurement from face-to-center of a fitting minus the socket depth. This is calculated because the pipe does not insert into the fitting completely to the center point.

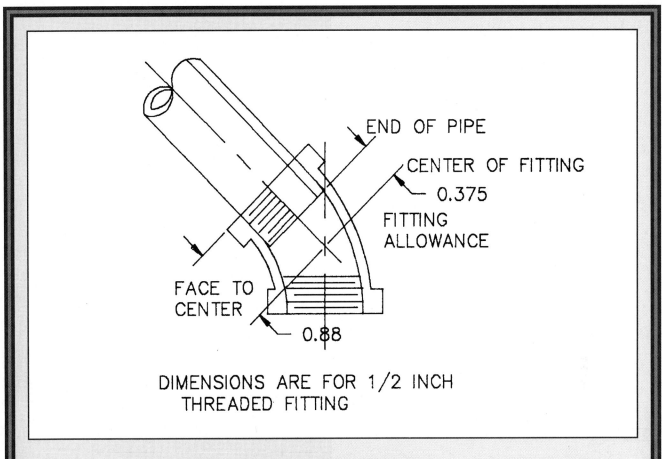

Figure 24-P – Fitting Allowance

The total end-to-end measurement of the pipe, then, is the calculated diagonal offset length minus the running length of the fittings. It is important to realize that the running length should be checked on the job as there are long and short pattern fittings, as well as occasional variations among manufacturers.

LESSON 25

FIRST AID EDUCATION

As specified in the safety requirements, first aid training is frequently handled by local authorities. First aid training is a serious business and we consider the hands-on approach to be the best method for learning first aid.

The best and the most well known organization to help in first aid training is the American Red Cross. In addition to publications, they offer courses in many localities. Information on courses, publications, and other services offered by the American Red Cross can be found at:

The American Red Cross National Headquarters
    2025 E. Washington Street, NW
    Washington, D.C. 20006
    202-303-4498
    http://www.redcross.org

Workplace - oriented programs include:

First Aid and Preparedness (1-1/2 hours)

Workplace Training: Standard First Aid (5-1/2 hours)

Adult CPR (3-3/4 hours)

Infant and Child CPR (5-1/2 hours)

AED Essentials (2 hours)

Bloodborne Pathogens Training: Preventing Disease Transmission (2 hours)

Injury Control Modules (1 hour each)
Ergonomics; Slips; Trips and Falls; Back Injury Prevention; Workplace Violence Awareness; Managing Stress; and Your Heart Matters

Oxygen Administration (2 hours)

Online resources also include:

Preparing Your Business for the Unthinkable
Guide to Business Continuity Planning CD-ROM
The Emergency Management Guide for Business and Industry

Publications available include:

First Aid Fast
HIV/AIDS Fact Book
Pet First Aid
Welcome Home
Community Disaster Education Materials

The Standard First Aid with AED Plus Infant and Child CPR Course Outline includes:

Standard First Aid with AED
Infant and Child CPR
Recognizing Emergencies
Protecting Yourself
Before Providing Care
Conscious Choking – Child
Prioritizing Care – Child
Rescue Breathing – Child
Cardiac Emergencies/Unconscious
Choking – Child
Review – Child
Written Examination and Closing – Child

It is recommended that you work with your teacher and your employer to become proficient in safety and first aid techniques.

LESSON 26

SAFETY ON THE JOB - OSHA - PHCC SAFETY MANUAL

To repeat what has been frequently stressed in this manual, safety is a most important concern of our profession. Accidents cost — the pain, suffering and personal losses of the individual and his or her family, time away from the job, worker compensation and insurance costs, and lost production on the job.

Safety does not just happen; a safe working attitude is not automatic. You must develop and hone your safety skills. YOU are ultimately responsible for your safety. No amount of safety gear — hard hats, protective clothing, rigging equipment, etc. — will protect a worker who fails to use the equipment properly, or who fails to be aware of what is going on.

A good mechanic develops a heightened sense for safety. He can anticipate the hazards in every job and will select and use the proper safety gear and practices to protect himself, his co-workers, and the public.

As indicated from the list of publications available from PHCC, safety does not only include accidents on the job. Safety also includes a drug-free workplace.

The PHCC Toolbox is a catalog of all PHCC publications. It is useful to obtain a copy of this catalog for your and your employer's use.

Other booklets and videos available from PHCC related to safety which may be of interest to you and your employer include:

The PHCC Safety Program

The PHCC Safety Program offers valuable guidelines for the development and/or expansion of safety training programs. Topics covered include Management/Leadership, Enforcement/Disciplinary Action, Hazard Assessment and Control, Communicating the Plan, Safety Planning/Training/Rules, Accident Investigation and Special Programs. Examples of typical safety forms are included throughout the manual.

Start-Up Safety Kit

"Lucky 13 Guidelines"
Use this packet as the backbone of your safety program. Kit contains 13 different rosters, posters and fliers to help you with safety training, emergency planning, and OSHA compliance.

Employees' Rights Poster

This four-section, 22"x17" poster alerts employees to their rights under the federal law - job safety, EEO Act, wage requirements, and employee polygraph protection.

Hazard Communication Employee Training Handbook

A 60-page employee handbook that describes chemical hazards encountered by p-h-c contractor employees.

Hazard Communication Employee Training Video

An overview of HazCom and the various rights and responsibilities under the rule. Helps train employees.

Hazard Communication Guide for PHC Contractors

A compliance guide to understanding OSHA's Final Rule 1910.1200 that requires employers to inform their employees about the dangers of using hazardous chemicals on the job.

Workplace Alcohol and Drug Testing

Easy employee workplace OraSure drug and alcohol testing is available through PHCC that does not require the worker to leave the job site.

Don't Drive Blindfolded! Video

Learn how you and your employees can drive more safely! Order "The Blindfold Effect," a PHCC video developed in partnership with Federated Insurance Companies. All PHCC members and their employees, whether they drive for work or personal reasons, can benefit from the message in the program. "The Blindfold Effect" can be used in a group or viewed individually.

"Heads Up!" for Safety Video and Training Program

This innovative safety training program presents a teamwork approach to maintaining safety on the job site. Brought to you by PHCC and Federated Insurance, the Heads Up! program addresses the p-h-c industry's most common safety hazards. The program features an "interactive" training video, a supervisor's train-the-trainer leadership guide and employee training manual. Additional leadership guides and employee training manuals are available.

What Construction Workers Should Know About Lead Safety

The lead exposure standard issued by OSHA requires that employers in the construction industry protect employees from excessive exposure to lead. Part One of the standard is an easy-to-read summary of the hazards of lead and preventing overexposure. Part Two is an in-depth presentation designed to meet the regulation's training requirements. Part Three contains a self-test and list of safety and health guidelines.

Supervisor's Safety Manual

Teaches crew leaders how to run safety meetings, make safety checklists, and how to prepare for unannounced inspections. Pocket-size with sample forms and tables.

PHCC Substance Abuse Program Manual

Provides guidelines on how to properly establish a substance abuse program for your company.

Supervisor's Substance Abuse Handbook

This PHCC manual will assist your company in setting up policies and procedures to keep employees free from the use, abuse and effects of chemical substances.

Employee Substance Abuse Handbook

Used with the PHCC Supervisor's Substance Abuse Handbook, employees are trained on the importance of a drug-free workplace. Pocket-size.

Emergency Response Safety Clean-Up Kit

Complete system of personal protection and biohazard spill clean-up. Available in a wall-mountable, durable, plastic case or disposable sealed plastic bag that also serves as a refill for the case. Kit contains: safety shield, apron, gloves, Nochar's A680 Bio-Haz Bond (solidifies liquid wastes), scoop with scraper, waste bag, germicidal disinfectant wipe, antimicrobial hand wipe, ID tag, and instructions.

Emergency Response Safety Kit

The OSHA Standard for Bloodborne Pathogens requires all p-h-c contractors to provide protective measures and training to all workers who have a risk of exposure to bloodborne pathogens and other infectious materials. This kit contains the protective equipment to supplement your first aid kit. Stored in a handy zipper poly bag. Recommended for each service vehicle.

Because safety training differs from state to state, we cannot address the safety regulations specific to the state in which you live. We can, however, identify a good general educational booklet for safety education: *PHCC SAFETY MANUAL*.

You should consult this reference guide for general safety guidelines. These guidelines will supplement your locally-required safety training. Take time to read the book and take time to be safe.

For your information, we have reproduced the *PHCC SAFETY MANUAL*. Additional copies of the booklet are available at reasonable cost, in 4" x 6½" size, from:

> Plumbing-Heating-Cooling Contractors--
> National Association
> 180 S. Washington Street
> P. O. Box 6808
> Falls Church, Virginia   22046-1148
> 703/237-8100
> http://www.phccweb.org

Note that the employee is expected to fill out the first page. This page may be kept by the employer as proof that the employer has educated the employee on safety.

Your employer may have in place a safety training program. Your employer's program should take precedence over the following information at all times.

PHCC SAFETY MANUAL

For Employees of:

_____

A service of the Plumbing-Heating-Cooling Contractors-National Association through its Safety Committee.

This is to certify that I, _____ (print name) have received and read the safety instructions and information in the PHCC Safety Manual. I further certify that I understand these instructions and that I shall observe and follow these instructions while I am employed by

_____
_____

(Company Name)

I understand that it is one of the requirements of my employment that in the event I am injured while in the course of my work, I shall report immediately to my supervisor or foreman, and obtain any first aid or treatment necessary.

In case of emergency, please notify:

Name:
_____

Address:
_____

Phone:
_____

Signed:
_____

Date:
_____

To be completed where applicable:

I, _____
(Supervisor's Printed Name), have instructed this employee in the fundamentals of safe working practices, including the instructions found in the PHCC Safety Manual, as well as the rules and practices which specifically apply to his or her job.

Project Location:
_____

Signed:
_____

Date:
_____

Return this Form for Filing to:

_____
_____
_____
_____
_____

Office Use:

Date Received:
_____

Signed:
_____

TABLE OF CONTENTS

OCCUPATIONAL SAFETY AND HEALTH ACT	264
GENERAL SAFETY REQUIREMENTS	265
SCAFFOLDING	267
GENERAL REQUIREMENTS	267
WELDED FRAME SCAFFOLDS	268
ROLLING SCAFFOLDS	269
NEEDLE BEAM SCAFFOLDS	269
FLOATS	270
TWO-POINT SUSPENDED SCAFFOLDS	271
EXCAVATIONS AND TRENCHES	272
TAGOUT/LOCKOUT PROCEDURE	276
PROTECTION OF PROTRUDING REINFORCING STEEL	276
PERSONAL PROTECTIVE EQUIPMENT	276
MATERIAL HANDLING	281
CRANES, MOTOR VEHICLES AND HEAVY EQUIPMENT	282
GENERAL	282
CRANES	283
MOTOR VEHICLES AND HEAVY EQUIPMENT	285
ELECTRICAL SAFETY	286
RIGGING	287
WEBBED NYLON SLINGS	290
GENERAL RIGGING PROCEDURES	291
WIRE ROPE CONNECTIONS	292
SAFE PRACTICES	294
CHAIN SLINGS	295
WIRE ROPE WEAR AND DAMAGE	296
WELDING AND BURNING OPERATIONS	299
VENTILATION AND PROTECTION	300
CYLINDER SAFETY	300
FIRST AID AND MEDICAL TREATMENT	300
LADDER SAFETY	301
HAND AND POWER TOOLS	303
GRINDING AND BUFFING	304
MACHINE SHOP SAFETY	305
FIRE PREVENTION	305
HOUSEKEEPING	306
FLOOR OPENINGS	306
USEFUL TABLES AND FORMULAS	312
STANDARD HAND SIGNALS	317
BARRICADES	323
SUPPLEMENTAL TRAINING AIDS FROM PHCC	324

We have selected this opportunity to express our appreciation to you for choosing this company as your employer. We feel confident your stay with us will be a safe and profitable experience.

The safety rules and procedures set forth in this manual are arranged to conduct you safely and effectively in the performance of your work. These rules are designed in compliance with all Federal, State and Local laws, coordinated with company doctrine; your personal safety is our prime concern. Your continued employment depends on your cooperation with our safety policies.

This manual was not prepared to be a substitute for OSHA requirements as listed in OSHA's Safety and Health Regulations for Construction, Part 1926 and General Industry Standards Identified as Applicable to Construction, Part 1910. Rather, this manual was prepared to increase your awareness in creating a safe working environment.

For specific information on OSHA regulations, consult the above referenced standards.

OCCUPATIONAL SAFETY AND HEALTH ACT
OSHA

It is the policy of the Company to be in voluntary compliance with the Occupational Safety and Health Act of 1970. We subscribe to the intent and purposes of the Act and endeavor to comply with the rules and regulations adopted by OSHA.

Section 5 of the Occupational Safety and Health Act outlines the duties and responsibilities of the employer and employee. We expect each employee to comply with Section 5(b) of the Act, which states:

"each employee shall comply with occupational safety and health standards and all rules, regulations, and orders issued pursuant to this Act which are applicable to his own actions and conduct."

GENERAL SAFETY REQUIREMENTS

1. Analyze your job from the standpoint of safety. Be sure you have all your protective equipment before starting a job. If you are in doubt about what you require, consult your supervisor. Any unsafe condition should be reported to the supervisor immediately.

2. All injuries must be reported to your supervisor who will notify the safety representative.

3. Absenteeism should be reported to your supervisor prior to shift start. If your absence is due to a work incurred injury or illness, a report must be filed with the insurance carrier upon your return. Consult your supervisor and/or safety representative in event of such occurrence.

4. Horseplay, mischief, or fighting will not be tolerated.

5. Employees are forbidden to enter Company property in possession of or under the influence of alcohol or unauthorized drugs. Contravention of this rule is ground for dismissal. Use of prescription or patent medicines which carry a warning against driving or operating machinery should be reported to your supervisor so that less demanding work can be assigned. The employee will comply with all "NO SMOKING" areas posted on Company property and customer worksites.

6. Maximum free hair length allowed in this shop is to the shoulder only.

7. Do not wear torn, loose, or frayed clothing; or scarves, ties, or jewelry which can catch in machinery or material, causing injuries. Remember, welding can cause arc burns; keep your face, neck and arms protected.

8. Each employee is responsible to his foreman for the tidiness of his work area. This means proper storage or return of tools, equipment, and material, as well as cleanup of scrap, dirt, and grease. Clear access must be maintained at all times to work areas and walkways. Keep the floors and loading areas clear of all slipping or tripping hazards.

9. Air hoses must not be kinked to control the flow of air. Horseplay with air hoses will not be tolerated. Compressed air injected into the body can cause a painful death to the recipient. Compressed air must not be used to clean clothing. If it is used to clean equipment, extreme care must be taken to avoid injury. Eye protection must be worn.

10. Defective and/or worn equipment should be reported to your supervisor with a request for repair or replacement. Defective tools should be returned to the tool room for replacement. Do not exceed the limits of any machine or tool, or use it for other than its intended purpose. Any person carrying out servicing and repairs must inform the supervisor and operators in the area. It is the responsibility of the operators to make sure maintenance or repairs are completed and guards on machinery replaced before resuming normal operations. Electrically controlled equipment must be locked out by qualified personnel only. If equipment has not been locked out, repairs must be done by a team of two.

11. Entry to electrical and compressor areas is restricted to authorized personnel performing specific assignments. Control panels should be left strictly alone. Only personnel designated by the foreman or other supervisors are permitted to operate controls.

12. Cranes and yard equipment must be operated only by personnel authorized to do so by their supervisors. Never stand or walk under a moving or suspended load.

13. Flammable substances, including paints, may not be stored inside the main buildings except as authorized. Thinners used for pipe or parts cleaning will be issued in quantities not exceeding one gallon in an approved container, and should be used in a well ventilated area to prevent buildup of explosive or toxic fumes. Combustible waste must be disposed of in metal trash containers and not left near welding or burning areas.

14. No smoking or open fires will be allowed during refueling. Ignitions must be turned off.

15. Fires must always be reported to the foreman, who will inform the Plant Superintendent and the Safety Coordinator. Fire extinguishers are to be used for fires of a minor nature. Know their location and how to operate them. Always return used or partially used extinguisher to the foreman's office (main shop) for recharging. Major fires will be handled by the Fire Department (Dial 911).

16. Radioactive areas will be barricaded and posted with a RADIATION HAZARD sign. Keep clear of these areas when X-ray work is in progress. Heed all warning signs; they are there for your protection.

17. For your protection, obey all warning signs such as "Keep Out," "No Smoking," "Eye Protection Required," and "Authorized Personnel Only."

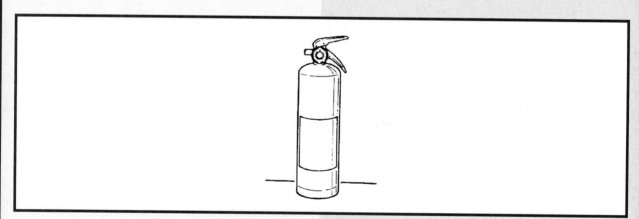

Fire Extinguisher

Hazard Signs

18. Sliding down ropes, cables, and guys, etc. is strictly forbidden.

19. Never jump from any elevated surface.

20. The handling of explosives is extremely dangerous. On all work of this nature consult your Foreman.

SCAFFOLDING

GENERAL REQUIREMENTS

1. Scaffolds should be substantially constructed to carry the loads imposed upon them and to provide a safe work platform. All scaffolds more than 10 feet high should have approved guardrails on all exposed ends and sides. Toeboards and screens should be provided on a scaffold if persons are required to pass under it.

2. Guardrails, midrails, and toeboards must be installed on all open sides of scaffolds 10 feet or more above the floor.

3. Only approved scaffolds should be used. Barrels, boxes and other makeshift substitutes for scaffolds should not be used.

4. Scaffold planks must be at least 2- x 10-inch full thickness lumber, scaffold grade, or equivalent.

5. Scaffold planks must be cleated and must extend over the end supports at least 6 inches — but not more than 12 inches.

6. All scaffolds must be at least two planks wide; no employee may work from a single plank.

7. Scaffold planks must be visually inspected before each use. Damaged scaffold planks must be destroyed immediately.

8. Adequate mud sills or other rigid footing, capable of withstanding the maximum intended load, must be provided.

9. Scaffolds must be tied on to the building or structure at intervals which do not exceed 30 feet horizontally and 26 feet vertically.

10. Do not overload scaffolds. Materials should be brought up as needed. Scaffolds must not be loaded in excess of one-fourth of their rated capability.

11. Where persons are required to work or pass under a scaffold, a screen of 18-gauge, ½-inch wire mesh is required between the toeboard and the guardrail.

12. Overhead protection is required if employees working on scaffolds are exposed to overhead hazards. Such protection must be a 2-inch plank or the equivalent.

WELDED FRAME SCAFFOLDS

1. Welded tubular scaffolds must be capable of supporting four times their intended load.

2. Overhead protection is required when exposed to overhead hazard.

3. A standard rail 2-inch by 4-inch or equivalent must be used on all exposed sides on any scaffold 10 feet or higher.

4. A minimum of 2 cleated 2-inch by 10-inch planks must be used for working surface.

5. All planking should be overlapped 12 inches minimum or secured from movement.

6. Planks should extend over end supports not less than 6 inches nor more than 12 inches.

7. Diagonal bracing must be used.

8. Scaffold must be on rigid footing.

9. Where persons are required to pass under scaffold, 18 gauge wire mesh must extend from toeboard to handrail.

10. Midrail 1-inch by 6-inch or equivalent must be present on all exposed sides.

11. Toeboards should be a minimum of 4 inches in height above scaffold floor.

12. Where height or length exceeds 25 feet, the scaffold should be secured at intervals not greater than 25 feet vertically and horizontally.

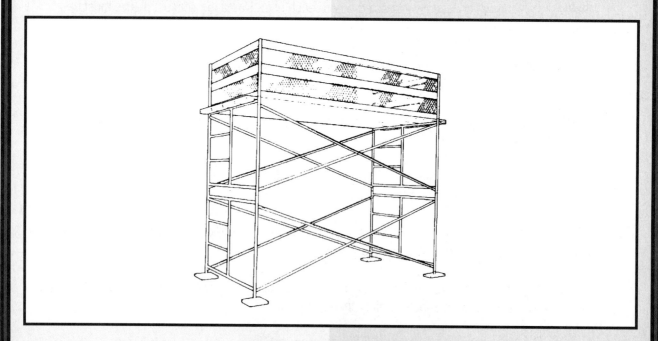

WELDED FRAME SCAFFOLD

ROLLING SCAFFOLDS

1. The height of rolling scaffolds must not exceed four times the minimum base dimension.

2. The work platforms must be planked tight for the full width of the scaffold. Block the underside of planks to prevent their movement.

3. Caster brakes must be locked when the scaffold is not in motion.

4. Get help when moving rolling scaffolds. Make certain that the route is clear. Watch for holes and overhead obstructions.

5. Secure or remove all loose materials and equipment before moving scaffold.

6. No one is allowed to ride on the rolling scaffold when it is being moved.

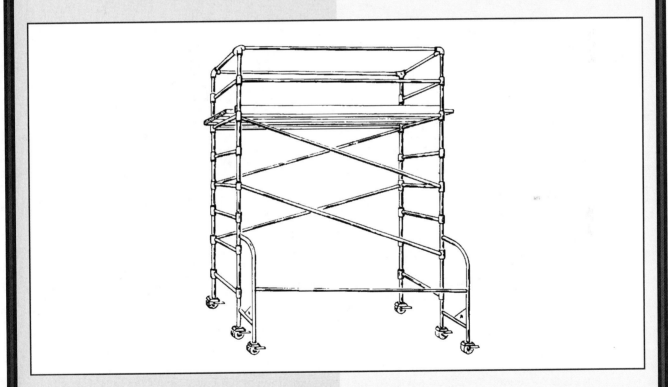

Rolling Scaffold

NEEDLE BEAM SCAFFOLDS

1. All needle beam scaffolds must be constructed to support the intended load with a safety factor of four.

2. All employees working from needle beam scaffolds must use safety belts and lifelines.

3. Needle beams must be at least 4 by 6 inches, and the span must not exceed 10 feet.

4. Rope supports must be at least 1-inch manila or larger. Attach with a scaffold hitch or eye splice, properly secured to prevent the beam from rolling or being displaced.

5. Needle beams suspended by wire rope must be secured with three wire-rope clamps, properly attached.

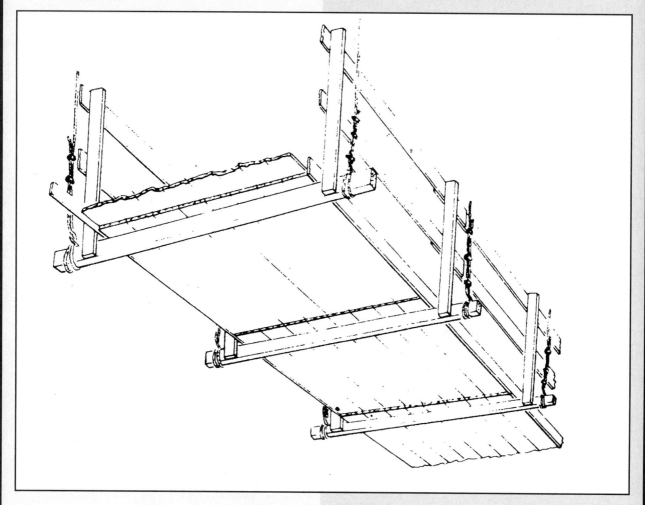

Needle Beam Scaffold

FLOATS

1. Floats are intended to support not more than three workmen and a few tools. They must be inspected carefully prior to each use.

2. The platform must be constructed from ¾-inch exterior plywood, Grade B-B or better. The minimum width must be 3 feet, and the minimum surface area must be 18 square feet.

3. The supporting beams must be 2- x 4-inch select lumber and must project at least 6 inches beyond each side of the platform.

4. A 1- x 2-inch edging must be placed on all sides of the platform to prevent tools from rolling off.

5. Supporting ropes must be 1-inch manila or equivalent in "as-new" condition and must be fastened so that the platform can't slip or shift.

6. When working from floats, you are required to wear a safety belt and to be tied off to the structure or to an independent lifeline.

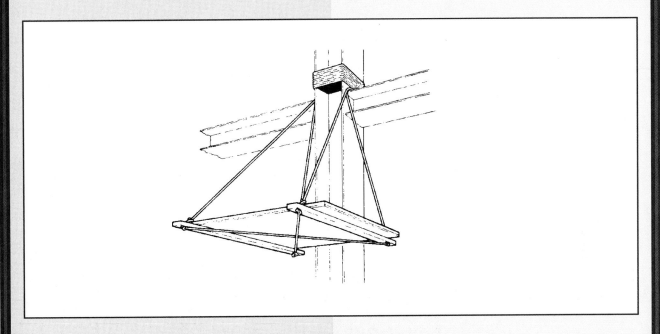

Float

TWO-POINT SUSPENDED SCAFFOLDS (SWINGING STAGES)

1. Each employee working from a two-point suspended scaffold must be tied off to an independent safety line.

2. Suspended scaffolds must not be less than 20, nor more than 36 inches wide.

3. Wire ropes used to suspend such scaffolds must have a safety factor six times the maximum intended load.

Two-Point Suspended Scaffold

4. Nonconductive insulating material must be placed over the suspension cables of each scaffold for protection when the chance of contact with an electrical circuit exists.

EXCAVATIONS AND TRENCHES

```
CAUTION!
CONTACT YOUR LOCAL OSHA REPRESENTATIVE FOR THE LATEST
TRENCHING AND SHORING REQUIREMENTS
```

All excavating and trenching operations and work done in them must conform with established standards.

1. All excavations must be sloped to the angle of repose, except in solid rock.

2. Materials must be placed 2 feet or more from the edge of the excavation. Precautions must be taken to prevent such materials from falling into the excavation.

3. Trenches 5 feet or deeper must be shored or sloped back to the angle of repose. Any excavation in unstable soil may require shoring or sloping.

4. Each excavation must be inspected daily by the responsible Superintendent. If evidence of cave-ins or slides is apparent, all work in the excavation must cease until necessary precautions have been taken to safeguard employees.

5. Where vehicles or equipment operate near excavations or trenches, the sides of the the excavation must be shored or braced as required to withstand the forces exerted by the superimposed load. Also, stop logs or other substantial barricades must be installed at the edges of such excavations.

6. Materials used for sheeting, shoring, or bracing must be in good condition. Timbers must be sounds, free of large or loose knots, and of adequate dimensions.

7. Employees working in bell-bottom pier holes must be protected by a substantial casing which extends the full depth of the shaft. When working in such holes, you must wear a shoulder harness secured to a lifeline which is tended full-time.

8. Safe access must be provided into all excavations by means of ladders, stairs, or ramps.

9. Trenches 4 feet or more in depth must have ladders spaced so that employees' lateral travel does not exceed 25 feet. Such ladders must extend at least 3 feet above grade level.

10. Walkways or bridges with standard guardrails must be provided where employees or equipment are required or permitted to cross over excavations or trenches.

11. In locations where oxygen deficiencies or concentrations of hazardous or explosive gases or dusts are possible, the atmosphere in the excavation must be tested prior to start of work and at intervals, as required. When such conditions exist or may develop, emergency rescue equipment must be kept readily available.

12. Sewers, ditches, underground utility compartments, pits, enclosed tanks, and other confined locations may contain hazardous vapors or insufficient oxygen. Protection against exposure to such hazards must be provided by ventilation (blowers, fans, etc.), by personal protective equipment (respirators) or by controlled exposure time (supervised exposure). Air samples may be required periodically and/or prior to entering the confined space. Check with your supervisor before entering. Never work alone under such conditions.

Trenches

Depth of trench	Kind of condition of earth	Uprights		Stringers		Size and spacing of members — Cross braces - Width of trench				Max spacing	
		Min Dim	Max Spac	Min Dim	Max Spac	Up to 3 ft	3 to 6 ft	6 to 9 ft	9 to 12 ft	Vert	Horiz
Feet		Inches	Feet	Inches	Feet	Inches	Inches	Inches	Inches	Feet	Feet
5 to 10	Hard, compact	3x4 or 2x6	8			2x6	4x4	4x6	6x6	4	6
	Likely to crack		3	4x6	4						
	Soft, sandy, or filled		Close sheeting	4x6		4x4	4x6	6x6	6x8		
	Hydrostatic pressure			6x8							
10 to 15	Hard		4	4x6							
	Likely to crack		2	4x6	4						
	Soft, sandy, or filled		Close sheeting	4x6		4x6	6x6	6x8	8x8		
	Hydrostatic Pressure	3x6		8x10							
15 to 20	All kinds of conditions		Close sheeting	4x12		4x12	6x8	8x8	8x10		
Over 20	All kinds of conditions			6x8			8x8	8x10	10x10		

Plumbing Apprentice Student Workbook Year One
Fourth Edition

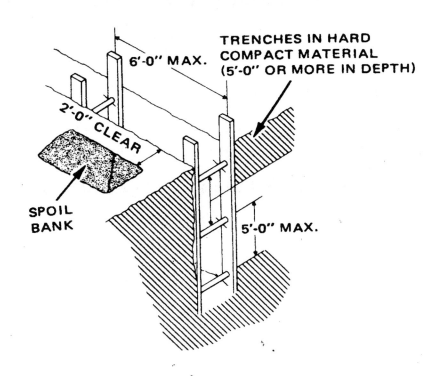

BRACING: SCREW JACKS OR TIMBERS SPACED NEVER GREATER THAN 5'-0" ON CENTER (ONE BRACE REQUIRED FOR EACH 4'-0" OF TRENCH DEPTH — NEVER FEWER THAN TWO BRACES)

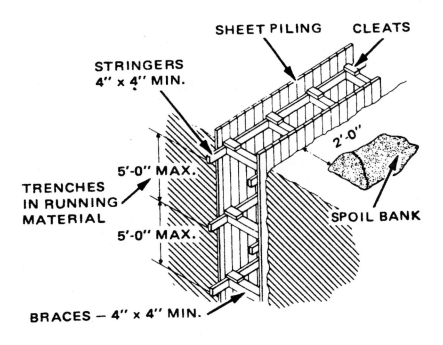

TAGOUT/LOCKOUT PROCEDURE

1. Familiarize yourself with the project procedures for tagging out and/or locking out electrical and mechanical equipment.

2. Do not remove tags, or alter the instructions.

3. Work should not begin on mechanical equipment or pressurized systems which could be energized or activated until it has been verified safe by your Supervisor and properly tagged and/or locked out.

PROTECTION OF PROTRUDING REINFORCING STEEL

1. Protruding rebar must be protected by:

- Installing plastic caps

- Protective trough as per sketch

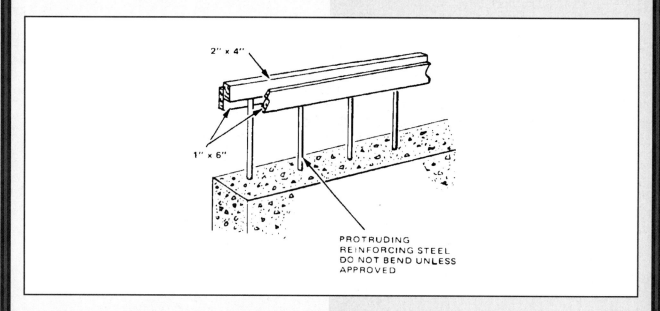

Protruding Reinforcing Steel

PERSONAL PROTECTIVE EQUIPMENT

1. All employees, visitors, and vendors must wear a hard hat in designated areas.

2. You must wear clothing suitable for the work you are doing. Minimum attire is long pants and a T-shirt.

3. Safety shoes with metatarsal protection are required.

4. Safety glasses will be worn in all areas. The company will supply non-prescription safety glasses.

Hard Hat

5. Hearing protective equipment is available upon request. You may be required to use it in designated areas or for specific jobs.

6. Respiratory equipment may be required in areas where health hazards exist due to accumulations of dust, fumes, mists, or vapors.

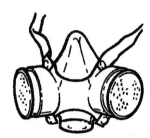

Respiratory Equipment

7. Safety belts and lifelines must be used when other safeguards, such as nets, planking, or scaffolding cannot be used. Be sure safety lines are independent of other rigging.

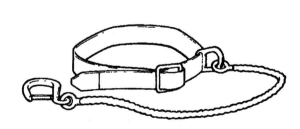

Safety Belts

Gloves

8. Wear gloves when handling objects or substances which could cut, tear, or burn the hands.

9. You must wear rubber boots when working in concrete, mud, or water.

10. Goggles, helmets, and shields that give maximum eye protection while welding and cutting should be worn by operators, welders, and their helpers. Goggles and face shields protect against corrosives, liquids, vapors, and solids. Shaded lenses of the required density are used to screen out harmful ultraviolet and infrared radiation.

11. Hearing protection shall be worn in the Shop due to excessive noise levels from grinding, buffing, chipping, and grit blasting activities. Continuous exposure to high noise levels can cause permanent hearing damage. Various types of hearing protection are available. Try them out to find which is most suitable for you. Contact your foreman for details.

12. Dust masks are available and shall be worn when there is a danger of inhaling particles. Excess facial hair which prevents a proper seal of mask to face shall be removed. A supplied air hood is available for working in confined areas where ventilation is inadequate. This apparatus is to be tested before each use. Portable air blowers are to be used by men working inside large O.D. pipe. The Worker should position himself and the blower so that particles are directed away from him.

13. Essential safety equipment is available to all personnel. The following table shows equipment required for specific tasks. Additional requirements may be published from time to time covering items not mentioned in the table.

| THIS EQUIPMENT IS <u>IN ADDITION</u> TO BASIC REQUIRED SAFETY EQUIPMENT ||
WORK PERFORMED	EQUIPMENT REQUIRED
Working in shop	Ear protection and safety glasses, hard hats, safety shoes.
Working outside shop	Hard hats, safety glasses and safety shoes
Burning	Burning goggles, gauntlet gloves and ear protection
Chipping and Chiselling	Chipping gloves
Grinding or working near grinder	Safety glasses, face shield and ear protection
Handling pipe, steel plate and shapes	Gloves, safety shoes and hard hats
Buffing pipe	Safety glasses and face shields
Handling scrap	Safety gloves and gauntlets or staple palm gloves and sturdy footwear
Using acids or other corrosives	Rubber apron, neoprene gloves, face shield, and respirator
Spray painting and blasting	Respirator or air supplied hood, ear protection, rubber apron, and safety glasses
Welding (all phases)	Welding hood, tinted lenses, gloves, screen to protect nearby workers, ear protection and safety glasses
Furnace area (exposure to heat)	Heat resistant clothing, safety glasses, and extra wide face shields
Machining	Safety glasses and gloves
Repairing or adjusting mechanical or electrical equipment	Safety tag and safety lockout

Eye protection must be worn when grinding with a portable or pedestal grinder. A shield worn over goggles or safety glasses is recommended.

REMEMBER, YOUR EYES CANNOT BE REPLACED!

Safety equipment is issued to you and is expensive. It is your responsibility to maintain this equipment in good condition. Defective equipment should be replaced immediately.

LENS SHADE NUMBERS FOR WELDING AND CUTTING OPERATIONS	
OPERATION	SHADE NUMBER
Resistance Welding (also protects against stray light from nearby welding and cutting if persons are out of the welding zone)	Clear filters or up to #2
Torch Brazing or Soldering	3 to 4
Light oxygen cutting and gas welding (to ⅛")	4 to 5
Oxygen cutting and medium gas welding (⅛" to ½") and arc welding up to 30 amps	5 to 6
Heavy Gas Welding (over ½") and arc welding and cutting from 30 to 75 amps	6 to 8
Arc welding and cutting from 75 to 100 amps	10
Arc welding and cutting from 200 to 400 amps	12
Arc welding and cutting exceeding 400 amps	14
NOTE: Safety glasses will be worn under all arc welding helmets, particularly for gas shielded metal arc welding.	

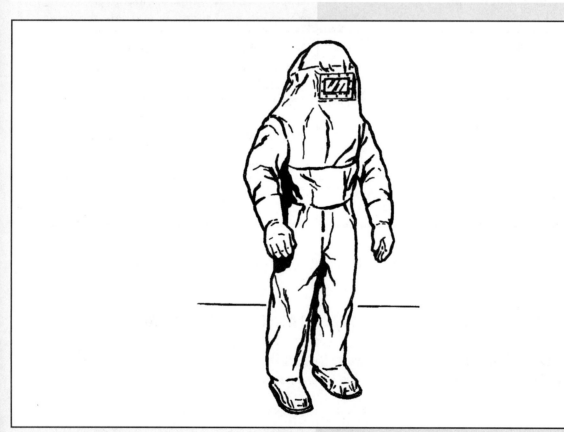

Protective Suit

MATERIAL HANDLING

1. BY LEARNING AND FOLLOWING SAFE PRACTICES FOR HANDLING MATERIAL YOU WILL HELP TO PREVENT INJURY AND WILL SAFEGUARD YOUR LIFE AND THAT OF YOUR FELLOW EMPLOYEES.

The materials and equipment moved in our operation involve weights which can kill or cause permanent disability. Check your equipment. If it shows any defects report the matter to your supervisor.

2. Know the approximate weight of your load and make certain your equipment is rated to handle it. All powered equipment and rigging is rated as to safe working load. This rating is posted. Never exceed the manufacturer's recommended safe working load.

3. Never stand, walk or work under suspended crane hooks or loads.

4. Riding a crane hook, yard equipment forks, or on a load is prohibited.

5. Never run hand trucks or mobile equipment over hoses, cables, or welding leads.

6. Employees working in railway cars must step out of the cars while they are switched.

7. Wear safety gloves for those operations specified in the table on protective equipment. Severe hand lacerations and punctures are a common hazard of our operations.

8. All material must be properly stacked and secured to prevent sliding, falling, or collapse. Aisles, stairs, and passageways must be kept clear to provide for the safe movement of employees and equipment and to provide access in emergencies.

9. Protruding nails must be bent or pulled when stripping forms or uncrating materials.

10. Pipe, conduit, and bar stock should be stored in racks or stacked and blocked to prevent movement.

11. Materials or scrap should never be dropped from elevated levels without trash chutes. Containers for this purpose will be provided.

12. Stored materials must not block any exit from a building.

13. Avoid moving loads by hand whenever possible. Use cranes and hoists for awkward or heavy lifts. When it is necessary to move anything by hand, always:

- Get help if you need it. Don't risk a hernia or a sprain.

- Squat, keeping your back straight. Don't bend from the waist when lifting or setting down the load.

- Hold the object close to the body and lift by straightening the legs.

- When changing direction, turn with your feet, not with your back.

- Be sure the load is secure against tripping, falling or blocking a walkway where you put it down.

- Protect your hands and fingers from rough edges, sharp corners, metal straps. Keep hands and fingers out of pinch points between the load and other objects.

Check the Mobile Equipment and Rigging sections for more information related to material handling.

Lifting

CRANES, MOTOR VEHICLES, AND HEAVY EQUIPMENT

GENERAL

ONLY PERSONNEL AUTHORIZED BY THE SUPERVISOR MAY OPERATE THE CRANES AND FORKLIFTS.

EQUIPMENT MUST NOT BE OPERATED BEYOND ITS RATED CAPACITY. RATING PLATES ON THE MACHINES MUST BE KEPT LEGIBLE

The equipment is built for safe and economical operation but it is only as safe as the operator.

1. All cranes, hoists, motor vehicles, elevators, and heavy equipment must be operated and maintained to conform with established standards.

2. Routine maintenance, fueling, or repairs must not be performed while the equipment is in use or the power on.

3. When handling or recharging batteries or using jumper cables, wear a face shield.

4. Rated load capacity charts, recommended operating speeds, special hazard warnings, and other essential information must be conspicuously posted in all cranes, hoists, and other equipment.

CRANES

1. Make a safety check of your equipment at the start of each shift. Check cables, chokers, hooks, shackles, braking, steering, linkages, warning signals. Report any defects affecting control or function of the equipment in writing in the log book provided and bring them to the attention of your supervisor immediately. The equipment must not be operated until defects presenting a hazard to workers are repaired. Operators must keep inspection records required by law.

2. Any repairs which affect the mast, boom, or suspension system must be certified in writing to be in accordance with the manufacturer's specifications and instructions. Such repairs are to be recorded in the log book, dated, and signed by the person responsible for maintenance or modification. The log book must be available for inspection at any time.

3. Before moving the crane, make sure both crane and load are clear of all men and other obstacles.

4. If equipment becomes faulty, stop immediately. When it is safe to do so, open the power switch. If the equipment fails to respond, turn control to OFF and report the matter to your supervisor for repairs. In the event of a power failure, switch controls of electrical equipment to OFF position until power is restored.

5. Yard speed limit for all vehicles is 10 miles per hour maximum.

6. Under no circumstances are riders permitted on the load or hook or in the crane cab. The only personnel authorized access to the top of a crane at any time are maintenance people so assigned.

7. To avoid confusion, only one worker should act as signalman to the crane operator, except in emergencies, when a STOP signal from any person shall be promptly obeyed. The signalman must use proper signals as shown on area bulletin boards and included within this booklet.

8. When approaching people, stop until they are clear of the path; then proceed.

9. Carry loads as close to ground level as possible, making sure that slings or load have sufficient clearance to avoid striking people or objects. Do not allow the load to swing, striking the boom.

10. The operator must keep the equipment under his control at all times and must not leave unattended any unsecured, suspended load. Operators must remain alert. Should you become ill during your shift, report to your supervisor. If it becomes necessary to leave the hoist or crane, turn controls to OFF position.

11. If a load comes to rest in an unstable position, do not attempt to lift it until workers are well clear of the area. No attempt will be made to pick up a DEAD LOAD with a crane without specific direction from a supervisor and in no such case will a chain be used. (A dead load is one which is fixed, snagged, or otherwise not free to be picked up directly.) Chains must not be used to carry bundles of scrap.

12. Oxygen, acetylene, or other high pressure gas bottles may be transported by crane only in safely constructed slings. Severe explosions have resulted from these bottles hitting each other or on the floor.

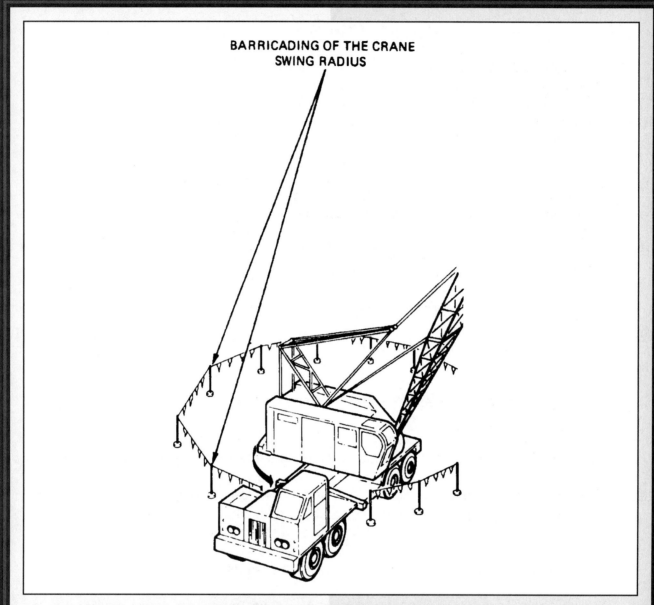

Barricading of the Crane Swing Radius

13. Accessible areas within the swing radius of all cranes must be barricaded to prevent employees from being injured by the counterweight.

14. All personnel are prohibited from riding the hook or load.

15. A fire extinguisher, rated at least 5 BC, should be located in the cab of each crane.

16. Safety latches are required on all crane hooks.

17. No crane or other equipment should be operated within 10 feet of energized electrical transmission or distribution lines.

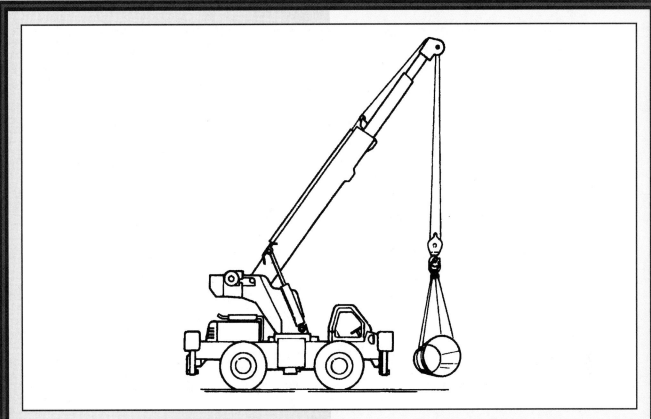

Crane

For lines rated over 50kV, the minimum clearance between the lines and any part of the crane or load must be 10 feet plus 0.4 inches for each 1kV over 50kV. Or use twice the length of the line insulator, but never less than 10 feet.

During transit with no load and the boom lowered, the minimum equipment clearance must be 4 feet for 50kV or less, 10 feet for 50kV to 345kV, and 16 feet for voltages up to 750kV.

A designated employee must observe clearance of the equipment and give timely warning for all operations where the operator's vision is obstructed.

Any overhead line must be considered energized unless a responsible client or utility company representative says that it is not energized.

MOTOR VEHICLES AND HEAVY EQUIPMENT

1. The parking brake must be set whenever the vehicle is parked. Equipment parked on an incline must have the wheels locked.

2. Where provided, seat belts must be used.

3. Do not ride in the bed of a truck containing materials which are not properly secured to prevent movement.

4. All personnel are prohibited from riding on loads, fenders, running boards, or tailgates, or with legs or arms dangling over the sides.

5. Drivers must not move vehicles until riders comply with all safety precautions.

6. Do not back up any vehicle or equipment when the view to the rear is obstructed unless:

- It is equipped with an operating backup alarm which is audible above the surrounding noise for a distance of 200 feet.

- Or an observer signals that it is safe to do so.

ELECTRICAL SAFETY

1. All temporary electrical wiring should be installed and maintained by qualified personnel in accordance with applicable codes.

2. All electrical tools and equipment must be grounded.

3. Damaged or defective electrical tools must be returned immediately to the tool room for repair.

4. Electricians are the only employees authorized to repair electrical equipment.

Grounding

5. When it is necessary to work on energized lines and equipment, rubber gloves, blankets, mats, and other protective equipment must be used.

6. Temporary electric cords must be covered or elevated. They must be kept clear of walkways and other locations where they may be exposed to damage or create tripping hazards.

7. Splices in electrical cords must retain the mechanical and dielectric strength of the original cable.

8. Temporary lighting must have guards over the bulbs. Broken and burned-out lamps must be replaced immediately.

9. Energized wiring in junction boxes, circuit breaker panels, and similar places must be covered at all times.

10. Hazardous areas must be barricaded and appropriate warning signs posted.

RIGGING

1. Most wire rope slings used in the shop have a Flemish Eye Splice. This type is more easily threaded into or removed from stacked pipe.

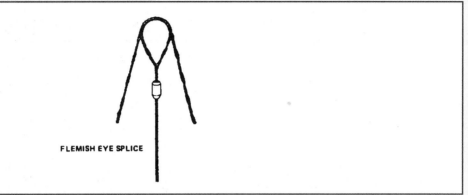

Flemish Eye Splice

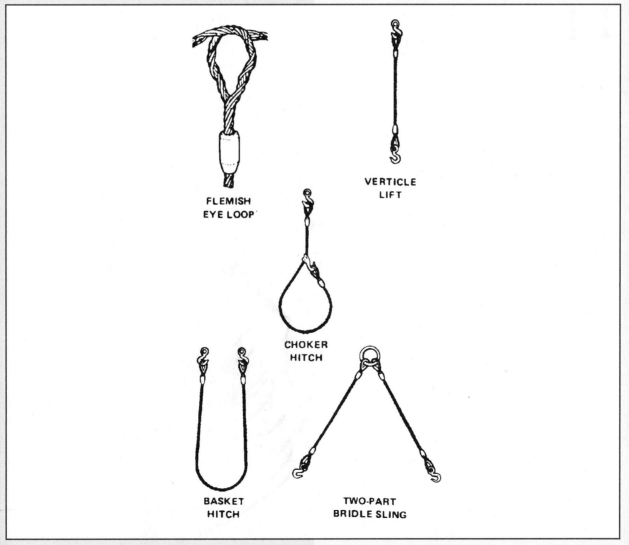

Slings

2. The most common wire rope diameters used in the Shop are ³⁄₈", ¹⁄₂", and ⁵⁄₈" inches. Sling lengths used are 3, 4, 6, 8, 12, and 20 feet.

3. Check slings before each use. Stretched or damaged slings will be replaced.

4. Do not exceed the recommended working load for the various types of hitches used as shown in the table following. This information is also posted at various locations in the Shop.

Safe working loads are calculated on a minimum curvature of ten (10) rope diameters at point of contact with the load.

LOAD CAPACITIES FOR VARIOUS HITCH TYPES

FLEMISH EYE LOOP ROPE SIZE		VERTICAL LIFT		CHOKER HITCH		BASKET HITCH		30°		60°		90°		120°	
Inches	mm	Pounds	Kg	Pounds	Kg	Pounds	Kg	Pounds	Kg	Pounds	Kg	Pounds	Kg	Pounds	Kg
1/4	6.4	920	417	700	318	1,840	835	1,780	807	1,600	726	1,300	590	920	417
3/8	9.5	2,020	916	1,520	690	4,040	1833	3,900	1769	3,500	1587	2,860	1297	2,020	916
1/2	12.7	3,740	1696	2,800	1270	7,480	3393	7,220	3275	6,480	2939	5,280	2395	3,740	1696
5/8	16.0	5,600	2540	4,200	1905	11,200	5080	10,840	4917	9,700	4400	7,920	3592	5,600	2540
3/4	19.0	8,080	3665	6,060	2748	16,160	7330	15,640	7094	14,020	6359	11,460	5198	8,080	3665
7/8	22.2	10,920	4953	8,180	3710	21,840	9906	21,100	9571	18,920	8582	15,480	7022	10,920	4953
1	25.4	14,180	6432	10,680	4844	28,360	12864	27,400	12430	24,600	11158	20,060	9099	14,180	6432
1-1/8	28.6	16,660	7557	12,500	5670	33,320	15114	32,300	14651	28,900	13109	23,500	10660	16,660	7557
1-1/4	31.8	20,740	9407	15,540	7049	41,480	18815	40,120	18198	36,000	16330	29,300	13290	20,740	9407
1-3/8	35.0	25,340	11500	19,000	8618	50,680	22988	49,000	22226	43,880	19904	35,840	16257	25,340	11494
1-1/2	38.0	30,620	13889	22,960	10415	61,240	27778	59,500	29989	53,040	24058	43,300	19640	30,620	13889
1-5/8	41.3	35,900	16284	26,920	12211	71,800	32568	70,000	31751	62,180	28204	50,760	23025	35,900	16284
1-3/4	44.5	41,160	18670	30,860	13998	82,320	37340	80,000	36287	71,280	32332	58,200	26400	41,160	18670
1-7/8	47.6	48,320	21917	36,240	16438	96,640	43833	93,600	42456	83,680	37956	68,320	30990	48,320	21918
2	50.8	52,760	23930	39,560	17944	105,520	47863	102,400	46448	91,380	41450	74,600	33840	52,760	23932

All calculated on the basis of 6: 1 Working Load Factor, and based on the use of Wire Rope Industries 6 x 19 Regular Lay Preformed Improved Plow Steel with Independent Wire Rope Centre for sizes from 1/4" diameter to 1" diameter inclusive, and Wire Rope Industries 6 x 37 Classification Regular Lay Preformed Improved Plow Steel with Independent Wire Rope Centre for diameters beyond 1"

WEBBED NYLON SLINGS

1. The 2-inch and 3-inch double thickness choker type slings are used in the shop. The length has no bearing on the load capacity.

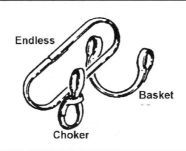

Choker

WEBBED SLING LOAD CAPACITIES (in lbs.)			
	VERTICAL	CHOKER	BASKET
2"	5,600	4,200	11,200
3"	7,100	5,300	14,200

2. Check before use for kinks, fraying or knots. A sling with a knot or two slings knotted together have only 25 percent of the rated capacity.

3. Hang slings up after use. Welding sparks or ground-in dust and grease can ruin nylon slings.

4. Nylon slings are primarily used for:

- Coated or wrapped pipe

- Stainless steel

- Handling material in spooling bays

If in doubt when to use a nylon sling, consult your supervisor.

VISUAL INDICATIONS OF DAMAGE TO WEBBING				
TYPE OF WEBBING	HEAT	CHEMICAL	MOLTEN METAL OR FLAME	PAINT & SOLVENTS
Nylon & Cordura	In excessive heat, nylon becomes brittle, has a shriveled, brownish appearance, fibers will break when flexed. Should not be used above 200°F.	Change in color usually appearing as a brownish smear or smudge. Transverse cracks when belt is bent over a mandrel. Loss of elasticity in belt.	Webbing strands fuse together. Hard shiny spots. Hard and brittle feel. Will not support combustion.	Paint which penetrates and dries restricts movement of fibers. Drying agents and solvents in some paints will appear as chemical damage.

QUALITIES OF NYLON SLINGS	
ADVANTAGES	DISADVANTAGES
Protects the load from dents, gouges, scrapes	Sharp edges on loads must be padded
Provides a wide, flat, non-slip bearing surface	Abrasion, dirt, and rough edges shorten sling life
Conforms to shape of load	Stretches under weight load
Lightweight, soft, easily handled	Lacks resistance to fire and high temperatures
Will not kink or slip in a choker hitch	Should not be used above 200°F (93°C)
Will not cause sparks	

GENERAL RIGGING PROCEDURES

1. All wire ropes, chains, hooks, sheaves and drums will be inspected periodically.

Check: Illustration following.

2. In chains, look for bent links, cracked welds, nicks and gouges.

3. When in doubt about the condition of any rigging, check with your supervisor.

4. Wear gloves to prevent injury. Keep your hands out of pinch points, and stay clear of the rigging as it tightens under the weight of the load.

5. Hands or feet will not be used for spooling cables onto drums.

6. If a slack line occurs, check the seating of the rope in sheave or drum before proceeding.

7. Note that increasing the angle from the vertical between legs of a two-leg sling increases the load stress on the legs.

8. The load must be properly set in the throat of the hook. Loading toward the point (except in grab hooks) leads to spreading of the hook.

EFFICIENCY OF WIRE ROPE CONNECTIONS
(As Compared to Safe Loads on Wire Rope)

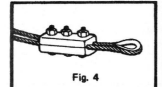
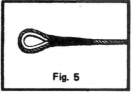

Figure	Type of Connection	Efficiency
1	Wire Rope Sockets — Zinc Type — properly attached	100% 100%
2	Wedge Sockets	70%
3	Clips	80%
4	Plate Clamp — Three Bolt Type	80%
5	Spliced Eye and Thimble:	
	1/4" and smaller	90%
	5/16"–7/16"	88%
	1/2"	86%
	5/8"	84%
	3/4"	82%
	7/8" and larger	80%

METHOD OF ATTACHING WIRE ROPE CLIPS

1. This illustrates the correct application of wire rope clips. All the saddles of the clips are in contact with the load end of the rope and the clips are correctly spaced.

2. The correct number of clips for safe application, and spacing distances, are shown in the table below.

3. Before ropes are placed under tension, the nuts on the clips should be tightened. It is advisable to tighten them again after the load is on the rope to take care of any reduction in the rope's diameter caused by the weight or tension of the load.

4. A wire rope thimble should be used in the loop eye to prevent kinking when wire rope clips are used.

CLIPS AND SPACING FOR SAFE APPLICATION			
Rope Diameter in.	Approximate Weight lb.	Minimum No. Clips for Each Rope End	Spacing of Drop Forged Clips in.
3/16	0.10	2	1⅛
¼	0.19	2	1½
5/16	0.29	2	1⅞
⅜	0.47	2	2¼
7/16	0.70	2	2⅝
½	0.78	3	3
⅝	1.06	3	3¾
¾	1.59	4	4½
⅞	2.40	4	5¼
1	2.72	5	6
1⅛	3.20	6	6¾
1¼	4.50	6	7½
1⅜	4.60	7	8¼
1½	5.80	7	9
1⅝	7.20	7	9¾
1¾	9.50	8	10½
2	12.50	9	12
2¼	15.50	9	13½
2½	18.00	9	15

SAFE RIGGING PRACTICES

1. Cable clips should be installed in accordance with standards in the preceding table.

2. The weight of the load should be determined to select the proper size of choker.

3. Sharp edges of the material to be rigged should be protected to prevent damage to the choker.

4. "Softeners" (i.e., pieces of lumber or the like) should be used to prevent material slippage.

5. Tag lines should be used when hoisting and rigging loads.

6. Material or equipment rigging should not be rigged from structural points which are unstable (such as unfinished work, handrail, or conduit).

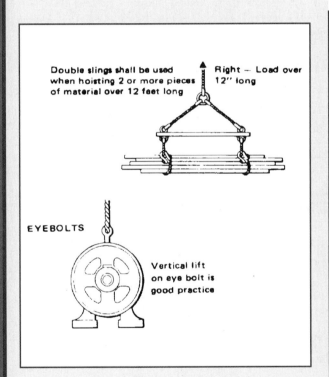

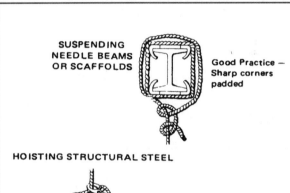

CHAIN SLINGS
Estimated Rating Capacity — Pounds Alloy Steel Chain*

Chain Size Inches	90°	60°	45°	30°	10°
9/32	3,250	5,600	4,600	3,250	1,190
3/8	6,600	11,400	9,300	6,600	2,300
1/2	11,250	19,500	15,900	11,250	3,800
5/8	16,500	28,600	23,300	16,500	5,700
3/4	23,000	39,800	32,500	23,000	8,000
7/8	28,750	49,800	40,700	28,750	10,000
1	38,750	67,100	54,800	38,750	13,400
1 1/4	57,500	99,600	81,300	57,500	20,000

Triple and Quadruple slings — Add 50% to Double Rating when used at same angle of inclination.

*Do not use these ratings for wrought iron chain.

FOLLOW THESE BASIC CAUTIONS

- Take up slack, then start load slowly.
- Keep chains free from twists, knots and kinks.
- Lift from center of hooks. Avoid lifting from the point.
- Distribute load evenly on all legs.
- Inspect chains regularly. Remember — elongation of links is a sign of overloading.

WIRE ROPE WEAR AND DAMAGE

The evidence in these illustrations will aid the inspector in determining the actual cause of wear or damage that he may find in any wire rope.

A wire rope which has been kinked. A kink is caused by pulling down a loop in a slack line during improper handling, installation, or operation. Note the distortion of the strands and individual wires. Early rope failure will undoubtedly occur at this point.

A typical failure of a rotary drill line with a poor cut-off practice. These wires have been subjected to excessive peening causing fatigue type failures. A predetermined, regularly scheduled, cut-off practice will go far toward eliminating this type of break.

A "bird-cage". Caused by sudden release of tension and resultant rebound of rope from overloaded condition. These strands and wires will not return to their original positions.

Localized wear over an equalizing sheave. The danger of this type wear is that it is not visible during operation of the rope. This emphasizes the need of regular inspections of this portion of an operating rope.

A single strand removed from a wire rope subjected to "strand nicking." This condition is the result of adjacent strands rubbing against one another and is usually caused by core failure due to continued operation of a rope under high tensile load. The ultimate result will be individual wire breaks in the valleys of the strands.

An example of a wire rope with a high strand — a condition in which one or two strands are worn before adjoining strands. This is caused by improper socketing or seizing, kinks or dogs legs. Picture A is a close-up of the concentration of wear and B shows how it recurs in every sixth strand (in a six strand rope).

An illustration of a wire which has broken under tensile load in excess of its strength. It is typically at the point of fracture. The necking down of the wire at the point of failure to form the cup and cone indicates that failure occurred while the wire retained its ductility.

A wire rope which has been subjected to repeated bending over sheaves under normal loads. This results in "fatigue" breaks in individual wires, these breaks being square and usually in the crown of the strands.

An example of "fatigue" failure of a wire rope which has been subjected to heavy loads over small sheaves. The usual crown breaks are accompanied by breaks in the valleys of the strands, these breaks being caused by "strand nicking" resulting from the heavy loads.

A close-up of a rope subjected to drum crushing. Note the distortion of the individual wires and displacement from their normal position. This is usually caused by the rope scuffing on itself.

A wire rope which has jumped a sheave. The rope itself is deformed into a "curl" as if bent around a round shaft. Close examination of the wires show two types of breaks — normal tensile "cup and cone" breaks and shear breaks which give the appearance of having been cut on an angle with a cold chisel.

An illustration of a wire which shows a fatigue break. It is recognized by the squared off ends perpendicular to the wire. This break was produced by a torsion machine, which is used to measure the ductility. This break is similar to wire failures in the field caused by excessive bending.

An example of a wire rope that has provided maximum service and is ready for replacement.

A fatigue break in a cable tool drill line caused by a tight kink developed in the rope during operation.

WELDING AND BURNING OPERATIONS

Welding and burning operations have a high potential for personnel injuries and fires. When doing either, you must follow these precautions:

1. Before starting to burn or weld, you must inspect your work area to ensure that sparks or molten metal won't fall on combustible materials. If you can't provide the necessary safeguards, check with your Supervisor.

2. You must not weld or burn in a hazardous area without obtaining written authorization from the responsible authority.

3. You must make certain that suitable fire-extinguishing equipment is available in your work area.

4. You are responsible for maintaining your burning or welding equipment in safe operating condition.

5. When burning or welding, you must wear approved eye protection, with suitable filter lenses.

6. Keep all welding leads and burning hoses off floors, walkways, and stairways. You are responsible for seeing that your equipment complies with safe practices at all times.

7. Never weld or burn on barrels, tanks, piping, or other systems which may have contained either combustible or unknown products without first obtaining approval from your Safety Representative or other responsible authority.

8. Wear gauntlet gloves and arm protection.

9. Ear plugs help keep sparks and slag out of the ear canal as well as providing protection from noise.

10. When welding, set up the screen provided to protect nearby workers. If screens are not sufficient, adjacent workers must wear suitable eye protection.

11. Appropriate containers must be used for electrode stubs and scrap. The floor must be clear of hazards.

12. Filter-type respirators are necessary when welding metal which give off toxic fumes. Check with your supervisor.

13. If you must work above the floor, use the approved stepstools for firm footing.

14. Do not use matches to light torches. Spark igniters must be used. Torches must not be used to light smoking materials.

15. When a crescent or special wrench is required to operate the acetylene cylinder

valve, the wrench must be kept in position on the valve.

16. The frames of all welding machines must be grounded.

VENTILATION AND PROTECTION

1. Welding, burning, and heating performed in confined spaces may require general mechanical or local exhaust ventilation to reduce the concentrations of smoke and fumes to acceptable levels. Your Safety Representative should be consulted prior to starting these operations.

2. If adequate ventilation cannot be provided, employees must be provided with, and required to use, air-supplied breathing apparatus.

3. When welding, cutting, or heating metals having toxic significance, such as zinc, lead, cadmium, or chromium-bearing metals, in the open air, you must wear filter-type respirators

CYLINDER SAFETY

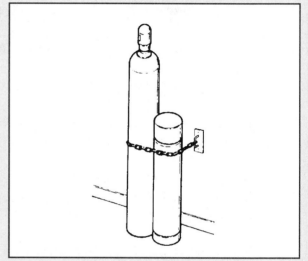

1. Cylinders must be secured during transport and use against tipping.

2. Cylinders may not be taken into confined spaces.

3. Levering against the valve to force open tight caps could result in valve damage. Use a wrench if the cap is too tight to open by hand. Caps must be replaced on gas cylinders not in use.

4. Arc welding electrode or ground leads shall not be hung over compressed gas cylinders.

5. Oxygen and acetylene (or other fuel gas) cylinders in storage must be separated from each other by 20 feet or by a 5-foot barrier which has a 1-hour fire rating.

FIRST AID AND MEDICAL TREATMENT

First aid facilities are provided for your safety. Qualified personnel are available to render treatment and to maintain required records.

1. Report all injuries IMMEDIATELY, no matter how minor, to your Supervisor and to First Aid. Treatment will be given, and the incident will be recorded. Should later medical care be needed, you will have fulfilled your obligations.

2. You must notify your Supervisor and First Aid prior to leaving the jobsite because of injury or illness,

whether personal or work-related.

3. If you get outside medical treatment (without clearing through First Aid) for a work-related injury or illness, you must notify First Aid at the start of the next scheduled work day. Failure to do so may result in disallowance of your claim and/or discharge.

4. Prior to returning to work after a disabling injury or illness, you must present a medical clearance to First Aid from the attending physician.

5. If you have a physical handicap, such as diabetes, impaired eyesight or hearing, back or heart trouble, hernia or aversion to heights, tell your Supervisor. You won't be expected to do a job which might result in injury to yourself or others.

6. Never move an injured or seriously ill person unless necessary to prevent further injury. Emergency steps for notifying First Aid are posted throughout the jobsite; familiarize yourself with them. First aid should not be administered by non-designated employees except in case of severe bleeding or cessation of breathing.

LADDER SAFETY

1. Inspect ladder before using. Ladders must be long enough and strong enough for the job. Ladders are for climbing. They are not to be used for levering, bracing, or any other purpose which might weaken the structure. If a ladder must be placed in a doorway or walkway, set up a barricade and warning sign.

2. Ladders must be braced against strong, immoveable objects, tied off at the top, and placed on firm ground at the base. The base of the ladder should be at least one-quarter of its supported length out from the support structure.

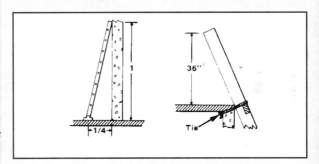

3. Tools or equipment should never be hung from or placed on the rungs of any ladder.

4. Do not carry anything in your hands while climbing. Hold the ladder with both hands. Tools and equipment can be drawn up on a hand line.

5. Ladders used for access to a floor or platform must extend at least 3 feet above the landing.

6. The areas around the top and base of ladders must be free of tripping hazards such as loose materials, trash, and electric cords.

7. You must face the ladder at all times when ascending or descending.

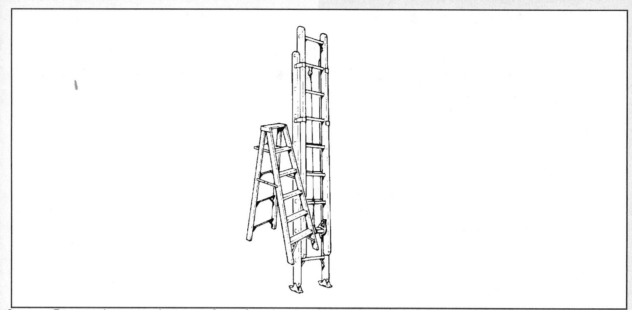

8. Be sure that your shoes are free of mud, grease, or other substances which could cause a slip or fall.

9. When working from a ladder, keep both feet on the rungs. Climb down and reposition the ladder as often as necessary.

10. No worker shall stand on either of the top two rungs of any single or extension ladder while working.

11. Stepstools are provided where a minor elevation is needed. They provide a firm footing. Makeshifts are dangerous.

12. Stepladders should be used only in the fully opened position. They should be set level with spreaders fully locked. Workers may work from the top platform of a stepladder only if it is equipped with a railing. Stepladders more than ten feet in height must be secured or held by another worker.

13. Metal ladders must not be used for electrical work or in areas where they could contact energized wiring. The use of metal ladders is restricted to special applications where the heavier wooden ladders are not practical.

14. Always move the ladder to avoid overreaching.

LESSON 27

PHCC SAFETY MANUAL (CONTINUED)

HAND AND POWER TOOLS

GENERAL

Only tools in safe working condition will be issued from the tool room.

Comply with all of the tool manufacturer's instructions, as well as the following safe practices:

1. Inspect your tools daily to ensure that they are in proper working order. Damaged or defective tools must be returned to the tool room immediately.

2. Power saws, grinders, and other power tools must have proper guards in place at all times.

3. Power tools should be hoisted or lowered by a hand line, never by the cord or hose.

4. Cords and hoses must be kept out of walk ways and off stairs and ladders. They must be placed so as not to create a tripping hazard for employees or to be subject to damage.

5. Electrically powered tools and equipment must be grounded when in use.

6. Hand tools should be used for their intended purpose only. The design capacity of hand tools should not be exceeded by unauthorized attachments.

7. When using the tools listed below, or when working near others using such tools, you must use the additional personal protective equipment specified. If you have questions about the protective equipment or safety rules, ask your Supervisor.

PNEUMATIC TOOLS

1. An approved safety check valve must be installed at the manifold outlet of each supply line for hand-held pneumatic tools.

2. All pneumatic hose connections must be fastened securely.

EQUIPMENT	PROTECTION
Jackhammers Tampers	Eye Protection Hearing Protection Foot Protection
Chipping Hammers Impact Wrenches Reamers	Eye Protection Hearing Protection
Cutting Torches Arc Welders	Eye Protection Hand Protection
Powder-Actuated Tools Grinders Hand-Held Chipping Hammers	Eye Protection Hearing Protection
Air Arcing	Eye Protection Hearing Protection

FUEL-POWERED TOOLS

1. All fuel-powered tools must be shut down while being refueled.

2. Smoking is prohibited during refueling operations. Other nearby sources of ignition, such as burning and welding, also must be halted during refueling operations.

POWDER-ACTUATED TOOLS

1. Only employees who possess valid credentials are permitted to use powder-actuated tools. The manufacturer's representative will conduct training classes at the jobsite upon request.

SAFE GRINDING AND BUFFING

1. Always wear approved eye, respiratory, and hand protection when working with or near grinders. Always visually inspect wheels for damage before mounting and before each use. Chipped or cracked wheels must be discarded.

2. Do not stand directly in line with a newly mounted wheel when starting up. Before grinding, always test run a newly mounted wheel at full speed for:

- thirty (30) seconds for reinforced discs.

- sixty (60) seconds for stand-mounted grinders.

3. The governor mechanism should be checked to make sure it is functioning. Make sure the RPM of the machine does not exceed the rated wheel speed.

4. Rests used on grinders shall not be more than one-eighth inch from the face, fastened securely, and must not be adjusted while the wheel is in motion.

5. Check mounting flanges for correct diameter

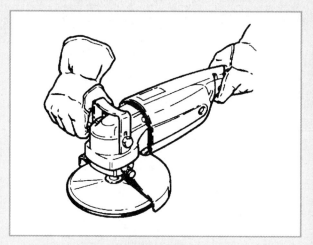

Grinder

(straight wheels at least one-third diameter of wheel) and for warpage. Do not use bent or dirty flanges.

6. All spindles, adapters, flanges, and other parts should be inspected periodically and maintained to size and verified to be in good condition.

7. Proper lubrication of air or electric motors and their bearings is essential.

8. Use blotters at least equal in size to the flanges when mounting the wheel. Do not over-tighten the mounting nut.

9. Use proper safety guards on grinders. Special guards are available for all grinders when working in confined areas. Make sure the guards are properly secured. Do not jam or poke the grinding wheel at the work. Grind only on the face of a straight wheel. Use disk wheels for side grinding. Light side grinding is permitted with a cup or saucer wheel. Make sure the wheel has stopped before putting the grinder down as it can travel, striking persons or equipment. Lay the machine down with the disk up.

10. Avoid dropping or bumping the wheel. Do not allow anything to strike a wheel which is not in use. Handle and store wheels carefully. Use suitable racks or bins according to

manufacturer's specifications.

MACHINE SHOP SAFETY

1. Preventive safety when working on a lathe includes periodic inspection of parts for wear or breakage. Injuries have been caused by broken parts ejected at high speed from a spinning lathe. Particular attention should be paid to checking of chucks and lathe tools.

2. Eye protection is essential.

3. Do not allow the cuttings to pile up under the lathe. Long coils can be picked up and may become snarled around or hit the employee, causing severe, even fatal injuries. Use proper hooks for removing these cuttings as they are formed.

4. The machine must come to a full stop before any adjustments are made as inserting blanks or removing finished work.

5. When operating a drill press, do not wear loose clothing or allow long hair to become entangled in revolving parts. Hair nets are available. Suitable eye protection must be worn.

6. Make certain the work is securely clamped to the drill table. This prevents it spinning out and striking the worker in the event a drill freezes or jams in the hole.

7. Do not use dull drills. They break easily and may cause injuries.

8. Clear chips from the drilled hole. Use a brush, not your hand, to sweep them out.

FIRE PREVENTION

1. Familiarize yourself with the location of all firefighting equipment in your work area.

2. Learn the classifications of fires:

- CLASS A — Ordinary combustible materials such as wood, coal, paper, or fabrics where wetting and cooling is the method of extinguishment.

- CLASS B — Flammable petroleum products or other flammable liquids where oxygen must be excluded for extinguishment.

- CLASS C — Fires in or near energized electrical equipment where, because use of water would be hazardous, a "nonconducting" extinguishing agent must be used.

3. Only approved solvents should be used for cleaning and degreasing. The use of gasoline and similar flammable products for this purpose is prohibited.

4. Keep the work area neat. An orderly jobsite reduces the fire and accident hazard.

5. When you must weld or burn near combustible materials, move them, cover them with fire-resistant fabric, or wet them down. When in doubt, consult your Supervisor.

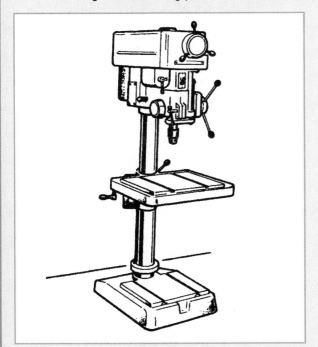

Drill Press

6. Flammable and combustible liquids must be handled only in approved, properly labeled safety cans.

7. Place oily rags in approved covered metal containers.

8. Do not attempt any work involving a source of ignition near a pit, sewer, drain, manhole, trench, or enclosed space where flammable gases may be present. Wait until tests have been made with an approved combustible gas indicator and the area has been declared safe for hot work.

9. The use of open fires is prohibited unless specifically authorized by the responsible Supervisor.

10. Do not weld or cut on a tank or in an enclosure that has contained gasoline or other flammable gas or liquid without specific instructions from your Supervisor.

HOUSEKEEPING

GENERAL

Good housekeeping on the job is mandatory, and every employee must do his part daily in this activity to keep the job clean for safety and efficiency.

1. Scrap materials and rubbish are fire and accident hazards. If an excess of these materials exists in your work area, ask your Supervisor to arrange for their removal.

2. You must use the trash barrels which are located throughout the jobsite. If you need one in your immediate work area, notify your Supervisor.

3. All stairways, corridors, ladders, catwalks, ramps, and passageways must be kept clear of loose material and trash.

4. Return all surplus materials to the stockpile at the completion of your job.

5. Do not leave tools and materials where they will create a hazard for others. Put them in the gang box or return them to the tool room.

6. Place oil rags in approved metal containers.

7. Wipe up spilled liquids immediately. If you can't handle the problem, notify your Supervisor so that he can arrange for the necessary cleanup.

8. Keep change rooms clean. Do not let soiled clothes, food scraps, and soft drink bottles accumulate. If drinking cups are used, deposit them in the containers provided. Also place sandwich wrappers, paper bags, and other trash in these containers.

9. Toilets and drinking fountains are provided for your convenience and comfort. Please help to keep them clean and sanitary.

FLOOR OPENINGS

1. Floor openings or holes should be protected by approved guardrails or covers. If covers are used, they should be strong enough to support the loads to be imposed upon them and should be secured to prevent accidental displacement.

2. The open edges of all floors six feet or more above the next floor or level should be guarded by an approved barricade secured to prevent accidental displacement.

3. Do not remove covers on floor openings without approval of your Supervisor. When a cover has been removed to bring in equipment or material, replace the cover immediately upon completion of material handling.

4. Do not remove warning signs or posters advising of floor openings or open covers.

METRIC EQUIVALENTS

The following information is presented to aid you in becoming "easy" with the magnitude of Metric values. As suggested several times in this book, the "Metric Conversion" has not happened as it was thought it would 20 years ago, but the authors believe that it is still necessary to be aware of the system.

COMMON METRIC EQUIVALENTS

(Approximate)
"Metric Conversion Act of 1972"
92nd Congress

1 *inch* = 25 *millimeters*

1 *foot* = 0.3 *meters*

1 *yard* = 0.9 *meter*

1 *mile* = 1.6 *kilometers*

1 *sq. inch* = 6.5 *sq. centimeters*

1 *sq. foot* = 0.09 *sq. meter*

1 *sq. yard* = 0.8 *sq. meter*

1 *cu. inch* = 16 *cu. centimeters*

1 *cu. foot* = 0.03 *cu. meter*

1 *cu. yard* = 0.7 *cu. meter*

1 *quart* = 0.001 *cubic meter*

1 *gallon* = 0.004 *cubic meter*

1 *ounce* = 28 *grams*

1 *pound* = 0.45 *kilogram*

1 *horsepower* = 0.75 *kilowatt*

1 *millimeter* = 0.04 *inch*

1 *meter* = 3.3 *feet*

1 *meter* = 1.1 *yards*

1 *kilometer* = 0.6 *miles*

1 *sq. centimeter* = 0.16 *square inch*

1 *sq. meter* = 11 *square feet*

1 *cu. centimeter* = 0.006 *cubic inch*

1 *cu. meter* = 1.3 *cubic yards*

1 *liter* = 1.06 *quarts*

1 *cu. meter* = 250 *gallons*

1 *gram* = 0.035 *ounces*

1 *kilogram* = 2.2 *pounds*

1 *kilowatt* = 1.3 *horsepower*

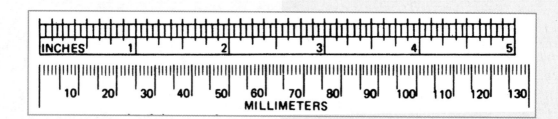

LESSON 28

PHCC SAFETY MANUAL CONTINUED

WEIGHT OF MATERIALS
National Safety Council

MATERIAL	WEIGHT PER CUBIC FOOT — POUNDS
Alcohol	49
Aluminum, cast-hammered	165
Aluminum bronze	481
Ammonia 27.9%	56
Asbestos	47
Ashes	43
Asphaltum	87
Babbitt metal	456
Bauxite	159
Benzine	46
Brass, cast-rolled	534
Brick, soft	100
Brick, common	112
Brick, hard	125
Brick, pressed	135
Brick, fire	145
Brick, sand-lime	136
Brickwork, mortar	100
Brickwork, cement	112
Bronze, 7.9 to 14% Sn	509
Cement, Portland	94
Chalk	137
Charcoal, pine	23
Clay	137
Coal, anthracite	97
Coal, bituminous	84
Coal, peat turf, dry	47
Coke	22-27
Concrete, plain	145
Concrete, reinforced	150
Copper, cast-rolled	556
Cork	15
Earth, dry, loose	75
Earth, dry, packed	93
Earth, moist, loose	81
Earth, moist, packed	100
Earth, mud, flowing	106
Earth, mud, packed	112
Emery	250
Feldspar	159
Gasoline	42
Glass, common	164
Glass, crystal	184
Glass, flint	188
Glass, plate	161
Gneiss	165
Granite	179
Granite, piled	96
Graphite	131
Gravel, dry, loose	90-105
Gravel, dry, packed	100-115
Gravel, wet	110
Gypsum	140
Hay and straw, in bales	20
Ice	55-57
Iron, cast pig	450
Iron, wrought	485
Iron, ferro silicon	437
Iron, ore, hematite	325
Iron, ore, hematite, in bank	160-180
Iron ore, hematite, loose	130-160
Iron ore, limonite	237
Iron ore, magnitite	315
Iron, slag	172
Kerosene	50
Lead	710
Lead ore, galena	465
Lime, quick, in bulk	55
Limestone	165
Linseed oil	58
Magnesite	187
Manganese	475
Manganese ore, pyrolusite	259

Material	Weight
Marble	170
Masonry, Ashlar, granite, syenite, gneiss	165
Masonry, Ashlar, limestone, marble	160
Masonry, Ashlar, sandstone, bluestone	140
Masonry, mortar rubble, granite, syenite, gneiss	155
Masonry, mortar rubble, limestone, marble	150
Masonry, mortar rubble, sandstone, bluestone	130
Masonry, dry rubble, granite, syenite, gneiss	130
Masonry, dry rubble, limestone, marble	125
Masonry, dry rubble, sandstone, bluestone	110
Masonry, brick, pressed brick	140
Masonry, brick, common brick	120
Masonry, brick, soft brick	100
Masonry, concrete, cement, stone, sand	144
Masonry, hollow concrete, cement, stone, sand	86
Masonry, hollow concrete, slag, etc.	78
Masonry, hollow concrete, cinder, etc.	60
Mica	175
Mortar	95
Mud	111
Naphtha	47
Nickel	537
Oils, mineral, lubricants	57
Paper	58
Paraffin	56
Peat	47
Petroleum	54
Petroleum, refined	50
Phosphate rock	200
Pitch	72
Plaster-of-Paris	103
Porcelain	250
Porphyry	172
Pumice	40
Quartz, flint	165
Riprap, limestone	80-85
Riprap, sandstone	90
Riprap, shale	105
Rubber, manufactured	95
Salt, granulated, piled	48
Saltpeter	67
Sand, gravel, dry, loose	90-105
Sand, gravel, dry, packed	100-115
Sandstone, bluestone	147
Shale, slate	175
Silver	655
Slag, bank	69
Slate	175
Snow, fresh fallen	8
Snow, compacted by rain	15-50
Soapstone, talc	169
Steel	489
Stone, crushed	100
Talc	169
Tar, bituminous	75
Terra cotta	119
Tile	115
Tin	459
Trap rock	185
Water, 4° C, maximum density	62.428
Water, 100° C	59.830
Water, sea	64
Wood, ash, white-red	40
Wood, cedar, white-red	22
Wood, cypress	30
Wood, fir, Douglas, spruce	32
Wood, fir, eastern	25
Wood, elm, white	45
Wood, hemlock	29
Wood, hickory	49
Wood, maple, hard	43
Wood, oak, white	46
Wood, pine, Oregon	32
Wood, pine, white	26
Wood, pine, red	30
Wood, pine, yellow, long-leaf	44
Wood, pine, yellow, short-leaf	38
Wood, redwood, California	26
Wood, spruce, white-black	27
Wood, walnut, black	38
Zinc, cast-rolled	440
Zinc ore, blends	253

GENERAL RULES

NAILS

Safe load lateral resistance in pounds equals 8 times the pennyweight.

    1   -   6d. nail = 8x 6, or 48 Lbs.
    1   -   8d. nail = 8x 8, or 64 Lbs.
    1   -   10d. nail = 8x10, or 80 Lbs.

MANILA ROPE

   1" rope = 1x1, or 1 Ton safe load.
   $1/2$" rope = $1/2$ x $1/2$, or $1/4$ Ton safe load
   For Sisal Rope, decrease safe loads by one-third.

PLOW-STEEL CABLE

Safe load in tons is 8 times the diameter in inches squared.   Thus, $1/2$" rope $1/2$ x $1/2$ = $1/4$ x 8 = 2 tons.

SHACKLE

Safe load in tons is diameter of pin in one-fourth inches (¼") squared and divided by three (3).

$$1/2" \text{ diameter} = 2 \text{ quarters}$$

$$\frac{2 \times 2}{3} = 1\,1/3 \text{ tons or 2,667 pounds}$$

WROUGHT IRON CHAINS (See Lesson 26 for alloy steel chain.)

Safe load in tons is six (6) times the diameter of chain stock in inches squared.

   $1/2$" diameter chain stock
   $1/2$ x $1/2$ x 6 = 1 $1/2$ tons or 3,000 pounds.

PLANK FOR SCAFFOLD

Two-inch by ten-inch (2"x10") or two-inch by twelve-inch (2"x12") plank, graded as suitable for fifteen hundred pounds per square inch (1,500 psi) bending stress, or better, may normally be used for a span in feet that is equal or less than the width in inches.

CHOCK SIZES FOR TANKS

Figure 28-A Chock Size

For every foot diameter of the tank use one inch of chock.

    For example: If tank is 2' in diameter, use 2" chock.
    For example: If tank is 3' in diameter, use 3" chock.
    For example: If tank is 6' in diameter, use 6" chock.

All chocks should be chamfered to fit the curvature of the tank.

USEFUL TABLES AND CHARTS

TEMPERATURE CONVERSION	
Celsius to Fahrenheit	Fahrenheit to Celsius
$C \times 1.8 + 32 = F$	$\dfrac{F - 32}{1.8} = C$

USEFUL APPROXIMATION OF PIPE WEIGHT
OD x t x 10 = WT. in LBS.
Outside Diameter Times Thickness Times 10 equals weight in pounds per foot

PIPE WALL THICKNESS AND WEIGHT PER FOOT

A.S.A. PIPE SCHEDULES

PIPE SIZE (INCHES)	O.D. (INCHES)	STANDARD WALL	EXTRA STRONG WALL	DOUBLE EXTRA STRONG	5	10	20	30	40	60	80	100	120	140	160
⅛	.405	.068 / .2447	0.95 / 0.3145		.035 / .1383	.049 / .1863			.068 / .2447		.095 / .3145				
¼	.540	.088 / .4248	.119 / .5351		.049 / .2570	.065 / .3297			.088 / .4248		.119 / .5351				
⅜	.675	.091 / .5676	.126 / .7388		.049 / .3276	.065 / .4235			.5876 / .109		.7388 / .147				
½	.840	.109 / .8510	.147 / 1.088	.294 / 1.714	.065 / .6383	.083 / .6710			.109 / .8510		.147 / 1.088				.187 / 1.304
¾	1.050	.113 / 1.131	.154 / 1.474	.308 / 2.441	.065 / .6838	.083 / .8572			.113 / 1.131		.154 / 1.474				.218 / 1.937
1	1.315	.133 / 1.679	.179 / 2.172	.358 / 3.659	.065 / .8678	.109 / 1.404			.133 / 1.679		.179 / 2.172				.250 / 2.844
1¼	1.660	.140 / 2.273	.191 / 2.997	.382 / 5.214	.065 / 1.107	.109 / 1.808			..140 / 2.273		.191 / 2.997				.250 / 3.765
1½	1.900	.145 / 2.718	.200 / 3.631	.400 / 6.408	.065 / 1.274	.109 / 2.805			.145 / 2.718		.200 / 3.631				.281 / 4.859
2	2.375	.154 / 3.653	.218 / 5.033	.436 / 9.029	.065 / 1.604	.109 / 2.638			.154 / 3.653		.218 / 5.022				.343 / 7.444
2½	2.875	.203 / 5.793	.276 / 7.661	.552 / 13.70	.083 / 2.475	.120 / 3.531			.203 / 5.793		.276 / 7.661				.375 / 10.01
3	3.5	.216 / 7.576	.300 / 10.25	.600 / 18.58	.083 / 3.029	.120 / 4.332			.216 / 7.576		.300 / 10.25				.438 / 14.32
3½	4.0	.226 / 9.109	.318 / 12.51	.636 / 22.85	.083 / 3.472	.120 / 4.973			.226 / 9.109		.318 / 12.51				
4	4.5	.237 / 10.79	.337 / 14.98	.674 / 27.54	.083 / 3.915	.120 / 5.613			.237 / 10.79	.281 / 12.66	.337 / 14.98		.437 / 19.01		.531 / 22.51
4½	5.0	.247 / 12.53	.355 / 17.61	.710 / 32.53											
5	5.563	.258 / 14.82	.375 / 20.78	.750 / 38.55	.109 / 6.349	.134 / 7.770			.258 / 14.62		.376 / 20.78		.500 / 27.04		.825 / 32.96

PIPE WALL THICKNESS AND WEIGHT PER FOOT

A.S.A. PIPE SCHEDULES

PIPE SIZE (INCHES)	O.D. (INCHES)	STANDARD WALL	EXTRA STRONG WALL	DOUBLE EXTRA STRONG	5	10	20	30	40	60	80	100	120	140	160
6	6.625	.280 / 18.97	.432 / 28.57	.864 / 53.16	.109 / 7.585	.134 / 9.289			.280 / 18.97		.432 / 28.57		.562 / 36.39		.718 / 45.30
7	7.625	.301 / 23.57	.500 / 38.05	.875 / 63.08											
8	8.025	.322 / 28.55	.500 / 43.39	.875 / 72.42	.109 / 9.914	.148 / 13.40	.250 / 22.36	.277 / 24.70	.322 / 28.55	.406 / 35.64	.500 / 43.39	.593 / 50.87	.718 / 60.93	.812 / 67.76	.906 / 74.69
9	9.625	.342 / 33.90	.500 / 48.72												
10	10.75	.365 / 40.48	.500 / 54.74	1.000 / 104.1	.134 / 15.19	.165 / 18.70	.250 / 28.04	.307 / 34.24	.365 / 40.48	.500 / 54.74	.593 / 64.33	.718 / 76.93	.843 / 89.20	1.000 / 104.1	1.125 / 115.6
11	11.75	.375 / 45.55	.500 / 60.07												
12	12.75	.375 / 49.56	.500 / 65.42	1.000 / 125.5	.165 / 22.18	.180 / 24.20	.250 / 33.38	.330 / 43.77	.406 / 53.53	.582 / 73.16	.687 / 88.51	.843 / 107.2	1.000 / 125.5	1.125 / 139.7	1.312 / 160.3
14	14.0	.375 / 54.57	.500 / 72.09			.250 / 36.71	.312 / 45.68	.375 / 54.57	.437 / 63.37	.593 / 84.91	.750 / 106.1	.937 / 130.7	1.093 / 150.7	1.250 / 170.2	1.406 / 189.1
16	16.0	.376 / 62.58	.500 / 82.77			.250 / 42.05	.312 / 52.36	.375 / 62.58	.500 / 82.77	.656 / 107.5	.843 / 136.5	1.031 / 164.8	1.218 / 192.3	1.437 / 223.5	1.593 / 245.1
18	18.0	.375 / 70.59	.500 / 93.45			.250 / 47.39	.312 / 59.03	.437 / 82.06	.582 / 104.8	.750 / 138.2	.937 / 170.8	1.158 / 208.0	1.375 / 244.1	1.562 / 274.2	1.781 / 308.5
20	20.0	.375 / 78.60	.500 / 104.1			.250 / 52.73	.375 / 78.60	.500 / 104.1	.593 / 122.9	.812 / 166.4	1.031 / 108.9	1.280 / 256.1	1.500 / 296.4	1.750 / 341.1	1.968 / 379.0
24	24.0	.375 / 94.62	.500 / 125.5			.250 / 63.41	.375 / 94.62	.562 / 140.8	.687 / 171.2	.968 / 238.1	1.218 / 296.4	1.531 / 367.4	1.812 / 429.4	2.062 / 483.1	2.343 / 541.9

WEIGHT PER FOOT FOR COMMON LARGER PIPE SIZES
(Standard Wall Pipe)

Pipe Size	O.D.	Wall Thickness	Weight per Foot (Lbs.)
		.250	63
24	24.000	.312	79
		.375	95
		.406	102
		.312	84
		.375	103
		.438	120
26	26.000	.500	136
		.625	169
		.750	202
		1.000	267
		1.500	393
		.312	99
		.375	119
		.406	130
		.438	138
30	30.000	.500	158
		.625	196
		.750	234
		1.000	310
		1.500	457

WEIGHT PER FOOT FOR COMMON LARGER PIPE SIZES
(Standard Wall Pipe)

Pipe Size	O.D.	Wall Thickness	Weight per Foot (Lbs.)
		.312	106
		.375	127
		.438	127
32	32.000	.500	168
		.750	250
		1.000	331
		1.500	489
		.250	96
		.312	119
		.375	143
36	36.000	.438	166
		.500	190
		.750	282
		1.000	374
		1.500	553
		.375	167
		.500	222
42	42.000	.750	331
		1.000	438
		1.500	649
		.375	191
		.500	254
48	48.000	.750	378
		1.000	502
		1.500	745

STANDARD HAND SIGNALS

NOTE: Tag lines are to be secure and of sufficient length prior to giving hand signal.

HOIST
With forearm vertical, forefinger pointing up, move hand in small horizontal circle.

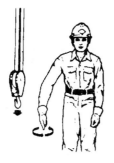

LOWER
With arm extended downward, forefinger pointing down, move hand in small horizontal circles

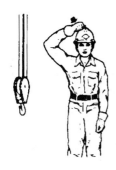

USE MAIN HOIST
Tap fist on head; then use regular signals.

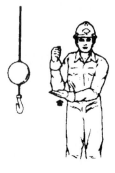

USE WHIPLINE
(Auxiliary Hoist)
Tap elbow with one hand, then use
regular signals.

RAISE BOOM
Arm extended, fingers closed,
thumb pointing upward.

LOWER BOOM
Arm extended, fingers closed,
thumb pointing downwards

MOVE SLOWLY

Use one hand to give any motion signal and place other hand motionless in front of hand giving the motion signal.
(Hoist slowly shown as example.)

RAISE THE BOOM AND LOWER THE LOAD

With arm extended, thumb pointing up, flex fingers in and out as long as load movement is desired.

LOWER THE BOOM AND RAISE THE LOAD

With arm extended, thumb pointing down, flex fingers in and out as long as load movement is desired.

SWING
Arm extended, point finger in direction of swing of boom.

STOP
Arm extended, palm down, hold position rigidly.

EMERGENCY STOP
Arm extended, palm down, move hand rapidly right and left.

TRAVEL
Arm extended forward, hand open and slightly raised, make pushing motion in direction of travel.

DOG EVERYTHING
Clasp hands in front of body.

TRAVEL (Both Tracks)
Use both fists in front of body, making a circular motion about each other, indicating direction of travel, forward or backward. (For crawler cranes only.)

TRAVEL (One Track)
Lock the track on side indicated by raised fist. Travel opposite track in direction indicated by circular motion of other fist, rotated vertically in front of body. (For crawler cranes only.)

EXTEND BOOM
(Telescoping Booms)
Both fists in front of body with thumbs pointing outward.

RETRACT BOOM
(Telescoping Booms)
Both fists in front of body with thumbs pointing toward each other.

EXTEND BOOM
(Telescoping Boom)
One Hand Signal. One fist in front of chest with thumb tapping chest.

RETRACT BOOM
(Telescoping Boom)
One Hand Signal. One fist in front of chest, thumb pointing outward and heel of fist tapping chest.

BARRICADES

Anyone who makes a hole or opening is responsible for having it barricaded.

NOTES

LESSON 29

TYPICAL PLUMBING FIXTURES

A fixture is a device that receives water from a branch supply and, using the water supply, discharges the waste materials into the drainage system.

This section of this lesson will study these fixture types:

- Water Closets
- Lavatories
- Urinals
- Drinking Fountains and Water Coolers
- Bidets
- Bathtubs
- Sinks
- Showers

WATER CLOSETS

A water closet is a fixture used to receive and remove human wastes from the bathroom or toilet room. These fixtures are made of vitreous china, plastic, or stainless steel. The china and plastic versions are available in many colors, depending on the manufacturer.

A basic requirement for these fixtures is that they must be emptied of waste materials after each use without employing any moving parts within the integral trapway. Further, they must cleanse the walls of the bowl, exchange the trap water, and replenish the trap seal. Water closets are emptied by introducing a water deluge into the fixture, and the deluge initiates an action that empties the bowl. This deluge effect is called a flush.

Since January 1, 1994, Federal law has required that water closets sold in interstate commerce for residential use be of the low consumption (1.6 gpf) type. Commercial use designs also had to be 1.6 gpf.

There are several methods for flushing a water closet, as described below. With the advent of the low consumption water closets which operate with 1.6 gallons (6 liters) or less per flush, some of the traditional bowl designs were modified to achieve the proper flushing action. A low consumption water closet is designed to discharge the contents of the trapway in approximately nine seconds. Such fixtures do not swirl as the traditional 3.5 gallons per flush (14 lpf) fixtures, but push the waste through the trapway. The trapway designs which we will study are of three basic designs:

1. Siphonic type which is similar to the designs called siphon jet or reverse trap.

2. Siphon Wash type which uses a deluge action to push the water from the bowl.

3. Blowout type which is a wall hung fixture for commercial use which is designed to remove any amount of materials.

Historically, the following designs have been used, but some will gradually be phased out.

SIPHON JET

In the siphon jet, a jet of water is directed into the upward leg of the fixture trap. This motion starts the water moving up the trap. When the water rises above the crown weir of the trap, a siphon is formed. This siphon is what empties the bowl. This bowl type has a large water surface and a deep water seal. Flushometer tank siphon-jet models are somewhat noisier than gravity style low consumption closets, but they may provide a more reliable flushing action as a trade-off for the noise. As discussed above, this concept will continue as the siphonic design.

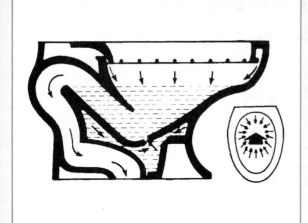

Figure 29-A – Siphon Jet Water Closet

REVERSE TRAP

The reverse trap type of device is similar to the siphon jet; however, the trapway and water seal are smaller in size. These are slightly smaller than the siphon jet and are somewhat less expensive. This style is expected to be discontinued as a design concept.

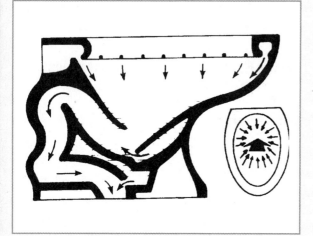

Figure 29-B – Reverse Trap Water Closet

BLOWOUT

Blowout type is a wall-hung fixture for non-residential use. It is designed to assure discharge of about anything that can be placed in a water closet bowl. This type uses 3.5 gallons per flush, and is permitted under an exception in the Federal rules, *but perhaps not under local rules*. The status of this water closet type should be verified in the jurisdiction where one or more is intended to be installed.

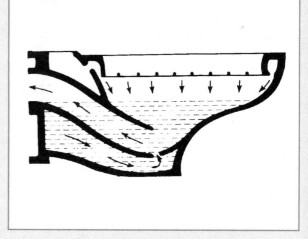

Figure 29-C – Blowout Water Closet

WASHDOWN

The washdown design water closet is highly effective within its limitations. This device is dependent upon the siphonic action in the cavern leg which is caused by the designed jet projection towards the leg. This design is smaller and less expensive, but it is not as effective in emptying the bowl. This design will be phased out.

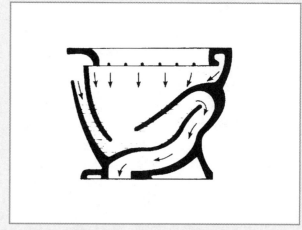

Figure 29-D – Washdown Water Closet

SIPHON VORTEX

The siphon vortex method is usually incorporated in one piece closets. A swirling action causes the bowl to evacuate its contents. These are effective, but they are more expensive than close coupled water closets, and are equipped with flushometer tanks or electric powered pumps to flush the fixture. This style will probably remain as a premium type fixture.

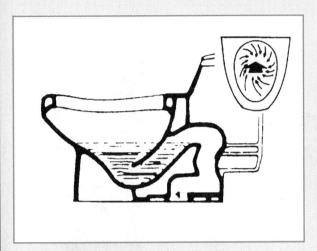

Figure 29-E – Siphon Vortex Water Closet

SIPHON WASH

This design requires only 1.6 gallons of water to flush. It has small surfaces and a small trapway. Even though the water amount is limited, the small bowl flushes through a deluge effect.

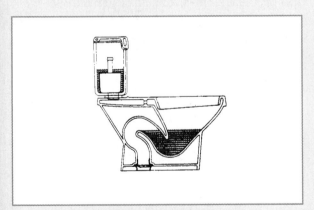

Figure 29-F – Siphon Wash Closet

FLUSHOMETER-TANK WATER CLOSET

This unit consists of a pressurized tank which uses a compressed-air-assisted discharge to flush the bowl. Approximately 1.6 gallons of water are used during the flushing cycle.

Note that the design is such that the air in the booster tank is compressed by the incoming water pressure — no external devices are needed for this operation.

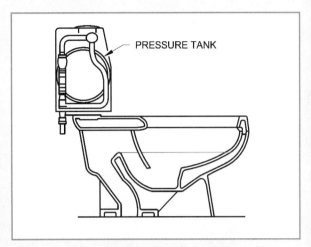

Figure 29-G – A Pressure-Assisted Water Closet With a Flushometer Tank

PNEUMATIC WATER CLOSET

In this design, a compressed air system assists the water to remove the contents of the bowl. The air is either provided by a compressor or a storage bottle.

Figure 29-H – Pneumatic Assist Water Closet

DUAL FLUSHING WATER CLOSET

A new product available is the dual flush water closet. This water closet provides the user the choice of 1.1 gallons/4 liters for liquid waste and 1.6 gallons/6 liters for discharging solid waste. According to the manufacturer, this water closet may save as much as 40% water over more typical designs.

GENERAL REMARKS

Each water closet type is designed to flush the

Therefore, the bowl construction is not complicated.

The bowls all incorporate a built-in trap to keep the sewer gases out of the room. Water in the trap is the barrier. As long as water remains in the trap, sewer gases will not enter the building. The com-

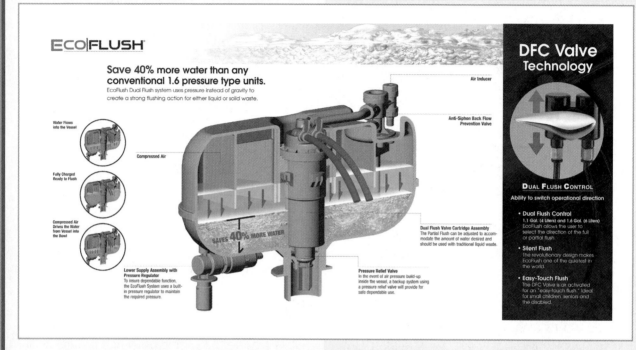

Figure 29-I – ECOFLUSH (WDI International, Inc.)

waste materials using a tank, a flushometer, or other pressurized flushing device. The flushing complexity of the bowl and the difficulty in manufacturing affect the cost of the bowl.

A one-piece closet is the most expensive water closet to make as the tank and bowl are formed as one unit, and a basic reality of china manufacturing is that the larger and more complex the product is, the more imperfect products will be produced in the manufacturing process.

A siphon jet closet is the next most expensive because it incorporates small tolerance details to achieve the high speed water jet to pull the waste material from the bowl.

The siphon wash closet is the most economical as it is designed to wash the waste material out merely through adding additional water to the bowl.

plete flushing cycle requires introducing the deluge, development of the siphon, minimum residual water as the siphon breaks, then refilling the trap to restore the seal against sewer gases.

ROUGH-IN

Floor-mounted water closets come in three rough-in measurements, that is, the center of discharge opening in the floor to the back wall:

```
10" — close
12" — standard
14" — overshot
```

When replacing a water closet, measure from the face of the *finished* wall behind the water closet to the center of the closet flange. The closet flange is that fitting which is connected to the closet bend and that holds the water closet in place with brass bolts.

When roughing-in for a new installation, get this dimension information from the supplier and install the closet flange so that the center of the opening will be the correct distance from the *finished* back wall surface for the fixture you intend to provide.

The standard rough-in dimension is 12", and many fixtures are only available for this dimension.

SIZE AND SHAPE

Water closet bowls are available in different sizes and shapes — be sure to check manufacturers' specifications and catalogs to obtain the style appropriate for your job.

The shapes of closet bowls and tanks are varied and include the following examples:

ONE-PIECE CLOSET

The bowl and the tank of this closet are formed in one piece. This is the most expensive style of water closet available for the reasons discussed above.

LOW-CONSUMPTION CLOSETS

Water closets considered to be water savers are now truly miserly — a design now has to use less than 1.6 gpf to meet federal requirements. Manufacturers are continuing the development of water closets that will use still less water for a satisfactory flush. When they exist in satisfactory numbers, we may expect to see that the fixture standards and plumbing codes will recognize them.

ELONGATED BOWL CLOSET

This type vitreous china bowl has an extended front which protrudes forward about two inches farther than a regular bowl. It has been used primarily in commercial buildings and institutions, but it is gaining popularity in residential applications.

WALL-HUNG CLOSET

The wall-hung closet is mounted to a bracket in the wall that is called a carrier or carriage bracket. The waste material is ejected out the rear outlet of the bowl rather than a base outlet. This type of closet allows for better cleaning and sanitation in the vicinity of the fixture.

WALL-HUNG TANK

This type of closet is an older design. The tank is mounted on the wall and the closet bowl is fed through a large diameter chrome-plated flush ell.

CORNER-TANK CLOSET

This type of specialty closet incorporates a standard bowl and a triangular tank. It is intended to fit in a corner, but has limited availability.

HANDICAPPED WATER CLOSETS AND SEATS

Free access for handicapped persons requires providing fixtures especially to serve their needs. These fixtures require a different rough-in if wall hung, and, in some cases, require special seats or closet designs for floor mounted water closets. Special seats come equipped with grab bars, or grab bars are required in the closet stall.

PRISON WATER CLOSETS

Prison water closets are usually made from corrosion resisting metal. All connections, brackets, and flushometers are concealed in a large chase behind the wall. This chase is inaccessible from the prison cell.

SPECIALTY WATER CLOSETS

There are many other specialty closets, including combinations of the types above, such as an elongated wall-mounted closet. Some type of specialty type water closets include clinic sinks and flushing rim service sinks which are normally installed in hospitals and other medical care facilities. These fixtures use water in the 4.0 to 6.0 gpf range.

FLUSH VOLUME

Most water closets are made to flush with a smaller amount of water than was formerly required. Variations in bowl and tank design usually reflect the anticipated water saving. Historically, water closets from the earliest years of the twentieth century used 7 or 8 gpf. The industry kept improving the products so that a few years ago, a water-saving design was 3.5 to 4 gallons per flush; low consumption models use 1.6 to 2 gallons per flush! As earlier stated, federal legislation now requires a maximum of 1.6 gallons per flush for most applications.

MATERIALS AND FINISHES

Although water closets have traditionally been made of vitreous china, other materials are now also being used:

Figure 29-J – Flushing Cycle at Rest

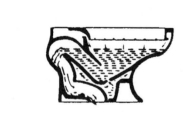

Figure 29-K – Flushing Cycle Beginning

1. Plastics such as polypropylene, PVC, and ABS

2. Cultured marble

3. Stainless steel or aluminum (for institutional uses)

The surface finish of water closets must be smooth and nonporous so that they can be easily cleaned and kept sanitary. These fixtures are available in a wide variety of colors to fit with any decorating scheme.

Figure 29-L – Flushing Cycle Full Siphon

FLUSHING CYCLE

The basic flushing cycle for the siphonic style water closet follows these steps:

Figure 29-M – Flushing Cycle Siphon Broken

1. The water in the bowl is at rest.

2. The flush begins, and as water rises in the trap, a fairly thin sheet of water seals off the downleg, preventing air from coming back from the discharge opening.

3. As slightly more water enters, the trap is filled with water and the siphon is established. The water coming from the flushing rim scours the entire bowl surface.

4. Soon the water supply stops and the siphon continues, emptying the bowl until

the siphon is broken by air entering the trap. The water level in the bowl is replenished by a slow after-filling feature built into the flushing mechanism. This flushing cycle is designed for a nine-second operation, even in gravity tank products.

FLUSHING DEVICES

The common flushing devices for water closets, as already mentioned, are flush tanks, flushometer valves, and flushometer tanks. Other special devices are available.

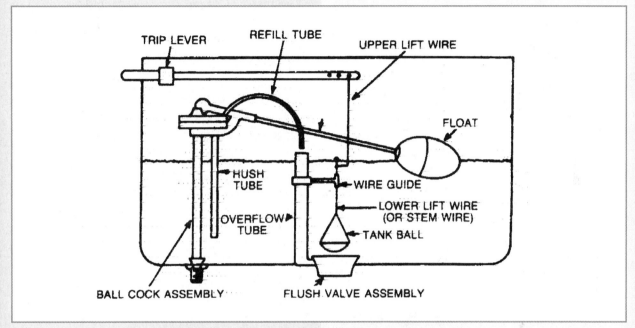

FIGURE 29-N – TYPICAL FLUSH TANK – LIFT WIRE MODEL

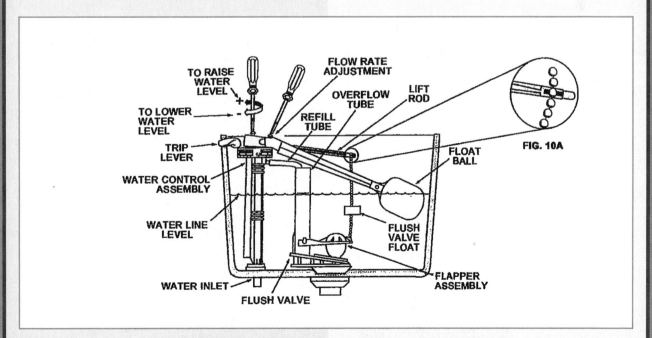

FIGURE 29-N(2) - TYPICAL FLUSH TANK - FLAPPER MODEL

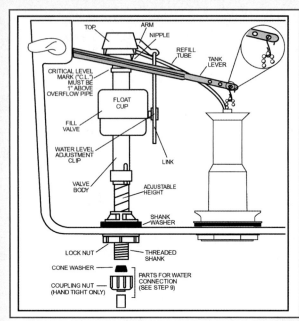

Figure 29-N(3) – Water Tower-2 piece WC American Standard

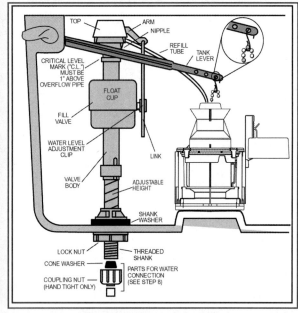

Figure 29-N(4) – Water Tower-1 piece WC American standard

GRAVITY-TYPE FLUSH TANK

A flush tank is a box-like vessel which contains devices which control the flow into and out of the tank. Parts in a gravity-type flush tank are shown in Figures 29-N and 29-N(2). Figure 29-N shows a lift wire model; 29-N(2) shows the version that uses a flapper and chain, rather than the lift wire type.

The flapper and chain model has been the normal trim in flush tanks for the last several years, but there are many tanks that use the lift wires. There are skilled repair plumbers that are on both sides of the question as to which is more reliable, but the parts costs of the flapper type is definitely less than buying all the items needed to replace the guide and wires in the first type.

The functions of the components in these more or less typical flush tanks are described below:

BALLCOCK OR FILL VALVE

The water control device (which should include an anti-siphon mechanism with a built-in siphon breaker) is designed to regulate the water height in the tank. It is operated by some type of float device or pressure sensing device which shuts off the water flow into the tank when the desired water level is reached. It is made of brass or plastic.

FLOAT ARM

This is the arm which connects the float device to the ballcock.

FLOAT

The float is a buoyant cylinder which is connected to the float arm and rises with the incoming water, applying pressure to the stem in the water control valve which closes the valve. Floats are made of styrofoam, copper, or other plastics. A saturated float will not rise properly, so the ballcock will not shut off the water entering the tank. Some floats and float arms are integral with the ballcock or fill valve.

OVERFLOW TUBE

This brass or plastic tube serves two purposes:

1. It provides an outlet for water to leave the tank if the ballcock fails to shut off.

2. It is the entrance line for the refill tube from the ballcock.

REFILL TUBE

This copper or plastic tube is attached to the ballcock and provides enough water to the bowl to replace the trap seal. Some new water closet designs do not incorporate a refill tube from the ballcock. The overflow is designed to replenish trap seal water.

The refill tube connects to the overflow tube through the use of a spring or clip fastener. The refill tube should not be allowed to extend into the overflow tube (for protection against backflow).

FLUSH VALVE

The flush valve is a brass or plastic circular opening through which the water leaves the tank and passes to the bowl. This opening is carefully machined to permit the tank ball to seat tightly against it.

TANK BALL OR FLAPPER

This rubber device seats on the flush valve when the flush is completed. There are at least two versions — one is essentially ball-shaped on the bottom, with a conical shape (which is connected to the lift wires) above; the other is hinged to the base of the overflow tube, and the closure member is raised by a chain to start the flush. Either of these products is hollow, so they float once they are raised from the flush valve rim.

LIFT WIRES (OR LIFT ASSEMBLY)

These brass or copper wires connect the tank ball to the arm of the flush trip lever. Flapper assemblies are operated by stainless steel or nylon chain which connect to the tank trip lever.

LIFT WIRE GUIDE

Where lift wires are used, this brass device is attached to the overflow tube and guides the lower lift wire to maintain the tank ball in alignment with the center of the flush valve opening.

TANK HANDLE

This plastic or brass device is the only external working part on the flush tank. It connects to the lift wires (or lift assembly) which lift the tank ball or flapper.

You may encounter variations on these part designs. As suggested above, flapper assemblies now use chains or nylon cords in lieu of liftwires.

IMPORTANT NOTE: LOW CONSUMPTION WATER CLOSETS ARE DESIGNED TO MAXIMIZE PERFORMANCE WITH MINIMUM WATER. IT IS IMPERATIVE THAT ORIGINAL DESIGN PARTS BE USED WHEN REPAIRS TO 1.6 GPF WATER CLOSETS ARE MADE. AN IMPROPERLY SELECTED PART (BALLCOCK OR FLAPPER ASSEMBLY) MAY CAUSE THE WATER CLOSET TO MALFUNCTION (IMPROPERLY FLUSH OR WASTE WATER). ALWAYS USE THE REPLACEMENT PARTS WHICH ARE RECOMMENDED BY THE MANUFACTURER.

Regardless of design or material, proper alignment is essential, or failures to reseat will be a frequent nuisance.

OPERATION OF THE FLUSH TANK

The flush tank works in this fashion:

1. When the handle is pushed down, the leverage of the handle pulls the flapper assembly (or lift wires and tank ball) upward. The tank ball must lift an inch to an inch-and-a-half, and the flapper must be past the 90° position (otherwise either will immediately drop onto the flush valve when the trip lever handle is released.

2. The water in the tank is thus released to flow through the flush valve and into the closet bowl. The water scours the flushing rim and starts the flushing action in the bowl.

3. The tank ball or flapper floats until the water level is almost down to the flush valve, when the moving water reseats the flapper (or tank ball) into the flush valve — thus resealing the tank.

4. As the water level in the tank goes down, the float drops, causing the ballcock to allow **new** water to enter the tank. This water starts to fill the tank once the flapper reseats itself on the flush valve seat.

5. As the ballcock fills the tank, it is also replacing the bowl water seal through the refill tube. (Not all water closet designs have a refill tube.)

6. The ballcock will shut off as soon as the water level is high enough in the tank.

FLUSHOMETER VALVES

The flushometer valve is another commonly used

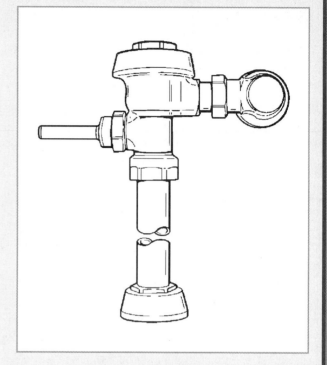

Figure 29-O – Sloan Royal Flushometer

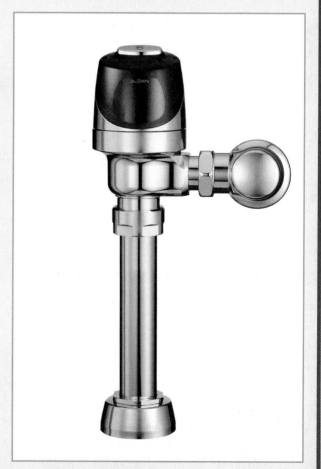

Figure 29-P – Sloan G2 Optima Plus Flushometer

Figure 29-Q – Sloan Royal Flushometer Installation Instructions

ROYAL I.I. — Rev. 3 (12/98)
Code No. 0816195

Royal® Flushometer
INSTALLATION INSTRUCTIONS FOR STANDARD EXPOSED WATER CLOSETS AND URINALS

NEW! Now featuring Sloan's exclusive
Dual Filtered Diaphragm ™

CSA® Certified   UPC® Listed by I.A.P.M.O

Made in the U.S.A.

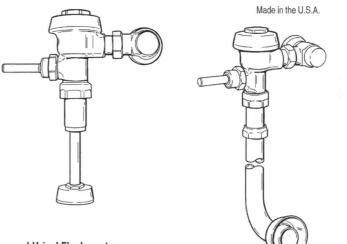

Exposed Closet Flushometer for 1-1/2" Top Spud
- MODEL 110/111
- MODEL 113
- MODEL 115
- MODEL 116

Exposed Service Sink Flushometer for 1-1/2" Top Spud
- MODEL 117

Exposed Urinal Flushometer for 1-1/4" Top Spud
- MODEL 180

Exposed Urinal Flushometer for 3/4" Top Spud
- MODEL 186

Exposed Closet Flushometer for 1-1/2" Back Spud
- MODEL 120
- MODEL 121
- MODEL 122

Installation of the Sloan Royal® Flushometer provides the quality, long life and water economy that makes Sloan Flushometers the most dependable ever. Sloan Flushometers provide outstanding water saving capabilities by precisely metering every flush. Royal Flushometers have been engineered to deliver a consistent, quiet flush, time after time. No internal adjustment of the Flushometer is required. Patented features provide unequaled performance on today's new generation of low consumption fixtures.

Royal Flushometers are designed for easy installation and maintenance and come complete with a metal oscillating ADA compliant Non-Hold-Open lever actuator, Bak-Chek® control stop with vandal resistant stop cap, adjustable tailpiece, vacuum breaker flush connection, spud coupling, sweat solder adapter kit, and wall and spud flanges.

The following instructions will serve as a guide when installing the Sloan Flushometer. As always, good safety practices and care are recommended when installing your new Flushometer. If further assistance is required, contact your nearest Sloan Representative office.

LIMITED WARRANTY

Sloan Valve Company warrants its Royal Flushometers to be made of first class materials, free from defects of material or workmanship under normal use and to perform the service for which they are intended in a thoroughly reliable and efficient manner when properly installed and serviced, for a period of three years (1 year for special finishes) from date of purchase. During this period, Sloan Valve Company will, at its option, repair or replace any part or parts which prove to be thus defective if returned to Sloan Valve Company, at customer's cost, and this shall be the sole remedy available under this warranty. No claims will be allowed for labor, transportation or other incidental costs. This warranty extends only to persons or organizations who purchase Sloan Valve Company's products directly from Sloan Valve Company for purpose of resale.

THERE ARE NO WARRANTIES WHICH EXTEND BEYOND THE DESCRIPTION ON THE FACE HEREOF. IN NO EVENT IS SLOAN VALVE COMPANY RESPONSIBLE FOR ANY CONSEQUENTIAL DAMAGES OF ANY MEASURE WHATSOEVER.

VALVE ROUGH-IN — Figure A

MODEL 110 — Water Saver, 3.5 gpf (13.2 Lpf)
MODEL 111 — Low Consumption, 1.6 gpf (6.0 Lpf)

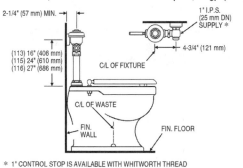

* 1" CONTROL STOP IS AVAILABLE WITH WHITWORTH THREAD

MODEL 113, 115 & 116 — Water Saver, 3.5 gpf (13.2 Lpf)
MODEL 113-1.6, 115-1.6 & 116-1.6 — Low Consumption, 1.6 gpf (6.0 Lpf)

* 1" CONTROL STOP IS AVAILABLE WITH WHITWORTH THREAD

MODEL 120, 121 & 122 — Water Saver, 3.5 gpf (13.2 Lpf)
MODEL 120-1.6, 121-1.6 & 122-1.6 — Low Consumption, 1.6 gpf (6.0 Lpf)

MODEL 180 — Standard, 3.5 gpf (13.2 Lpf)
MODEL 180-1.5 — Water Saver, 1.5 gpf (5.7 Lpf)
MODEL 180-1 — Low Consumption, 1 gpf (3.8 Lpf)

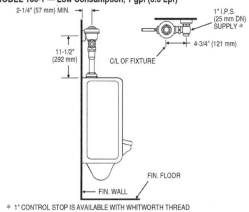

* 1" CONTROL STOP IS AVAILABLE WITH WHITWORTH THREAD

* 1" CONTROL STOP IS AVAILABLE WITH WHITWORTH THREAD

MODEL 186 — Water Saver, 1.5 gpf (5.7 Lpf)
MODEL 186-1 — Low Consumption, 1 gpf (3.8 Lpf)
MODEL 186-0.5 — 0.5 gpf (1.9 Lpf)

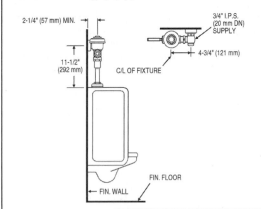

MODEL 117 — Standard, 6.5 gpf (24.6 Lpf)

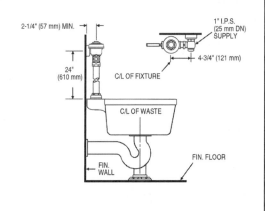

Figure A

NOTE: Water Closet Valves with "-2.4" Model Designation Deliver 2.4 gpf (9.0 Lpf)

PRIOR TO INSTALLATION

Before you install the Sloan Flushometer, be sure the items listed below are installed (see Figure A).

- Closet/urinal fixture
- Drain line
- Water supply line

Important:

- *ALL PLUMBING SHOULD BE INSTALLED IN ACCORDANCE WITH APPLICABLE CODES AND REGULATIONS.*
- *WATER SUPPLY LINES MUST BE SIZED TO PROVIDE AN ADEQUATE VOLUME OF WATER FOR EACH FIXTURE.*
- *FLUSH ALL WATER LINES PRIOR TO MAKING CONNECTIONS.*

The Sloan ROYAL® is designed to operate with 10 to 100 psi (69 to 689 kPa) of water pressure. THE MINIMUM PRESSURE REQUIRED TO THE VALVE IS DETERMINED BY THE TYPE OF FIXTURE SELECTED. Consult fixture manufacturer for minimum pressure requirements.

Most Low Consumption water closets (1.6 gpf/6 Lpf) require a minimum flowing pressure of 25 psi (172 kPa).

TOOLS REQUIRED FOR INSTALLATION

- Slotted screwdriver
- Sloan A-50 "Super-Wrench™" or smooth jawed spud wrench

INSTALLATION

If an existing Control Stop is being used in this installation, skip ahead to Step 4 for Vacuum Breaker installation.

Step 1 — Install Optional Sweat Solder Adapter (Figure 1)

Install the Sweat Solder Adapter only if your supply pipe does not have threaded ends. If your installation includes a supply line with a threaded iron pipe nipple, skip ahead to Step 2.

Measure distance from finished wall to centerline of Fixture Spud. Cut water supply pipe 1-1/4" (32 mm) shorter than this measurement. Chamfer O.D. and I.D. of water supply pipe.

Slide Sweat Solder Adapter onto water supply pipe until end of pipe rests against shoulder of Adapter. Sweat solder Adapter to water supply pipe.

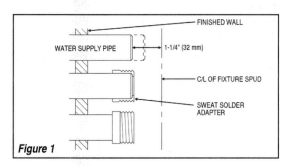

Figure 1

Step 2 — Install Wall Flange and Covering Tube (Figure 2)

Determine length of Covering Tube by measuring distance from finished wall to the first thread of Adapter (dimension "X" in Figure 2). Cut Covering Tube to this length.

Slide Covering Tube onto water supply pipe. Slide Wall Flange over Covering Tube until it rests against the finished wall.

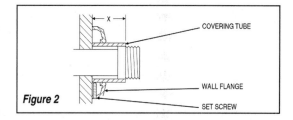

Figure 2

Step 3 — Install Control Stop (Figure 3)

Install the Sloan Bak-Chek® Control Stop to the water supply line with the outlet positioned as required. Tighten the Control Stop coupling with a wrench. DO NOT install the Vandal Resistant Stop Cap at this time. Secure the Wall Flange and the Covering Tube with the Set Screw. Tighten with a 1/16" hex wrench.

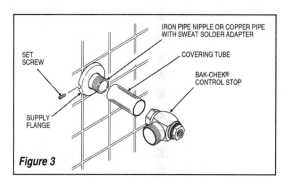

Figure 3

Step 4 — Install Vacuum Breaker Flush Connection (Figure 4)

Note: *Follow the instructions on the bag of your High Back Pressure Vacuum Breaker Kit included with your Flushometer: insert Baffle into groove of Vacuum Breaker Sack; insert Sack in Vacuum Breaker Tube; and place Friction Ring on top of Sack.*

Slide the Spud Coupling, Nylon Slip Gasket, Rubber Gasket and Spud Flange over the Vacuum Breaker Tube and insert tube into Fixture Spud. Tighten the Spud Coupling onto the Fixture Spud by hand.

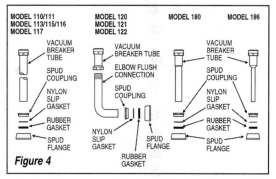

Figure 4

Step 5 — Install Flushometer (Figures 5A, 5B and 5C)

SLOAN ADJUSTABLE TAILPIECE (Figure 5A)

The Sloan Adjustable Tailpiece compensates for "off-center" roughing-in on the

IMPORTANT NOTES

With the exception of the Control Stop inlet, DO NOT USE pipe sealant or plumbing grease on any valve component or coupling!

Protect the chrome or special finish of Sloan Flushometers — DO NOT USE toothed tools to install or service these valves.

Use a Sloan A-50 "Super-Wrench™" or smooth jawed spud wrench to secure all couplings. Also see "Care and Cleaning" section of this manual.

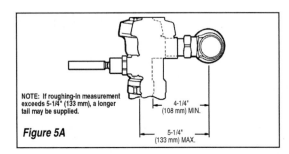

Figure 5A

job. Maximum adjustment is 1/2" (13 mm) IN or OUT from the standard 4-3/4" (121 mm) (centerline of Flushometer to centerline of Control Stop).

SLOAN FLUSHOMETER BODY (Figure 5B)

Lubricate the tailpiece O-ring with water. Insert Adjustable Tailpiece into Control Stop. Tighten the Tailpiece Coupling by hand.

Align the Flushometer directly above the Vacuum Breaker Flush Connection. Tighten the Vacuum Breaker Coupling by hand.

Align the Flushometer Body and securely tighten first the Tailpiece Coupling (1), then the Vacuum Breaker Coupling (2), and finally the Spud Coupling (3). Use a wrench to tighten these couplings in the order shown.

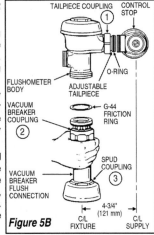

Figure 5B

SLOAN TRIPLE SEAL HANDLE ASSEMBLY (Figure 5C)

Sloan's triple-sealed Flushometer Handle is ADA-complaint.

Install the red A-31 Handle Gasket on the Handle Assembly. Insert the Handle Assembly into the Handle opening in the Flushometer Body. Securely tighten the Handle coupling with a wrench.

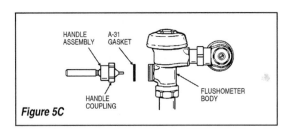

Figure 5C

Step 6 — Flush Out Supply Line (Figures 6A and 6B)

If your Control Stop has a Vandal Resistant Stop Cap in place, follow the Vandal Resistant Control Stop Cap Removal Instructions before proceeding.

Shut off Control Stop. Use the Sloan "SuperWrench™" to loosen the Outside Cover. Remove the Outside Cover and the Inside Cover. Lift out the Inside Parts Assembly (see Figure 6A).

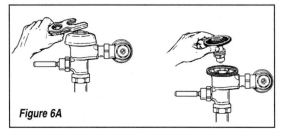

Figure 6A

Reinstall the Outside Cover. Tighten with wrench. Open the Control Stop. Turn on the water supply to flush any dirt, debris, or sediment from the line. Then shut off the Control Stop. Remove the Outside Cover.

Inspect the Inside Parts Assembly featuring our Dual Filtered Diaphragm™. The upper ring is the secondary filter and includes the metered bypass. It is kept in position by three (3) barbs that lock it into the top of the rubber diaphragm. The lower ring is the primary filter and slips into the lip on the lower side of the diaphragm. These two (2) plastic filtering rings should be in position on the upper and lower sides of the Permex® diaphragm. Insert the Inside Parts Assembly into the Flushometer Body. Install the Inside and Outside Covers. Tighten the Outside Cover with a wrench. Open the Control Stop and press the Handle to activate.

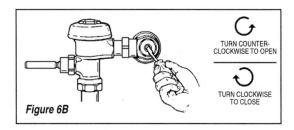

Figure 6B

Step 7 — Adjust Control Stop (Figure 7)

Adjust Control Stop to meet the flow rate required for the proper cleansing of the fixture. Open Control Stop COUNTERCLOCKWISE one full turn from the closed position. Activate Flushometer. Adjust Control Stop after each flush until the rate of flow delivered properly cleanses the fixture.

Important: *The Sloan Royal® Flushometer is engineered for quiet operation. Excessive water flow creates noise, while too little water flow may not satisfy the needs of the fixture. Proper adjustment is made when the plumbing fixture is cleansed after each flush without splashing water out from the lip AND a quiet flushing cycle is achieved.*

After adjustment, install the Vandal Resistant Control Stop Cap to the Control Stop.

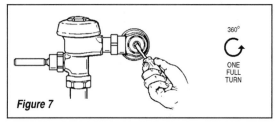

Figure 7

IMPORTANT NOTES
With the exception of the Control Stop inlet, DO NOT USE pipe sealant or plumbing grease on any valve component or coupling!
Protect the chrome or special finish of Sloan Flushometers — DO NOT USE toothed tools to install or service these valves.
Use a Sloan A-50 "Super-Wrench™" or smooth jawed spud wrench to secure all couplings. Also see "Care and Cleaning" section of this manual.

4

Step 8 — Install Vandal Resistant Stop Cap (Figure 8)

Thread the Plastic Sleeve onto the Stop Bonnet until it is snug (tighten only by hand; do not use pliers or a wrench).

Place the metal Control Stop Cap over the plastic Sleeve and use the palm of the hand to push or "pop" the Cap over the fingers of the Plastic Sleeve. The Cap should spin freely.

Important: *DO NOT install the Cap onto the Sleeve unless the Sleeve has been threaded onto the Control Stop Bonnet. If the Sleeve and Cap are assembled off of the Control Stop, the Sleeve WILL NOT come apart from the Cap.*

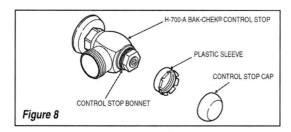
Figure 8

VANDAL RESISTANT CONTROL STOP CAP REMOVAL INSTRUCTIONS (Figure 9)

Use a large flat screwdriver as a lever to remove the Cap from the Control Stop. Insert the screwdriver blade between the bottom edge of the Cap and the flat surface of the Control Stop body as shown in Figure 9. Push the screwdriver handle straight back toward the wall to gently lift the Cap. If necessary, work the screwdriver around the diameter of the Cap until you can grasp the Cap and lift it completely off the Sleeve. The Sleeve should remain attached to the bonnet of the Control Stop.

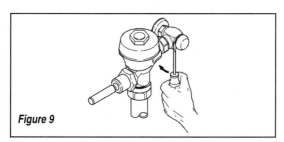
Figure 9

CARE AND CLEANING OF CHROME AND SPECIAL FINISHES

DO NOT use abrasive or chemical cleaners to clean Flushometers that may dull the luster and attack the chrome or special decorative finishes. Use ONLY soap and water, and then wipe dry with a clean cloth or towel.

While cleaning the bathroom tile, protect the Flushometer from any splattering of cleaner. Acids and cleaning fluids can discolor or remove chrome plating.

TROUBLESHOOTING GUIDE

I. **Flushometer does not function (no flush).**
   A. Control Stop or Main Valve is Closed. Open Control Stop or Main Valve.
   B. Handle Assembly is damaged. Replace Handle (B-73-A) or install Handle Repair Kit (B-51-A).
   C. Relief Valve is damaged. Replace Royal Performance Kit.

II. **Volume of water is not sufficient to siphon fixture.**
   A. Control Stop is not open wide enough. Adjust Control Stop for desired delivery of water volume.
   B. Dual Filtered Diaphragm Assembly is damaged. Replace Royal Performance Kit.
   C. Incorrect Dual Filtered Diaphragm Assembly is installed in Flushometer; for instance, Urinal assembly inside a Closet Flushometer, or Low Consumption assembly inside a higher consumption fixture. Determine the flush volume required by the fixture and replace Royal Performance Kit. Use valve label and markings on fixture for reference.
   D. Water supply volume or pressure is inadequate. If no gauges are available to properly measure supply pressure or volume of water at the Flushometer, then remove the Relief Valve from the Dual Filtered Diaphragm Assembly, reassemble the Flushometer and completely open the Control Stop.
   - If the fixture siphons, more water volume is required. Install a higher flushing volume Royal Performance Kit. **IMPORTANT — Laws and Regulations requiring Low Consumption Fixtures (1.6 gpf/6 Lpf Water Closets and 1.0 gpf/3.8 Lpf Urinals) prohibit the use of higher flushing volumes.**
   - If the fixture does not siphon or if a Low Consumption flush is required, steps must be taken to increase the water supply pressure and/or volume. Contact the fixture manufacturer for minimum water supply requirements of the fixture.

III. **Length of flush is too short (short flush).**
   A. Dual Filtered Diaphragm Assembly is damaged. Replace Royal Performance Kit.
   B. Handle Assembly is damaged. Replace Handle (B-73-A) or install Handle Repair Kit (B-51-A).
   C. Incorrect Dual Filtered Diaphragm Assembly is installed in Flushometer; for instance, Urinal assembly inside a Closet Flushometer, or Low Consumption assembly inside a higher consumption fixture. Determine the flush volume required by the fixture and replace Royal Performance Kit. Use valve label and markings on fixture for reference.

IV. **Length of flush is too long (long flush) or continuous.**
   A. Metering by-pass hole in Diaphragm is clogged. Remove the Dual Filtered Diaphragm Assembly. Remove the Primary and Secondary Filter Rings from the Diaphragm and wash under running water. Replace Royal Performance Kit if cleaning does not correct the problem.
   B. Diaphragm or Relief Valve is damaged. Replace Royal Performance Kit.
   C. Incorrect Dual Filtered Diaphragm Assembly is installed in Flushometer; for instance, Closet assembly inside a Urinal Flushometer, or Water Saver assembly inside a Low Consumption Flushometer. Determine the flush volume required by the fixture and replace the Royal Performance Kit. Use valve label and markings on fixture for reference.
   D. Inside Cover is damaged. Replace Inside Cover (A-71).
   E. Supply line water pressure has dropped and is not sufficient to close the valve. Close Control Stop until pressure is restored.

V. **Chattering noise is heard during flush.**
   A. Inside Cover is damaged. Replace Inside Cover (A-71).

VI. **Handle Leaks.**
   A. Handle Seal or Assembly is damaged. Replace Handle (B-73-A) or install Handle Repair Kit (B-51-A).

Refer to the Royal Flushometer Maintenance Guide for additional Troubleshooting and Repair Part information. If further assistance is required, please contact the Sloan Valve Company Installation Engineering Department at 800/982-5839.

IMPORTANT NOTES

With the exception of the Control Stop inlet, DO NOT USE pipe sealant or plumbing grease on any valve component or coupling!

Protect the chrome or special finish of Sloan Flushometers — DO NOT USE toothed tools to install or service these valves.

Use a Sloan A-50 "Super-Wrench™" or smooth jawed spud wrench to secure all couplings. Also see "Care and Cleaning" section of this manual.

PARTS LIST

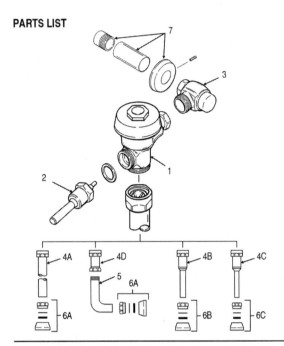

Item No.	Part No.	Description
1	†	Valve Assembly
2	B-73-A	ADA Compliant Handle Assembly
3	H-700-A	Bak-Chek® Control Stop
4A	V-600-AA	1-1/2" (38 mm) x 9" (229 mm) Vacuum Breaker Assembly ‡
4B	V-600-AA	1-1/4" (32 mm) x 9" (229 mm) Vacuum Breaker Assembly
4C	V-600-AA	3/4" (19 mm) x 9" (229 mm) Vacuum Breaker Assembly
4D	V-600-A	Vacuum Breaker Assembly
5	F-109	1-1/2" (38 mm) Elbow Flush Connection ‡
6A	F-5-A	1-1/2" (38 mm) Spud Coupling Assembly
6B	F-5-A	1-1/4" (32 mm) Spud Coupling Assembly
6C	F-5-A	3/4" (19 mm) Spud Coupling Assembly
7	H-633-AA	1" (25 mm) Sweat Solder Kit with Cast Set Screw Flange
	H-636-AA	3/4" (19 mm) Sweat Solder Kit with Cast Set Screw Flange

† Part number varies with valve model variation; consult factory.

‡ Length varies with valve model variation; consult factory.

For a complete listing of Flushometer Valve components and Repair Kits, see the Royal Maintenance Guide or consult your nearest Plumbing Wholesaler.

For optimum water conservation and Flushometer performance, use *only* Genuine Sloan Parts.

Manufactured in the U.S.A. by Sloan Valve Company under one or more of the following patents: U.S. Pats. 5,295,655; 5,505,427; 5,542,718; 5,558,120; 5,564,460; 5,649,686; 5,730,415. Other Pats. Pending. BAK-CHEK®, PARA-FLO®, PERMEX®, TURBO-FLO®, DUAL FILTERED DIAPHRAGM™.

Sloan's New Dual Filtered Diaphragm™ Assembly has been an exclusive feature of our Royal Flushometer since January 1998. Our design eliminates valve "run on" by capturing dirt and debris particles in a series of patented filters before they ever reach the metering bypass hole. Our Dual Filtered Diaphragm saves water and reduces flushometer valve maintenance.

Our Dual Filtered Diaphragm Assembly is factory assembled to assure accurate flush delivery and is sold only in Royal Performance Kits. These Royal Performance Kits also include our B-51-A Handle Repair Kit with factory lubricated Triple Seal and our V-651-A High Back Pressure Vacuum Breaker Repair Kit. All of these components have been engineered to meet the performance demands of today's plumbing systems and are exclusive to the Royal Flushometer, the standard for use with Low Consumption fixtures.

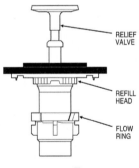

Royal Performance™ Kits §

Royal Performance Kits can be used to upgrade older Royal, Regal and similar design Diaphragm Flushometer Valves. To identify the flush volume of the Dual Filtered Diaphragm Assembly, look at the shape and color of the relief valve, refill head, and flow ring(s).

Kit Number	Flush Volume	Fixture Application	Relief Valve	Refill Head ‡	Flow Ring(s)
A-1101-A	1.6 gpf/6.0 Lpf	Water Closets, Low Consumption	Green	Gray	Smooth
A-1102-A	3.5 gpf/13.2 Lpf	Water Closets, Water Saver	White	Gray	Smooth
A-1103-A	2.4 gpf/9.0 Lpf	Water Closets, 9 Liter European	Blue	Gray	Smooth
A-1106-A	0.5 gpf/1.9 Lpf	Urinals, Wash Down	Green	Black	Use two (2): Smooth & Slotted
A-1107-A	1.0 gpf/3.8 Lpf	Urinals, Low Consumption	Green	Black	Slotted
A-1108-A	1.5 gpf/5.7 Lpf	Urinals, Water Saver	Black	Black	Smooth

§ Kit includes:
 A — Dual Filtered Diaphragm Assembly
 B — Handle Repair Kit with Triple Seal Packing
 C — High Back Pressure Vacuum Breaker Repair Kit
 D — Tailpiece "O" Ring

‡ Water Closet Refill Heads (gray) have larger slots than Urinal Refill Heads (black).

SPECIAL APPLICATIONS

For old style blowout urinals that require 3.5 gpf (13.2 Lpf), use Kit No. A-1102-A. For wash down urinals that require 1.0 gpf (3.8 Lpf), use Kit No. A-1107-A. For older water closets that require 4.5 gpf (17.0 Lpf), use 3.5 gpf flush kit No. A-1102-A and remove the Flow Ring.

For Service Sinks that require 6.5 gpf (24.6 Lpf), use Sloan Repair Kit No. A-36-A and remove the Flow Ring before installing.

NOTICE: The information contained in this document is subject to change without notice.

SLOAN VALVE COMPANY • 10500 Seymour Avenue • Franklin Park, IL 60131 • Ph: 800/982-5839 • Fax: 847/671-4380

http://www.sloanvalve.com
Printed in U.S.A. 12-98

device to control the flow of water into a bowl to discharge fecal and liquid waste material.

The flushometer valve produces a quick scouring flush rate. The proper operation of this device is dependent upon sufficient water volume and pressure. Consult the manufacturer's specifications for minimum working pressure and pipe size to supply the valves.

There are two types of flushometer valves: piston and diaphragm.

The basic working parts of a diaphragm flushometer are the following:

HANDLE

The device which upsets the plunger and opens the small valve in the center of the diaphragm.

PLUNGER

The plunger is connected to the diaphragm center valve.

DIAPHRAGM

This device is the heart of the flushometer valve. It is a flexible barrier between the low pressure and high pressure regions in the valve.

VACUUM BREAKER AND TAILPIECE

This device prevents back-siphonage from the fixture into the water supply.

FLUSHOMETER VALVE OPERATION

The diaphragm flushometer valve works as follows:

1. When the handle is activated, a small center valve in the diaphragm is opened, dumping the water above the diaphragm down the flush tailpiece.

2. A low pressure region is thus formed above the diaphragm. The diaphragm is pushed upward off its seat by supply pressure in the region below the diaphragm, allowing water to enter the fixture, and the small valve in the diaphragm closes if the handle has been released.

3. A small orifice in the diaphragm allows the pressures above and below the diaphragm to equalize after a few seconds, which forces the diaphragm to reseat and stop the water flow.

4. This total cycle lasts about ten seconds in time.

The piston flushometer valve type works on the same principle as the diaphragm type.

The flushometer valve is the most-often used flushing device for commercial water closets and urinals because the fixture is ready for the next user sooner than other flushing types. They are also easily maintained and are vandalism resistant.

Battery-operated flushometer mechanisms are commonly utilized in high-traffic areas or where some fixture users will not operate the valve.

FLUSHOMETER TANK

A third closet flushing device which has gained popularity in recent years is the flushometer tank. This tank, when used with a mating bowl designed for use with the product produces highly reliable flushing operation within the federal limit guidelines.

The product works as follows:

1. As water enters the unit, it must first pass through the backflow prevention devices, a pressure reducing device, and an air aspirator.

2. The aspirator draws air in with the water

which later becomes compressed in the top of the tank as the tank fills.

3. When the pressure in the tank is equal to the line pressure, the tank shuts off.

4. Activation of the flush button releases the *turbo charged* tank water through the bowl at approximately 35 gallons per minute.

Other air assisted flushing devices are becoming more prevalent in today's markets. These devices use compressed air to accelerate the siphonic flushing in the bowl, thereby using less water.

Consult manufacturer's data for additional information.

PUMP DEVICES

Some 1.6 gpf water closets of the dual flush design utilize a hydraulic pump in the tank to operate the fixture. The tank of this fixture is refilled with water using a ballcock. However, the hydraulic pump is used rather than a flush valve and flapper assembly to control the flow through the water closet bowl.

This technology will likely be the more predominantly applied system for future water closet designs, for many reasons:

1. A single pump device can be easily matched to the hydraulic demands of the bowl, with the placement of the proper electronic microchip in the pump control assembly.

2. With the elimination of the flapper, there is no leakage from the tank to the bowl.

3. Great water savings are realized with dual flush selection.

WATER CLOSET ACCESSORIES

Special accessories for water closets include the following:

CLOSET BOLT

Corrosion-resisting closet bolts are used to attach the closet to the floor flange or wall carrier.

WAX RING, PUTTY, GASKET, OR CLOSET-TO-DRAIN CONNECTOR

These are alternate materials used as a seal between the floor-flange and the china fixture to assure a water- and gas-tight connection. The newest of these devices, the closet-to-drain connector was developed to fill the need for a better seal where the water closet affixes to the closet flange. The connector is comprised of a sleeve which has o-rings on its outer perimeter and a flexible membrane on its inlet. The connector is inserted into the open drain (once adjusted to the proper pipe diameter by moving the outer perimeter o-ring to the correct position on the connector) and the water closet outlet is nested into the connector inlet. Once seated on the flange, the assembly is capable of withstanding a 10 foot head pressure test.

CLOSET SEAT

The closet seat can be made of laminated wood, molded plastic, or other special manufacture. It bolts to the bowl of the closet. It is usually selected in the same color and rim contour as the bowl.

Closet seats are available with or without covers and with closed or open fronts.

The standard closet seat for commercial use is elongated, open front, without cover. The standard closet seat for residential use is standard size, closed front, with a cover.

Seats for use on accessible fixtures are presently manufactured with a higher seat and with grab bars for use by handicapped or elderly people.

Special hinge assemblies and mounting devices

are also available.

Some *high tech* water closet seats have special

Figure 29-R– Kohler Bardon™ Superior Urinal

features such as bidet sprays, heaters, and even blood pressure monitors!

TANK LOCKS

The tank lock is a device used to lock the lid down on commercial, tank-type closets. This type of vandal-proofing device prevents tampering with the internal parts of the water closet.

URINAL

The urinal is a fixture which serves only to receive and remove urine.

Formerly confined to use in commercial or institutional facilities, residential urinals are now being marketed. Most codes now require that a urinal has a liquid seal integral trap in the fixture. There are urinals made for both male and female usage, but most installations are for males.
Urinals are available in colors as well as white.

The most common materials used to make urinals are vitreous china, enameled cast-iron, stainless steel, and plastic.

Urinals are designed in a variety of flushing and mounting methods:

PEDESTAL

This type of urinal is floor-mounted and is flushed by a driving jet of water.

It should be noted that pedestal urinals are not made anymore. Code prohibitions, concern for conservation, and cleaning considerations now favor wall hung products.

STALL, WASHDOWN

This fixture washes the large stall wall and dilutes the urine in the trap. Some localities require all urinals to be stall type, and other areas prohibit them.

WALL HUNG

The wall hung urinal is mounted directly on the wall in order to facilitate cleaning the wall and floor area around the fixture. The trapway is either a washdown, siphon jet, or blowout design. It may or may not have privacy shields.

CIRCULAR, WASHDOWN

This type of urinal is circular in shape with a circular stream tower used to wash out the waste water.

TROUGH URINAL

The trough urinal is an extended collector, floor or wall-mounted, which can be used by more than one person at a time. Most codes prohibit trough urinals, except for temporary facilities.

METHODS OS FLUSHING URINALS

The standard method of flushing a urinal is through the use of a flushometer valve, but groups of urinals have been arranged to be flushed with tanks. The tank is supplied with water through a small pipe, and the tank contents are discharged when the water level in the tank is at the highest point. This system has been used in primary and secondary schools, because a significant number of students will not operate flushometer valves. Such tanks should be fitted with timers and solenoids to prevent water waste during periods of non-use of the building. Most manufacturers have discontinued tanks due to water conservation requirements.

Battery-operated flushometers have largely replaced such tank systems.

WATERLESS URINALS

One of the newest offerings to the fixture industry is the waterless urinal. Developed over 100 years ago, the modern version of this product uses a special trap which is filled with a liquid which is lighter than water (or urine). The urine travels down the side walls of the urinal, through the liquid, and safely into the drainline.

This fixture requires scheduled maintenance and the liquid does require replenishment, based upon traffic through the restroom.

Each fixture set can save up to 45,000 gallons of water per year in very high-traffic toilet rooms.

F-1000

WATERFREE URINAL
VITREOUS CHINA

FALCON WATERFREE TECHNOLOGIES

Falcon waterfree urinals reduce water and sewer costs, maintenance and repair bills, and create more hygienic, odor-free restrooms. A patented, sealed cartridge eliminates the need for water, typically conserving 40,000 gallons per unit per year. Purchasing and installing Falcon waterfree urinals is less expensive than manual and automatic flush units because flush valves and associated piping are not required. Maintenance costs and vandalism problems associated with flush valves are also eliminated.

FEATURES

- Touch-free operation
- Uses no water
- Mechanical-free design
- Patented, sealed locking cartridge
- Smooth, non-porous surfaces
- ADA compliant

BENEFITS

- Improved hygiene and safety
- Reduced water and sewer costs
- No costly flush valve or sensor repairs
- Odor-free, vandal proof
- Minimal care and easy cleaning

Patented, sealed cartridge uses a biodegradable sealant liquid to control odors.

     ASME A112.19.2M

Figure 29-S– Falcon F-1000 Waterfree Urinal

F-1000
WATERFREE URINAL
VITREOUS CHINA

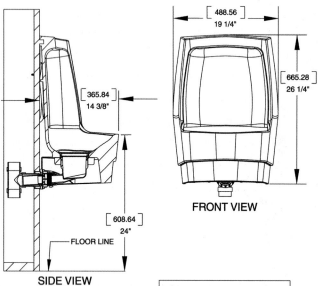

SIDE VIEW

FRONT VIEW

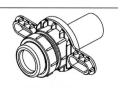

UNI-COUPLER

THIS ADAPTIVE COUPLER DESIGN ENABLES THE INSTALLER TO CONNECT THE UNI-COUPLER DIRECTLY ONTO THE SPUD FLANGE OR IN-WALL NIPPLE.

FALCON HANGER BRACKETS

EXISTING SPUD FLANGE

SPECIALLY DESIGNED FALCON HANGER BRACKETS PROVIDE ADDED SWING RADIUS FOR EASE OF INSTALLATION

These dimensions are for informational purposes only. For specific rough-in dimensions, please refer to specification sheets available on our web site or by contacting us directly.

INCLUDES

- Fixture—available in white and colors
- H1 Cartridge housing (pre-installed)
- 1 Cartridge kit*
- 2 Wall hangers with anchors
- Uni-coupler
- Instruction sheet

*Additional cartridges sold separately

COMPLIANCE CERTIFICATIONS:
Meets ANSI/ASME A112.19.2.M-1998 and A117.1 (section 605.2) for Vitreous China Fixtures. In compliance with IAPMO 1GC 161-2000 and ANSI Z124.9-94.

 MEETS THE AMERICAN DISABILITIES ACT GUIDELINES AND ANSI A117.1 ACCESSIBLE AND USABLE BUILDINGS AND FACILITIES - CHECK LOCAL CODES.

Falcon Waterfree Technologies, LLC
1593 Galbraith Avenue SE
Grand Rapids, MI 49546
Tel: 866.275.3718 (toll-free)
Fax: 616.954.3579
E-mail: info@falconwaterfree.com

www.falconwaterfree.com

Specifications subject to change.
© Copyright 2005 Falcon Waterfree Technologies, LLC. All rights reserved.   003 (2/05)

Figure 29-S– Falcon F-1000 Waterfree Urinal (continued)

BIDET

A bidet is a vitreous or semi-vitreous fixture used to wash the perineal parts of the body for hygienic purposes. It can also be used as a foot bath. The word bidet is pronounced be-day. This fixture closely resembles a water-closet in design, color options, and height.

A bidet fixture fitting is either comprised of an air gap designed fitting or a submerged sprayer mechanism, which is located in the base of the bidet bowl compartment. If a submerged sprayer is used, a vacuum breaker assembly is required. Where this vacuum breaker was once required to be located 6" above the flood level

Figure 29-T – Kohler Portrait® Vitreous China Bidet

rim of the bidet, improvements in vacuum breaker designs now permit units to be mounted at 1" above the rim (if the vacuum breaker is approved and listed for this installation).

A bidet is equipped with a hot and cold water control which regulates the temperature and pressure of the spray device, located in the base or on the rim of the bidet bowl. The bidet is also equipped with a pop-up waste plug to retain or drain the water after use. Upon completion of use, the bidet is drained and flushed through the flushing rim.

SINKS

The following definition of sink is reprinted from the *Plumbing Dictionary, Fifth Edition,* with permission from the American Society of Sanitary Engineering.

1. *a shallow fixture that is commonly used in the kitchen; in connection with the preparation of food, for laboratory purposes, and for certain industrial processes. There are many types of special sinks, the purpose of which is indicated by the name prefixed before the word sink, such as slop sink, vegetable sink, etc.*

2. *a stationary basin or a cabinet connected with a drain and usually a water supply for washing and drainage.*

Of course, not all sinks are shallow, as we shall see in this section.

Sink is a term which is often confused with lavatory: a **lavatory** is a personal washing basin; a sink is a functional work fixture.

Different types of sink uses include the following examples:

KITCHEN SINKS, RESIDENTIAL

Usually made of stainless steel, enameled cast-iron, enameled steel composites, or plastic, these sinks are formed in one, two, or three compartments and are available in colors (if enameled or plastic types are used).

SERVICE SINKS (ALSO CALLED SLOP SINK OR JANITOR SINK)

The service sink is used for janitorial applications. Service sinks are usually heavy, white, enameled cast-iron, vitreous china, or plastic type with a 2" or 3" drain. Service sinks usually have a deep bowl to accommodate the filling of a bucket. One variation is a floor-mounted plastic composition unit which is called a mop sink.

BEAUTY OR SHAMPOO SINKS

This special class of sink is designed with special rim features to facilitate hair washing. When provided with spray hoses (nearly always), a vacuum breaker is required at the point of connection of the hose.

BAR SINK

The bar sink is usually a smaller variation of the kitchen type sink with a 2" or 3½" waste outlet. A gooseneck faucet is used to provide clearance for cleaning glasses and filling pitchers.

SCULLERY SINK

This sink is for commercial food preparation. The walls, legs, and bowls are made of stainless steel. Stainless drainboards used for food preparation are on one or both sides of the sink. A deep dish compartment used mainly for a food waste grinder may be built into this sink. Older scullery sinks were made of enameled cast-iron.

CHEMICAL SINK

Made of porcelain enamel, cast-iron, plastic, stainless steel, or other chemical-resisting materials, the chemical sink is usually very large with continually circulating water to prevent chemicals from remaining in the sink for a prolonged time. Water is circulated to dilute the chemicals to protect the drain lines.

LAUNDRY SINK

This is a deep-bowl single- or double-compartment sink used mainly to soak or dye clothing. The laundry sink, usually called a laundry tray or laundry tub, was originally manufactured from soapstone, a heavy water-resistant cement mixture formed into a deep compartment; most such products now are made of plastic.

MEDICAL SINK

Specialty sinks for the medical profession are designed according to use. For example, plaster casting sinks are available for use in mixing and applying casts. Special strainers and interceptors are available to prevent plaster from entering the drainage piping.

SINK MOUNTING

Originally, sinks that mount on counter tops were installed with a separate rim. Sink manufacturers then developed designs wherein the functions of the mounting rim were accomplished with an extension of the sink edge. These sinks are called self-rimming, since they do not need a separate item to attach to the counter top. In either case, a series of clamping devices holds the sink firmly in place in the counter top; caulking is placed under the rim to keep water from leaking around the sink into the cabinet below. Undermount sinks are also becoming very popular. Kitchen sinks with tiling edges are also available.

SINK FAUCETS

Faucets are available for the different sink types in many variations. The proper faucet must be selected for the use intended. For example, a single faucet to serve the needs of a three-compartment scullery sink must have a swing spout long enough to reach each compartment. When sinks are spec-

Figure 29-U – Kohler Bon Vivant® Self-Rimming Kitchen Sink

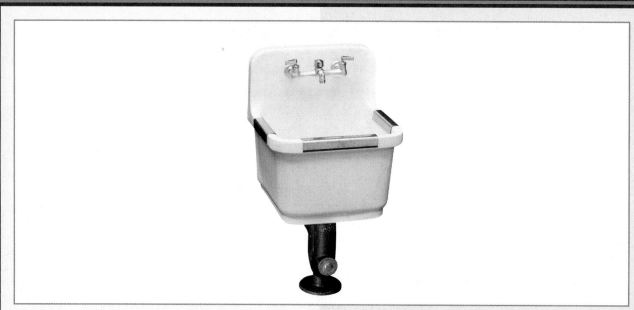

Figure 29-V – Kohler Sudbury™ Service Sink

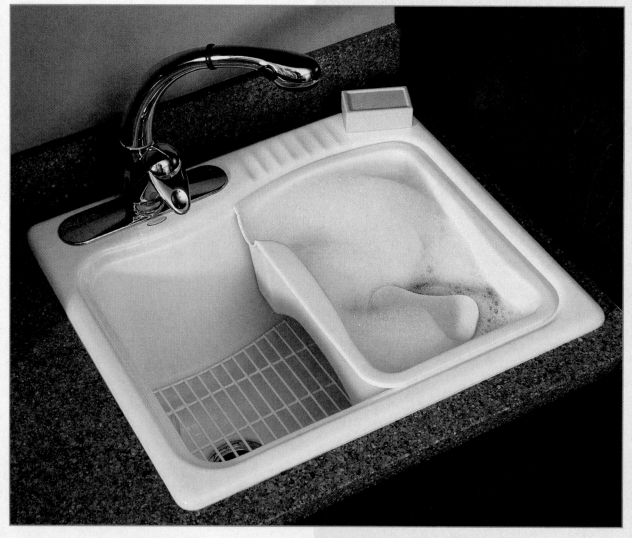

Figure 29-W — Kohler River Falls™ Self-Rimming Laundry Sink

ified, you must indicate the number of holes (if any) in the surface to accommodate faucet(s), air gap fittings, instantaneous heaters, soap dispensers, or side sprays.

SINK MATERIALS

Many materials are used to manufacture sinks. With the increase in luxury or design fixtures, the choice of material is often ruled by the style of the sink to be installed. Materials include these examples:

ENAMELED STEEL/COMPOSITE

Enameled on rough grained steel with a composite backing, this material gives the appearance of an enameled-iron fixture, but with great durability and reduced weight.

ENAMELED CAST-IRON

Enameled on rough grained cast-iron, this material is heavy and very durable.

STAINLESS STEEL

Stainless steels are alloys of nickel and iron which are light-weight, strong, impact resisting, and very easily maintained.

PLASTIC

Plastic is a material easily formed into sink shapes. Traditionally, this material was used mainly for mobile home and chemical sinks, but now it is used for laundry, kitchen, and bar sinks as well. Stringent standards have been written to assure durability of these products to be equivalent to other traditionally used materials. Some counter tops and sinks of the same plastic material are fabricated and joined with such precision that the total assembly appears to be homogeneous.

VITREOUS CHINA

Older type service sinks were made of china. Care must be taken to protect this sink from chipping or cracking. A dropped glass may shatter the china in unusual circumstances, but normally this material is very rugged.

GENERAL CONSIDERATIONS

When ordering a sink for a specific installation, certain items must be considered. If the sink does not fit the space and use for which it was designed, extra time and cost must be expended to force a fit. This usually means that no one, including the customer or your employer, is really happy with the job.

The placement and number of faucet and accessory holes must be ascertained before installation. Many new sinks have holes for accessories such as sprayers and soap dispensers. It is also possible to install single-handle or double-handle faucets. The center-to-center dimension of the faucet elements must be taken into consideration also.

The size, number, and location of drain openings must be planned for ahead of time. One, two, and three compartment sinks are available. Some sinks have a special compartment for the food waste grinder. Bar sinks may need a different size drain fitting than a standard kitchen sink. Local codes must be consulted for proper installation.

You must find out what the sink will be used for in order to install the appropriate material, size, and accessories for the sink. It may seem reasonable to sell something you have in stock because you have it, but if it is not appropriate to the job you may end up with many callbacks.

There is a wide price range available in sinks. Consultation with your customer is necessary to arrive at a fixture appropriate for the job at a cost the customer agrees to.

There are many possible faucets to go with each sink. Again, consultation with the customer is necessary before installation is started. Trips to a supply house to pick up a different faucet are expensive and wasteful.

LAVATORY

The following definition of lavatory is reprinted from the Plumbing Dictionary, Fifth Edition, with permission from the American Society of Sanitary Engineering.

1. a basin or vessel for washing. 2. a plumbing fixture, as above, especially placed for use in personal hygiene and not principally for laundry purposes and not ever for food preparation or utensils in food services. 3. a fixture designed for the washing of the hands and face.

Lavatories come in different styles and colors to suit a variety of tastes. Different types are described below:

WALL-HUNG LAVATORY

This lavatory hangs on a metal bracket connected to the wall or on a concealed arm carrier. The rim height is normally 31" from the finish floor, although some installations may be higher for accessible use or for convenience.

This type is the most popular, and among the least expensive.

VANITY LAVATORY

Vanity actually refers to the cabinet under the counter top into which the lavatory fits. The cabinet may be used with a molded stone, vitreous china, marble, or plasticized bowl. Vanity cabinets, as well as lavatories and sinks, must be selected to fit the available space. The measurement should be made on the top, as it usually extends beyond the vanity cabinet by a small amount. Vanity lavs are available in various shapes. See Figure 29-X.

SELF-RIMMING OR UNDERMOUNT LAVATORY

These fixtures mount on top of or beneath the counter top, respectively. They are available in many shapes and styles to accommodate the many types of bathrooms, wash rooms, and other locations where lavatories are installed.

HANDICAPPED (OR WHEELCHAIR) LAVATORY

Precisely speaking, the handicapped lavatory is a function of its installation, not its design as a fixture. There are requirements for the maximum drain dimension to the wall (to allow wheelchair movement), for padding the piping under the lavatory, and for a faucet mechanism that is easy to operate and that does not require closing the hand in a tight grip. Electronic faucets are preferable in these installations.

The padding serves two purposes: because many wheelchair-bound persons have no sensations of feeling in their legs - the padding protects against bruising and from burns (from hot piping). See Figure 29-Y.

The padding is normally colored white, with the material opaque, but it is also available in transparent material. This version was developed because the opaque type has been used as a hiding place for contraband materials.

PEDESTAL LAVATORY

An older style bowl which is mounted on an enameled-iron or china pedestal, this lavatory has regained popularity in recent years.

MEDICAL LAVATORY

The medical lavatory has special water and cleansing soap controls. Automatic faucets are available to eliminate the need for medical personnel to touch the faucet after hands have been washed. It is usually deep-welled for soaking.

Figure 29-X – Kohler Pennington™ Countertop Lavatory

DENTAL LAVATORY

This is a small, wall-hung lavatory, intended for serving a person scrubbing his or her teeth.

MATERIALS OF CONSTRUCTION

Lavatories are made of materials similar to those used in the manufacture of sinks. Marble and cultured marble are also available.

The same considerations as for sink trim must be made for lavatories. For example, most faucet holes molded or cut in a lavatory are at 4" center-to-center measurements. If an 8" faucet is desired, the fixture must be ordered with an 8" center drilling.

DRINKING FOUNTAIN AND CUSPIDORS

A drinking fountain is a water fountain device that combines a stream for drinking and a drain opening for removal of waste. It is made of enameled cast-iron, vitreous china, brass, galvanized, or stainless steel.

The nozzle directs a stream of tap water upward in an arc for ease in drinking without requiring the use of a glass or cup. The nozzle is a special design which provides for air gap protection, while minimizing germ transmission with the protected bubbler.

Cuspidors are special fixtures which are installed in conjunction with drinking fountains in some institutional buildings. The cuspidor is for expectoration, and therefore has flowing water, without a nozzle for drinking.

Figure 29-Y – Plumberex Trap Gear™ Undersink Protector

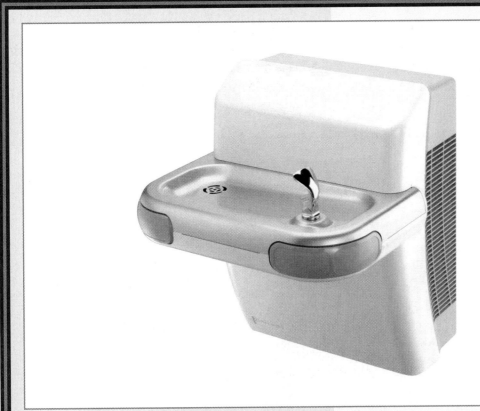

Figure 29-Z – Halsey Taylor Contour™ Barrier-Free Fountain With Back Panel

WATER COOLERS

Water coolers differ from drinking fountains in that they cool the tap water supply with mechanical refrigeration. The tops of these devices are made of vitreous china, brass, or stainless steel, and the cabinets are usually steel. The casings of newer models are made of patterned design stainless steel which is easily maintained.

GENERAL

Characteristics of sinks, lavatories, and fountains point out the progress made in the development of fixtures. Modern fixtures are smooth, colorful, and easy to care for with functional surfaces made to drain properly to the drain outlet. Fixtures are now available in a variety of styles and sizes to fit many size requirements and applications. Installations are now easier and quicker as the fixtures are lighter with assembly hardware simple to operate.

RELATED TRIM

When installing sinks, lavatories, and drinking fountains, there are related trim items used for the drain installations and connections. Such items used in connection with these fixtures include the following:

BASKET STRAINER

Usually chrome-plated or finished with a color to match the fixture, the basket strainer is a 3½" x 1½" reducer fitting used to adapt a kitchen sink opening to fit the 1½" drain size. Some basket strainers are cable operated to open and close the drain from a sink-top control.

P.O. (PATENTED OVERFLOW) PLUG ASSEMBLY

This is a plug and chain stopper used for lavatory and laundry tray waste connections. It has an opening below the cone face that is against the bottom of the fixture. This opening is connected to an overflow passage built into the fixture, so that if the stopper is in place (at the fixture outlet) water entering the overflow passage can still exit via the tailpiece.

DUPLEX STRAINER

This is a small drain adapter for residential laundry tub, and some sink connections.

Figure 29-AA – Halsey Taylor Barrier-Free Cooler

Figure 29-BB – Kohler Memoirs® Bath

POP-UP WASTE

This is a drain connection which attaches to a lavatory. It includes a stopper which is controlled by an arm built into the faucet which, when operated, lifts or lowers the drain stopper in the drain assembly.

TAILPIECE

The tailpiece is a chromed-brass or plastic drain tube which extends from the strainer, P.O. assembly, or pop-up assembly, to the trap. It may be purchased with or without locknuts and washers. They are also available with a wye branch inlet to receive condensate drainage or dishwasher discharges from adjacent fixtures.

Note that trim or accessory orders must conform to the size, type, color, and other specification of the fixture. Consult manufacturers' specifications for exact sizes.

BATHTUBS

A bathtub is a fixture large enough for most of the body to be submerged in water. Bathtubs are used in homes and apartments, residential institutions, and hospital installations. They are described in terms of design configuration, length, drain location, material, color, and depth.

Tubs are available in the following designs and arrangements.

RECESSED TUB

A recessed tub is usually a rectangular tub, open and unfinished on the back and ends. The front side has a finished apron which extends to the floor. This tub is meant to sit in a recess, with the back and side walls of the recess tiled or otherwise waterproofed above the tub. This is the most popular bathtub in use today.

CORNER TUB

Usually a rectangular (sometimes square) tub, unfinished on two sides; it has an extended apron sweeping its length and width. It is used where one side and end of the tub are open to the room with the back and other end wall tiled in. These tubs are specials in the present fixture market.

LEG TUB

The leg tub is usually in an oval shape. It was prevalent in early plumbing, and has recently come back into vogue. This tub stands on four cast-iron feet about three inches off the floor. None of the sides of the older type leg tubs was finished, but the newer types have a more finished, sleek appearance and come in decorator colors and styles.

WHIRLPOOL TUB

Whirlpool tubs are of various shapes and sizes. They incorporate a system by which the water is circulated by a pump so that the water swirls around

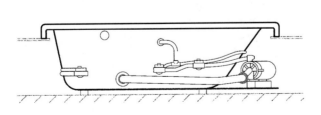

Figure 29-CC – A Whirlpool Tub

the bather. They are intended for filling and draining for each use.

For maximum safety, the fixtures should be designed and constructed in accordance with national standards. The circulation system should be self-draining.

The fixture should be equipped with listed suction fittings which have been tested for resistance to hair and body entrapment.

SOAK TUB

Usually a square tub with a deep circular or oval soaking well formed in the middle, it is much deeper than the standard well. They are usually made without finished aprons, since they are placed in a structural opening in an island setting.

HOT TUB AND SPAS

These products are intended for recreational use rather than actual bathing. They are equipped with filtration and heater systems to control water quality and temperature so they do not have to be filled and drained after each use.

OTHER CONSIDERATIONS FOR TUBS

There are other things involved with tub selection. Manufacturer's specifications should be consulted to insure proper application of a particular tub.

SIZE

Size is a critical factor when ordering a tub. With many variations possible, absolute adherence to job specifications is necessary. Tubs come in lengths of 4', 4'6", 5' (standard length), 5'6", and 6'. Length refers to the long side of the rectangular shape. The width of most standard tubs is 30" to 32".

The depth of the bathtub wells varies with the manufacturer. Apron height dimensions range from 14" to 16" to 32" (soak tubs).

Drain placement must be specified when ordering a tub. Facing the tub apron, a left-handed tub has the outlet on the left side; a right-handed tub has the outlet on the right.

The tub bottom is designed to slope towards the drain opening.

Slip resisting surfaces are available in all tub materials and should be specified.

MATERIALS OF CONSTRUCTION FOR TUBS

Bathtubs are made of many different materials; the three most commonly used materials are listed below.

ENAMELED STEEL

This is an inexpensive, light-weight tub made of porcelain-on-steel. The enameled steel may be reinforced with a composite backing for added durability. Caution must be taken not to drop anything on the surface of non-composite backed tubs, as it may crack the porcelain and expose the steel. If this happens, rusting to the point of failure can occur in a short time.

ENAMELED CAST-IRON

Cast-iron is more expensive than an enameled steel unit. It is also heavier, stronger, and will retain heat longer than the steel tub.

PLASTIC

These tubs are made of a plastic such as fiberglass or acrylic (which describes the finished surface), and are available as a tub-shower, a tub-only, or shower-only models.

Bathtubs are available in various colors in the enameled style, and in the plastic-formed versions.

BATHTUB FIXTURE STANDARDS

Every fixture made must conform to certain mandatory specifications. Those for bathtubs include the following elements:

1. All surfaces of the tub must be non-porous; that is, the surface must not provide a place where organisms can reside or alien material (like soap scum) can be retained. Any such matter would constitute a health hazard. The roughness of non-slip surfaces are not considered to be violations of this requirement.

2. Every tub must have waste opening and overflow opening to accommodate a drainage device called a waste and overflow fitting. This fitting is designed to receive water from either the waste line or the overflow hole to prevent overfilling and spillage from the tub. Water reaching the height of the overflow will begin to drain into the tub waste and overflow tubing, thus preventing overflow onto bathroom floor surfaces.

BATHTUB OPTIONAL ACCESSORIES

Optional accessories for bathtubs include:

1. Grab bars for support (these require the installation of backing boards in the wall)

2. Molded or fold down seats

3. Soap dishes and shelves

4. Molded railings

5. Built in pop-up drain

6. Slip-resisting surface on the bottom of the tub

7. Whirlpool circulation piping and pumps

8. Accessible designs to accommodate the physically challenged

BATHTUB INSTALLATION CONSIDERATIONS

It is important to remember when setting a tub that it must be continuously supported along the back and ends. If these edges are not level, the tub may not drain properly.

If the supports are not full length, the tub will rock as the bather changes positions during use. Such movement can lead to improper draining, failure of the drain connection, and possibly structural failure of the tub (if made of steel or plastic).

Improperly supported plastic tubs may crack or *spider web*.

SHOWERS

A shower has a somewhat similar function to a bathtub, but it must not be confused with a bathtub-shower unit. A shower is used solely for the purpose of washing the body in a standing or seated position, and not for soaking.

TUB-SHOWERS

A tub-shower is a combination of a bathtub and a shower, where either bathing or showering can occur. The preferred version of this type is made in one piece, but these are almost completely restricted to new construction. The product is large enough that very few existing homes have doorways and passages to the bathroom that are large enough to permit the movement into place of the one-piece models. Fixtures are available that are assembled from a few pieces for remodel application.

Tub-showers are usually used in residential applications and may be either a recessed enamel tub with tile walls or a preformed plastic or enameled steel unit.

A bathtub can be converted to a tub-shower by providing means for deflection of the spraying water back into the tub, for proper drainage and for the protection of the building or structural components.

OPTIONS

Showers can be bought as a pre-molded unit complete with a drain pan, or they can be fabricated to any size, using tile for the floor and wall surfaces and a seepage liner for additional protection. The shower, like the bathtub, must be made of non-porous material; a water-absorbing surface would harbor dirt and germs.

As with tubs, showers can be described in terms of shape, design, size, drain placement, material, color, and depth. The shape and design of a shower depends upon the purpose it serves. If a small, personal, residential shower is needed, a molded fiberglass unit may be the best investment. For a school with 800 students, small individual units may

Figure 29-DD – Kohler Sonata® 60" Shower Module

be impractical; therefore, community showers are fabricated for shower rooms.

The shower must have waterproof walls and a waterproof floor, with a drain (or drains) large enough to accommodate the water flow. Floors should slope towards the drain.

RANGE OF SIZES OF SHOWERS

The size of the shower is dependent upon its use. The common residential shower sizes range from 30" x 30" (square fiberglass multi-piece stalls) to 3'0" x 5'0" (one-piece molded plastic units) to any convenient size (for built-up tiled units). The shower wall height is usually 6'0". Be sure to consult your local codes as there are minimum area requirements for showers, especially when accommodating local accessibility requirements.

The depth of the base for a shower ranges from four to eight inches. The threshold lip of the shower base must be high enough to keep the water in, yet must not be too high to make entry hazardous. Lower thresholds are used for showers to accommodate the physically challenged.

Drain placement is important when ordering a molded unit or when fabricating a large unit. Small shower stalls (30" x 30" or 32" x 32") usually have the drain located in the center of the molded pan. Large molded one-piece units (3' x 4' or 3' x 5') may have to be ordered with a specified right-hand or left-hand drainage connection

When fabricating a large shower, drain placement is important to consider, as it is necessary to drain the shower floor as efficiently as possible with a minimum flow distance to the drain. One drain serving a shower 10' x 12' would be impractical as a long run may make it difficult to wash dirt and soap suds to the drain opening. Consult manufacturers' specifications for the manufactured showers, and consult local code authorities prior to fabricating a shower. For showers for more than one person, the drains must be placed so that the water from one bather does not flow across the bathing area of another bather.

MATERIALS OF CONSTRUCTION FOR SHOWERS

Materials used in the manufacture of showers are mainly these products:

PLASTICS

Fiberglass or other plastics are used for the drain pans and the shower walls.

PAINTED OR ENAMELED STEEL

This is a very inexpensive material used in prefabricated stall units for the walls and floor. Steel has lost most of the market to plastic shower stalls due to problems with rust.

SITE-FABRICATED SHOWERS

This type of shower is made at the jobsite. The plumber is responsible for placing a waterproof membrane under the shower floor pan. This membrane is either lead, copper, plastic, or an elastomeric material such as PVC or CPE. The shower floor, which can be made of soapstone, marble, tile, or concrete, is then fabricated on top of this membrane. Prefabricated bases are also available. These bases offer an economical alternative to a built-up base. When installing plastic sheeting as the secondary protective membrane, never tar coat the plastic. It will destroy the material.

OPTIONAL ACCESSORIES FOR SHOWERS

Optional accessories and features presently manufactured include the following:

Shower colors vary by the units ordered. The choice is nearly limitless with a tiled shower, but color choices are also available with the prefabricated bases or with complete units.

Grab bars (don't forget the backing boards)

Seats

Doors

Curtains

Multiple shower heads

Pulsating flow shower heads

Slip resisting floor surface

LESSON 30

FIXTURE FITTINGS (FAUCETS)

Fixture fittings, also called faucets, as traditionally defined, are **valves for drawing a liquid from a cask, pipe, or other vessel**. Faucets are fittings which deliver water to fixtures. A faucet or fixture fitting is designed to perform the following functions:

1. Provide an inlet space (or spaces) for the water supply

2. Control the flow rate

3. Provide complete shut-off

4. Deliver water through a desired spout or outlet

Factors to be taken into consideration when selecting a fixture fitting (faucet) for a particular job include:

1. Application — light duty (residential), moderate (commercial), or heavy (most institutional and industrial uses)

2. Cost — usually, the cost would relate to the severity of the application

3. Life expectancy required — usually, but not always, relates to the severity of the application

4. Safety — strong enough to withstand surges likely to occur, made so that it will not come apart after moderate wear, capable of tight shutoff after moderate wear

FAUCET PARTS

The parts of a conventional screw stem, deck-mount faucet are shown in Figure 30-A and described below:

STEM

The stem is the flow control part which positions the washer onto the seat or lifts it from the seat. It usually has a holding (or retainer) cup which holds the washer firmly in place. The stem position is controlled by a threaded section that engages the housing.

WASHER

The washer is a soft-material, laminated fiber, ceramic, or rubber disc which screws or snaps onto the stem and closes the opening in the seat to stop flow.

SEAT

The seat is the mating surface for the washer; also, the supply inlet. The standards for the faucet require replaceable seats in most faucet designs. Seats may be made of plastic, brass, ceramic, or chrome-plated brass.

BODY

The body or housing is the machined, cast, or extruded section of the fixture fitting to which all mating parts assemble. The body also houses the mixing chambers in faucets in which the hot and cold water mix. By federal regulation, the body is limited to a lead content of 8% maximum if made from a copper (brass) alloy.

HANDLE

The handle is the device which is attached to the stem which facilitates stem positioning.

SPOUT

The spout is the outlet for the water.

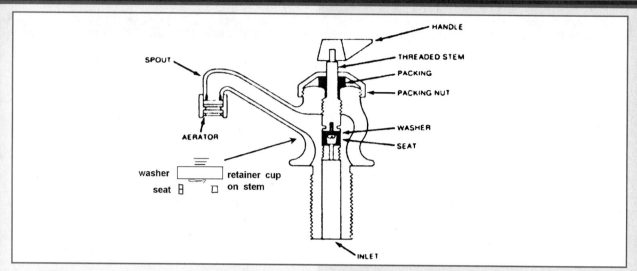
Figure 30-A – Typical Single-Bibb Faucet

SUPPLY INLET

The supply inlet is the inlet side to a fitting which can be connected to the water supply by means of a soldered, threaded, or ground-joint assembly.

PACKING

Packing is the material which prevents water from leaking around the top of the stem.

PACKING NUT OR COLLAR

This is the part through which the stem passes, and which restrains the packing to maintain the seal against the stem.

AERATOR

The aerator is a device which attaches to the spout outlet and introduces air into the water as it leaves the faucet. Most flow rate control is performed with flow restricting aerators. The aerator may include a spray device. Since the basic aerator function entrains air into the water stream, aerators are usually prohibited in hospitals, because any air-borne bacteria in the air will be inserted into the water as it leaves the faucet.

SINK AND LAVATORY FIXTURE FITTINGS

Many manufacturers make two-handle mixing fittings with variations of the conventional design. Although the conventional **seat and washer** may not be seen, the concept of water control is the same. Some less expensive washer type faucets use a snap-on washer instead of a screw-retained type washer, although some chatter may occur with the snap-on device. **Plastic disc** faucets use two plastic discs to control the water flow.

Some faucets are available with single lever controls. These faucets use front-rear (or up-down) lever position to control volume, and left-right lever position to control the mix of hot and cold water. Their internal mechanisms control water by various methods, such as the following:

1. The **ball** type fitting uses a machined brass, stainless steel, or extruded plastic ball to control the flow by varying the alignments of passageways through the ball with rubber seals within the valve body.

2. The **cartridge** type fitting uses a machined or molded cartridge and O-rings to regulate flow.

3. The **spring-force** type fitting uses a spring washer and rubberized seal to control the

water flow.

4. The **ceramic disc** fitting uses polished ceramic parts to control water flow.

Most industrial and commercial installations require **self-closing** or **self-metering** fittings These faucet types are selected to reduce water waste.

A **self-closing** fitting is similar to a standard type except that the stem device is spring-loaded to close immediately after release of the handle.

A **self-metering** fitting will close a short time after the handle is released.

OPTIONAL ACCESSORIES FOR FIXTURE FITTINGS

There are many different options available for every faucet as was discussed briefly in the section on sinks and lavatories. It is important to take into account these optional features when choosing a faucet for a job. Options include the following:

1. Mounting choices

    - Deck mount
    - Wall mount

2. Number, spacing, and types of connections

    - 8" center-to-center is standard for kitchen sink faucets
    - 6" or 8" center-to-center valves are common on bar sinks and upscale lavatories
    - 4" centersets are common on typical lavatories and laundry trays
    - Solder, threaded, or ground joint inlet connections

3. Spray hose devices

4. Different spout lengths

5. Flexible aerators

6. Dual handled or single lever

7. Foot and knee controls (hospitals, medical and dental offices)

8. Electronically activated

9. Handle materials (chrome, stainless steel, acrylics, etc.)

10. Body materials (chrome, stainless steel, plastics, etc.)

11. Blade handles for handicap style faucets

12. Soap dispensers

Most faucets do not have any options for flow rates. Federal standards for faucets have set a maximum of 2.5 gallons per minute on lavatory, sink faucets, and showerheads for normal private use. The maximum flow for public-use lavatories is 0.5 gpm or 0.25 gallon per cycle.

BATH AND SHOWER FAUCETS

Bath and shower head fittings are called mixing valves. The internal mechanisms of mixing valves are designed with water control devices like the lavatory and sink faucets — either two stem (one hot, one cold) or single lever (which blends the hot and cold water). Older tub-shower mixing valves may have three stems — one for hot, one for cold, and one to divert the water from the tub to the shower and vice-versa.

For bathtub or shower applications, most codes now require fitting types that guard against sudden temperature change. There are two common methods of accomplishing this temperature protection:

1. Thermostatic mixers sense leaving water temperature directly and cause the hot and cold controls to reposition themselves if the

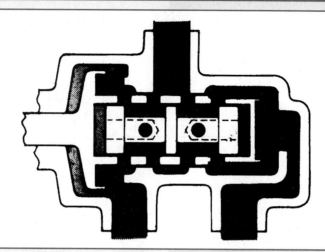

Figure 30-B – HydroGuard®T/P Series e700

discharge water temperature changes. While this approach seems to be the best, such mixing valves are more costly and require more maintenance than regular types. They are also subject to time lag from the time of the abnormal temperature leaving the faucet until the internal parts can reposition themselves. Thus, the bather may still experience brief temperature changes that may be startling or even painful.

2. Pressure balance mixing valves are said to be less costly and require less maintenance. These valves operate by sensing inlet pressure on the hot and cold supplies. If there is a sudden, significant change in the pressure of either, the valve adjusts the interior mechanisms to either shut off both sides, or at least, approximate the same percent flow through the two sides. Since most sudden temperature changes are the result of inlet pressure changes, this valve is fairly satisfactory in most cases, and may even give superior performance, because the usual cause of temperature

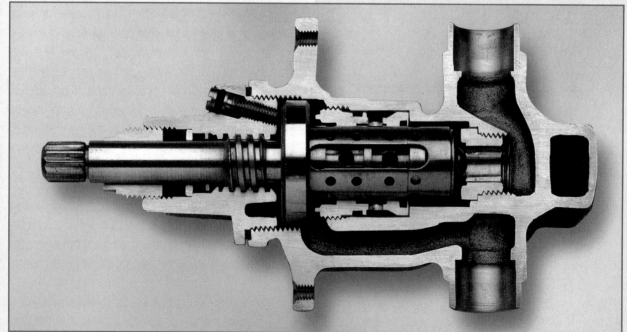

Figure 30-B(2) – Symmons Pressure-Balancing Shower Valve

change is corrected more quickly than with thermostatic models. See Figure 30-B. Combination pressure-balancing and thermostatic mixing valves are also available.

You should realize, however, that other conditions can cause sudden temperature change, so these valves do not provide a total solution. Most manufacturers provide a limit stop on the hot water side of the pressure balance valve to limit the flow of hot water at all times. The installer must adjust this limit stop at the time of installation.

The problem of temperature changes in showers is very serious and a substantial cause of injury. Most discussion is about scalding, but a greater problem is falls. With sudden temperature change, hot or cold, a person in normal health will act reflexively (i.e., quickly) to move out of the water stream. Because the shower or tub bottom surface can be rather slippery, especially when soapy water is present, the person might fall while trying to move quickly.

The products available may not be totally satisfactory if not properly maintained, and, of course, there are many, many existing installations where there is no protection whatsoever. Total safety involves a combination of a slip-resistant surface, grab bars (which are screwed to the structure), properly maintained scald-preventing valves, and consumer consciousness about safe bathing procedures.

One further caution about these balanced pressure or thermostatic valves. At least some are an arrangement of parts that connect hot and cold supplies to a mixing chamber through throttling ports, and then through an on-off port to the shower head or bathtub spout. This type can shut off tightly (no drip to the shower head or bath spout), and yet there is an open path from the hot water to the cold water supply pipes. Depending on piping details, this path can produce strange results like flushing a water closet with hot water, hot water appearing at cold faucets (and vice versa), or complete inability to deliver the correct water to faucets in the vicinity of the shower valve. If not factory equipped, consider installing check valves at each mixing valve unless you are sure that this cross-over problem cannot occur.

OPTIONAL ACCESSORIES FOR MIXING FIXTURE FITTINGS

Optional accessories for bathtub, shower head, and tub-shower mixing fittings include the following:

1. Body sprays

2. Two or three handles

3. Shower head styles, including side spray (telephone-type showers)

4. Anti-scald buttons which immediately stop the hot water supply

5. Screw-driver-controlled repair *stops* which can be closed to cut off the water supply so that the valve can be repaired

6. Handle materials (various metals, finishes, and plastics)

DIVERTER SPOUTS

Diverter is a term used to describe one type of spout serving faucets used to supply bathtubs and shower heads. The spout is equipped with a closure plate and rod that raises the closure plate into place to block the spout. When the spout opening is covered, water is diverted up to the pipe which supplies the shower head. In this way, the diverter is used to permit water to flow from the spout or to be directed to the shower head.

Lift spout diverters are not positive shut off devices. They leak a small amount of water to the tub during the showering process, in order for them to reopen when the pressure is off. If they were positive shut-off, the head pressure in the shower arm might keep the lift gate in the diverter position. When the

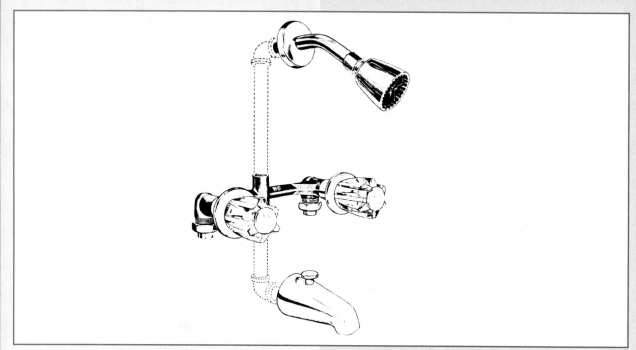

Figure 30-C – Typical Shower Diverter Installation (Wolverine Brass)

water is turned on again, the bather might get a surprise spray from the shower head.

Once the water pressure is turned off, the diverter falls back to the at-rest position, so that at each new use of the faucet, water comes out the spout, rather than from the shower head. The use of full flow fittings to the spout is required; some types of restricted flow fittings may cause the water to unintentionally flow to the shower head.

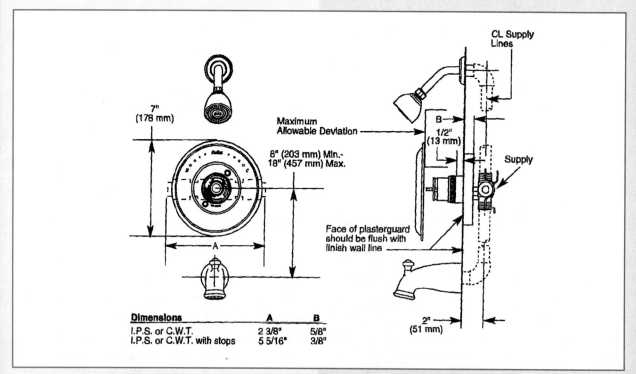

Figure 30-C(2) – Delta Bath Mixing Valve Single Handle

OTHER FAUCETS

BOILER DRAIN

A boiler drain is a screw stem valve which is used to drain tanks, water heaters, etc. They are provided with a hose thread on the outlet, and are not normally equipped with a backflow device.

SILL COCKS AND HYDRANTS

These devices are faucets piped to the outside of buildings. There are many types, all of which are provided with hose threads. The device should be provided with either integral or add-on backflow protection.

NON-FREEZE WALL HYDRANT

In this faucet the seat and washer assembly is actually inside the building. The stem and body are available in various lengths to accommodate different wall thicknesses. See Figure 30-D.

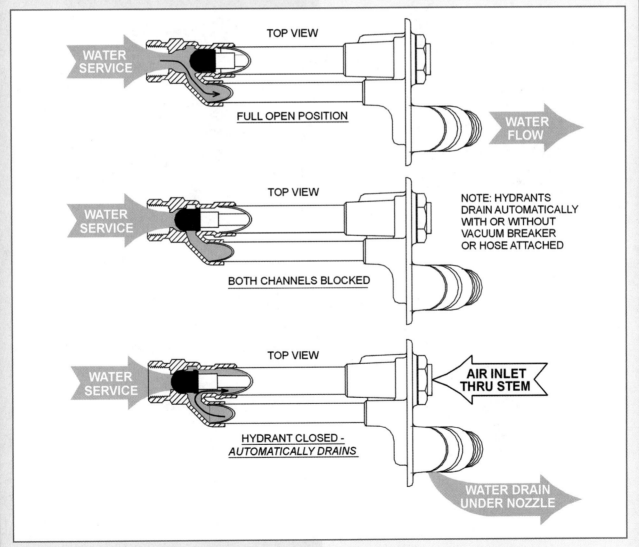

Figure 30-D – Woodford Model 65 Freezeless Wall Hydrant Flow Diagram

POST OR SANITARY YARD HYDRANT

This frostproof faucet has the washer and seating assembly below the ground. The small weep hole located at the base of this device must be free from obstruction so that the contents of the post will seep out once the faucet is turned off. See Figure 30-E. Such designs may constitute a cross-connection by some codes.

Preferred types of post or sanitary yard hydrants have a storage vessel which is located in the ground as part of the assembly. When the water is shut off at the hydrant, the water in the post drains into the reservoir. When the water is turned on again, a venturi fitting in the hydrant assembly siphons the water from the reservoir, in order to prevent a stagnant water condition in the assembly.

WASHING MACHINE VALVE

A single-handle-operated 90° turn dual ball valve is

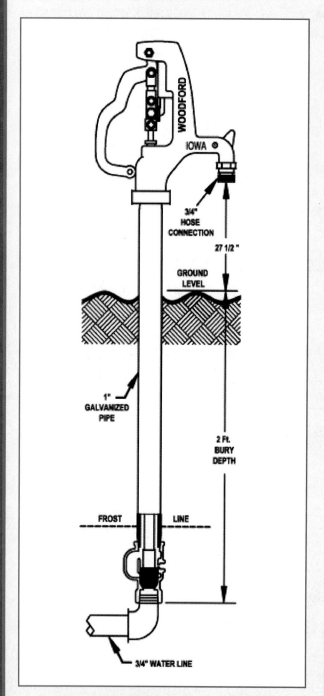

Figure 30-E – Woodford Yard Hydrant

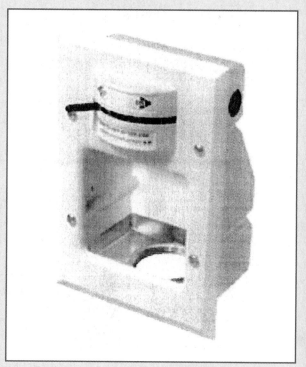

Figure 30-F – Symmons Laundry-Mate W-600

provided to simplify shutting off (or turning on) the water supply to a washing machine. Since most washing machines are equipped with integral air gap backflow protection, separate vacuum breakers are not normally placed on these assemblies.

LESSON 31

VALVES

A valve, as defined in the Fifth Edition of the *ASSE Plumbing Dictionary*, is a

device by which the flow may be started, stopped, or regulated by a moveable part which opens or obstructs a passage.

A valve is used to control the flow of liquids, gases, or vapors.

Valves are available in many sizes, shapes, and applications. They may be used frequently, intermittently, or only in emergencies (as in the case of safety valves). They may be operated manually, by external mechanisms (electric motor, hydraulic actuators, pneumatic power, etc.), or in response to system conditions (over- or under-pressure, over-limit temperature, etc.).

Valves hold the interior moving parts in proper relationship by some sort of external assembly which also incorporates the piping connections. Flow control is accomplished by washers, plugs, balls, gates, or plates.

VALVE PARTS

The following elements are common to all types of manually-operated valves:

BODY

The valve body is used to contain the fluid, mount the other parts, and is attached to the piping.

CLOSURE SYSTEM

The closure system opens or closes the fluid pathway through the valve.

STEM

The stem extends out from the body to control the closure system.

HANDLE

The handle manipulates the stem.

PACKING

Packing prevents the fluid from leaking around the stem.

TYPES OF VALVES

Valves can be placed in various kinds of categories — the following descriptions are based on the functions of the valves. Many other types of characterization are possible.

OPEN-CLOSE

Open-close type valves are designed either to be fully opened or totally closed. The desired feature of this type of valve is minimum restriction to flow. Tight shut-off is essential in some applications using these valves, but it may not be a critical reason for choosing this valve.

These valves are selected because they are high-capacity, low pressure-drop types; therefore, they are not effective for throttling (i.e., restricting) flow until they are nearly closed. If these valves are operated in this near-closed position, wear is likely and valve deterioration results.

Examples of this valve type are the following:

    Gate
    Butterfly
    Plug
    Ball

(Plug valves are suitable for both on/off and throttling).

GATE VALVE

Gate valves are used to completely close a line so that piping or equipment can be isolated for repair, inspection, or replacement. It is not uncommon that these valves do not shut off tight, but they are nearly equivalent to straight pipe when they are wide open. They are ineffective for throttling until almost completely closed — at this point wear will take place, degrading the ability of the valve to develop good close-off. The valves also tend to be noisy when severely throttled.

The valve mechanism is made up of a gate which slides in a channel formed in the valve body, with the gate being attached to the stem. Stem packing and handle complete the basic valve.

The valve closes by having the gate close tightly against the channel circular faces.

In order that the gate does not drag against these faces throughout its travel, the gate is made wedge shaped so that it is free of the seating surfaces as soon as it begins to open. Because of this shape, the gate is sometimes referred to as the wedge.

A variation on the one-piece wedge is to divide it into two discs. The stem end is made so that the two discs are forced apart at the point of closure to try to improve the shut-off characteristic of the gate valve and to prevent sticking in the closed position.

There are two stem designs:

Outside Screw and Yoke

Inside Screw - with two variations

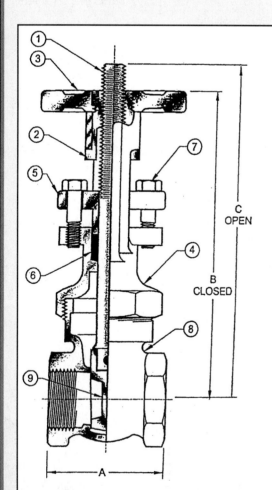

MATERIAL LIST

NO.	PART	MATERIAL	SPECIFICATION
1	Stem	Brass	CDA-360
2	Yoke Bushing	Brass	
3	Handwheel	Ductile Iron	
4	Bonnet	Brass	CDA-844
5	Packing Gland	Brass	
6	Packing		
7	Gland Follower Bolt	Stainless Steel	
8	Body	Brass	CDA-844
9	Wedge	Brass	

Figure 31-A – Hammond 118-FP Brass Gate

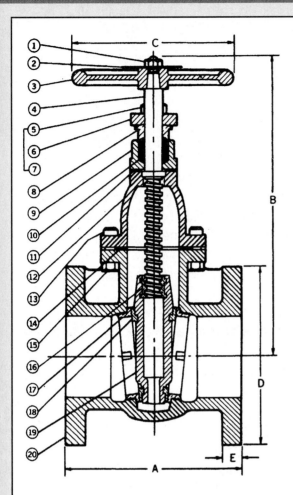

	IR1144-HI SPECIFICATIONS		
MSS	MSS SP-70, Type 1, Class 125, Flanged Ends, Iron Trim		
ANSI	ANSI B16.1 for Class 125 Flanges		
	ANSI B16.10 for Face to Face Dimensions		
MATERIAL LIST			
NO.	PART	MATERIAL	SPECIFICATIONS
1	Handwheel Nut	Ductile Iron	ASTM A536
2	Identification Plate	Aluminum	
3	Handwheel	Cast Iron	ASTM A 126, Class B
4	Stem	Steel	ASTM A108-Gr. 1020
5	Gland Follower Nut	Steel	ASTM A563-B
6	Gland Follower	Ductile Iron	ASTM A536
7	Gland Follower Bolt	Steel	ASTM A307-B
8	Packing Gland	Cast Iron	ASTM A 126, Class B
9	Stuffing Box	Cast Iron	ASTM A 126, Class B
10	Packing	Non-Asbestos	Graphite
11	Stuffing Box Gasket	Non-Asbestos	Graphite
12	Bonnet	Cast Iron	ASTM A126-Class B
13	Body Bolt	Steel	ASTM A307-A
14	Body Gasket	Non-Asbestos	Graphite
15	Body Bolt Nut	Steel	ASTM A563-B
16	Wedge Bushing	Cast Iron	ASTM A126-Class B
17	Seat Ring	Steel	ASTM A570-Gr. 33
18	Wedge Face Ring	Cast Iron	ASTM A126-Class B
19	Wedge	Cast Iron	ASTM A126-Class B
20	Body	Cast Iron	ASTM A126-Class B
21	Stuffing Box Nut (not shown)	Steel	ASTM A563-B

Suffix HI – Signifies International valves.

Figure 31-B – Hammond Gate Valve

Rising stem
Non-rising stem

The outside screw and yoke (OS&Y) design has the thread external to the valve so it is not subject to corrosive attack or fouling by the pipe line fluid. This is a rising stem design, so space must be allowed for the stem extension. OS&Y valves are usually specified for critical applications, such as for fire protection systems. See Figure 31-A.

Inside screw valves have the threads in contact with the pipe contents, so they are not suitable for corrosive or fouling fluids. If the gate is threaded to the stem, a shoulder is placed on the stem to restrain it in the bonnet; when it is turned, the gate is retracted from or extended into the flow path. This design is the non-rising stem version.

If the thread is in the bonnet just below the packing, and the gate is connected to the stem with a swivel joint, the stem and gate extend or retract together. This design is the rising stem version. Be sure you provide room for the extended-stem position. See Figure 31-B for a non-rising stem model.

BALL VALVES

Ball valves are similar in general design to plug valves, except that instead of a tapered plug, the moveable part is a ball. The ball is set in a Teflon™ or similar liner, so the ball is always relatively easy to turn. See Figure 31-C.

These valves are made of hardened parts, so they are long-lasting but they should not be used for throttling applications. A 90° turn of the handle will completely open or close the ball valve.

Some ball valves are designed with a body and ball that is relatively oversized, so that the area of the opening through the ball is the same as the cross-sectional area of the pipe. These are referred to as *full port* valves, and they are required for some applications — usually on the water service.

Ball valves are less expensive than gate valves and usually produce tight shut off. Therefore, they have become the valve of choice rather than gate valves in many applications, especially applications involving backflow preventers.

BUTTERFLY VALVE

Butterfly valves are low pressure-drop valves that are suitable for full-open, full-closed application. The valve consists of a body, stem, and closure plate. Where a gate valve retracts the gate from the flow path, the butterfly valve rotates the closure plate to close the flow path or to be aligned parallel with it. Thus, when the valve is fully opened, the closure plate is still in the flow path, but it produces very little flow restriction in the parallel position.

The butterfly valve is a less complicated valve than a gate valve to manufacture, and therefore they are

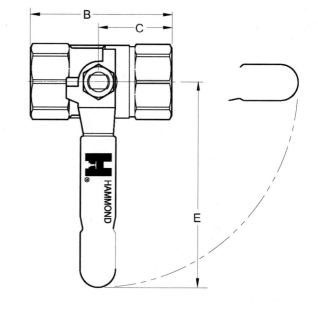

MATERIALS LIST

ITEM	PART	MATERIALS	ASTM SPEC.
1	Body	Brass, Forged	B283
2	Tailpiece	Brass, Forged	B283
3	Ball	Brass	B16
		Brass, Cast	B584
4	Seat	PTFE	
5	Stem	Brass	B16
6	Thrust Washer	PTFE	
7	Gland Nut	Brass	B16
8	Gland Packing	PTFE	
9	Handle	Steel w/ Zinc Plating	B633
10	Handle Nut	Brass	B16

* Please consult factory for more information regarding steam applications.

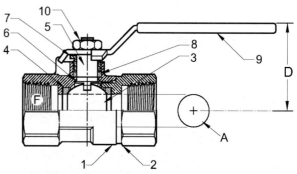

Figure 31-C – Hammond 8201 Forged Brass Ball Valve

less expensive, especially in larger sizes.

The valve handle is installed parallel to the closure plate, and the body has stops on it to restrict the angle to a 90° rotation — thus, when the handle is parallel with the pipe, the valve is wide open, and at 90° to the pipe, the valve is closed. See Figure 31-D.

mal pressures. These valves have the advantage that scale or other solids in the valve will not keep it from closing (such solids in the gate valve seats can keep it partly open) because the sliding action of the plug and body will shear off any such foreign material.

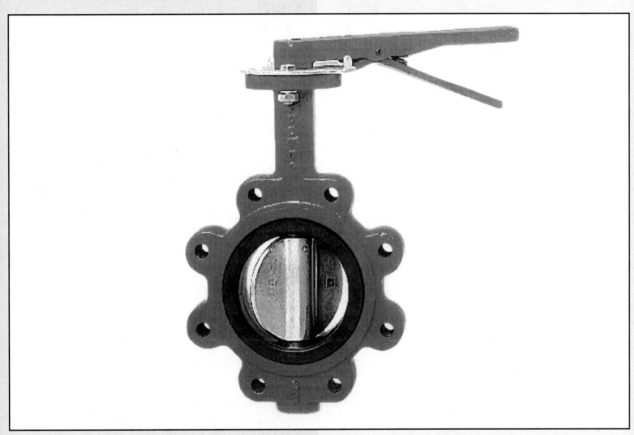

Figure 31-D – Watts Resilient Seated Butterfly Valve

PLUG VALVES

Plug valves (also called cocks) are used where quick shut off is required. Like the butterfly valve, they are operated with a 90° turn of the handle.

This valve type is an extension of a wine cask stop, probably one of the earliest valve types. The construction is very simple: a tapered plug with a hole drilled through it is placed in a matching tapered opening in a body. The body has connections for the pipe line. This design works on very low pressure applications, but they are difficult to use at nor-

An improved version is the lubricated plug cock, where the plug has internal passages machined into it, and lubricant is forced into these passageways to make the plug easier to turn. The end of the plug is fitted with a lubricating fitting so that the plug can have lubricant added whenever necessary (Figure 31-E). This type valve is effective at regulating flow.

VALVES USED FOR FLOW REGULATION

Certain valve types are designed to regulate the fluid flow as needed by the fixture or device being supplied. These valves have a higher pressure-

Figure 31-E – Lubricated Plug Valve

drop than open-close valves.

Flow restriction is obtained by the following arrangements:

1. Change of direction of flow through the valve

2. Introduction of a restriction system in the flow path

3. A combination of both

Examples of this type valve are:

- Globe
- Needle
- Angle
- Plug

GLOBE VALVES

Globe valves acquired their name because the body is globe-shaped.

The globe valve can be used to regulate fluid flow or to develop tight shut off. The inlet path makes an abrupt right angle turn to approach the disc and seat orifice. The flow path includes two more abrupt right angle turns to reach the outlet. These abrupt right angle turns and the orifice size are the reason for high pressure drop in this design. Figure 31-F shows a typical valve.

The disc and seat of the globe valve may be replaceable, depending on the size of the valve and its use. Globe valves that are usually used residentially have non-renewable seats and composition-type discs to seal the flow. The composition-type disc is called a washer and is usually made of a rubber or laminated fiber.

Globe valves are manufactured in two styles:

- Straight pattern (Figure 31-F)
- Angle pattern (Figure 31-G)

The washer position relative to the seat is controlled by a threaded stem, fitted with a handle on the outside, and packing assembly to prevent stem leakage.

Globe valves are available with threaded, flanged, or welding ends for steel pipe, and solder-cup ends for copper tubing. Globe valves for compression joining are now on the market.

Valves are marked as to the direction of the flow.

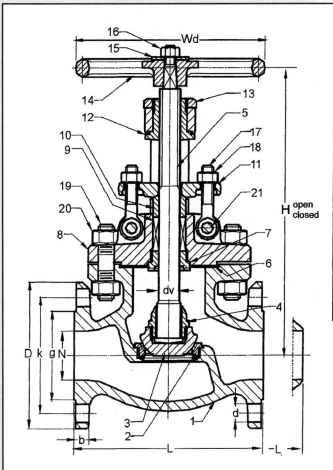

Figure 31-F – Hammond 1560CB2 Flanged Ends Globe Class 150

The direction of flow should be through the seat and against the washer. With this arrangement, when the valve is shut off, the packing can be replaced because there would be no pressure in that part of the valve.

NEEDLE VALVE

Needle valves are essentially globe valves with a cone-shaped closure element instead of a flat disc element. The seat is shaped to the same contour as the cone-shaped member. As the closure element starts to lift as the valve begins to open, the area of the flow path is very small so the valve provides a fine adjustment of low flow volumes to control instrument, meter line, or gauge piping. They are also convenient for high pressure and/or high temperature control. See Figure 31-H.

Trim assembly is a tapered "needle" which fits very snugly into its female socket. Thread pitch of the stem assembly is quite fine; therefore, the stem can be adjusted precisely, and the flow regulated closely.

Figure 31-G – Hammond Bronze Angle Globe Valve

CHECK VALVES

Check valves are designed to prevent reverse flow in fluid lines. They should not be confused with the more sophisticated backflow preventers, which will be studied in later lessons.

This valve is available in various types. The main check valve types are the following:

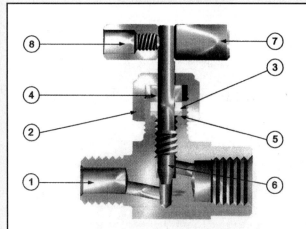

Model Shown: 4F4M-SN6LR-SS

Materials of Construction

Item #	Description	Material
1	Body	ASTM A 276 Type 316
2	Packing Nut	ASTM A 479 Type 316
3	Packing*	PTFE
4	Packing Gland	ASTM A 479 Type 316
5	Packing Washer	Stainless Steel
6	Stem (R-Stem)	ASTM A 276 Type 316
6	Stem (K-Stem)	ASTM A 276 Type 316, with PCTFE
7	Handle**	Aluminum
8	Handle Screw	Stainless Steel

* Optional elastomeric O-ring stem seals and Grafoil® packing are available - See How to Order
** Handles for Grafoil® packed valves and valves with R stem types are stainless steel T-bars
Lubrication: Graphite filled hydrocarbon

Figure 31-H – Parker SN6 Series Needle Valve

LIFT

In the lift check valve, a tapered piston disc sits on a mating tapered seat. Pressure from the inlet side pushes the piston disc off the seat. Reverse flow forces the plug into the seat, thus stopping the reverse flow. See Figure 31-I.

SWING

This check valve, by far the most popular, is designed like a door with an overhead hinge. Forward flow opens the door; reverse flow closes the door. In one model, closure disk normal position is 90° to the flow path. A second type places the disk at 45° to the flow path. See Figure 31-J for a swing check.

BALL

In a ball check valve, the ball sits on a tapered seat and is lifted off the seat when inlet pressure is exerted. Reverse flow reseats the ball. See Figure 31-K.

NOTE: ALL THE CHECK VALVES DESCRIBED ABOVE MUST BE INSTALLED IN A HORIZONTAL LINE!

Exception: The 90° swing check will function in a vertical line that has flow upward..

SPRING-LOADED

In a spring-loaded check valve, the closing disc is mounted on a center post and spring loaded closed so that forward flow moves the disc clear of the seat, and the spring moves the disc against the seat when flow stops, before reverse flow can occur. These valves do not produce water hammer when they close like swing check valves can in some circumstances. See Figure 31-L.

CHECK VALVE SEATING

In general, check valves do not close off tightly

against reverse flow. It may seem odd, but closure is better with higher pressure lines, because the higher pressure differential produces more closure force.

Improved valve seating may be realized by specifying *soft seat* check valves. These valves have a rubber (or other type "soft" material) disc attached to the closure mechanism. Soft seat check valves are often installed upstream of certain types of backflow preventers to prevent undesirable leakage from relief ports. These valves are more likely to stick closed with some types of water or other fluid that they are controlling.

Care must be taken to install check valves in the proper manner. Horizontal check valves should be installed in a horizontal position; vertical checks in a vertical position. The valves must be installed to conform with the desired direction of flow.

Ball and lift check valves should be installed in a horizontal line with the moving part placed upward.

Swing checks must be installed so that gravity will

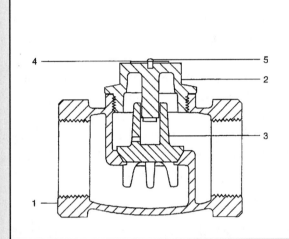

Materials of Construction

NO.	DESCRIPTION		MATERIAL	ASTM SPEC.
1	Body		Bronze	B-62 C83600
2	Cap	(½")	Brass	B-124 C67500
		(¾"- 2")	Bronze	B-62 C83600
3	Disc	(½")	Brass	B-124 C67500
		(¾"- 2")	Bronze	B-62 C83600
4	ID Plate		Aluminum	
5	Drive Screw		Steel	Electro Brassed

Figure 31-I – Stockham Class 125 Bronze Lift Check Valve

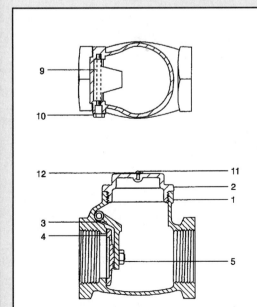

Materials of Construction

NO.	DESCRIPTION		MATERIAL	ASTM SPEC.
1	Body		Bronze	B-62 C83600
2	Cap		Bronze	B-62 C83600
3	Hinge		Bronze	B-62 C83600
4	Disc	1/4-1	Brass	B-16 C36000
		1 1/4-3	Bronze	B-62 C83600
5	Nut		Brass	B-16 C36000
9	Pin		Brass	B-16 C36000
10	Screw		Brass	B-16 C36000
11	Drive Pin		Mild Steel	
12	Name Plate		Aluminum	

Complies with MSS-SP-80, Type 3

Figure 31-J – Stockham Class 125 Bronze Swing Check Valve

assure that the flapper has seated properly. Spring-loaded checks may be installed in almost any position, except flow downward.

AUTOMATIC REDUCING VALVES

Automatic pressure reducing valves or regulator valves are designed to receive high pressure at the inlet and to automatically maintain a constant lower pressure at the outlet.

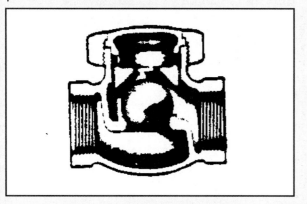

Figure 31-K – Ball Check Valve Flow from Left to Right

High pressure fluids are contained by a diaphragm-operated disc and seat assembly. The diaphragm tension is maintained by an adjustment spring. The outlet pressure acts to close the diaphragm, and is balanced by the spring force. If outlet pressure falls, the disc is opened from the seat, increasing flow (and pressure).

If pressure rises above the balance point, disc opening is restricted, reducing flow (and pressure). See Figure 31-M.

The operation of these valves may be assisted by an external mechanism, such as compressed air.

This valve should be installed on a horizontal line, sitting in a vertical position. In some larger installations, a bypass valve system should be installed in case replacement or repair of this valve is necessary. Pressure gauges are often installed upstream and downstream of these devices in more sophisticated systems.

There are two possible reasons for using these valves — one is to reduce an unacceptable high pressure to a lower pressure that the down-stream equipment can handle; the second is to provide an essentially constant downstream pressure with a variable (or varying) upstream pressure.

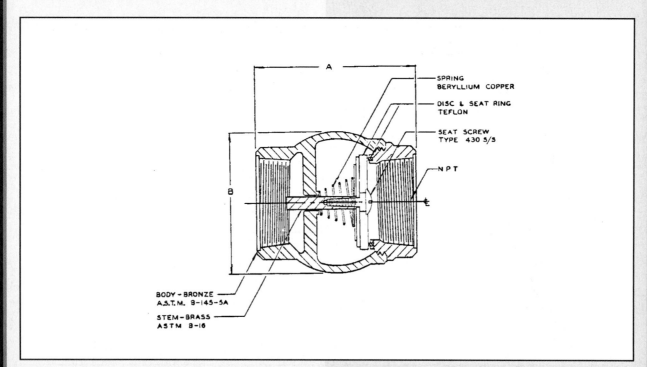

Figure 31-L – Typical Spring Check Valve Flow from Right to Left

Figure 31-M – Watts Series 25AUB-Z3 Water Pressure Reducing Valve

RELIEF

Relief valves are intended to operate in emergencies, that is, when the normal operating controls of the system malfunction in some manner. They are designed to relieve excessive temperature or pressure that may occur in a system.

Pressure types are installed wherever an unsafe pressure can occur as the result of firing a burner or electric heater, operating a compressor or pump in a closed system, having a closed vessel warmed up (even by ambient heating), or having a pressure reducing valve stick open and deliver excessive pressure. See Figure 31-N.

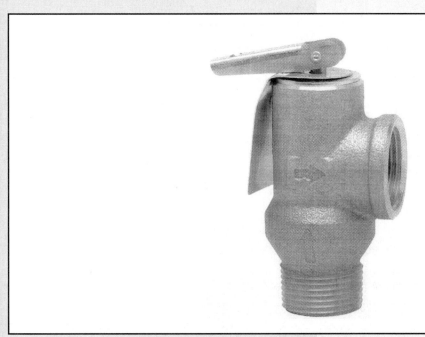

Figure 31-N – Watts Series 530C Calibrated Pressure Relief Valve

Temperature types are used to limit temperatures in water heaters and boilers.

Vacuum types are used to admit air into a system in the event that a hazardous less-than-atmospheric pressure is developed in the system.

The standard relief valve is a spring loaded device which begins to open at a set pressure, and it requires a further slight pressure increase to become fully opened. It reseats when the pressure falls below the initial operating pressure by a slight amount. The result of this operating characteristic is that at pressures near the trip point, the forces holding the valve closed are small, and right at the trip point, the small flow through the valve tends to **wire draw** the seat. It is not unusual for this type valve to leak slightly after only a few operating cycles.

Temperature relief valves are made to open at a certain temperature. Some reset closed on temperature fall, while others have a plug that melts at the set temperature, so reclosing is not possible.

Another version uses a pressure disk that ruptures when the system pressure exceeds the rating of the rupture disk.

Combination temperature and pressure relief valves are available in various sizes and ratings.

The spring-loaded device remains closed until the vessel pressure exceeds (by a small amount) the rated pressure of the valve. The valve then opens and relieves the internal pressure of the tank. When the vessel pressure is less than the spring pressure (by about 5%), the valve recloses. The pressure rating of the valve is dependent upon the force set in the spring and the area of the valve seat.

Temperature-activated relief valves are actually pressure relief valves except that the pressure that opens them is developed by the temperature of the fluid in the sensing bulb, rather than the pressure that exists in a closed vessel or system.

Figure 31-O shows a combination temperature and pressure relief valve that is installed on water heaters.

The rupture disc device remains intact until the rated pressure is reached and then breaks out completely, allowing full flow of fluid. This flow continues until

Figure 31-O – Watts Series 10L Temperature and Pressure Relief Valve

the system is shut down and the rupture disk is replaced with a new one. Disc devices are used only on rare occasions, as the convenience and reduced spillage qualities favor the spring-loaded valve.

SPECIALTY

Specialty valves are designed to perform specific functions.

Examples are:

Tempering

Solenoid

Diaphragm

The **tempering valve** is designed to mix hot and cold fluids in order to achieve a specific mixed temperature. These valves should **not** be used as a scald-prevention control for a shower. A scald preventive valve (designed in accordance with the American Society of Sanitary Engineering Standard #1016) is required to maintain a temperature setting within ±3° and to respond to temperature or pressure changes instantly. Tempering valves are not designed with this degree of sensitivity, and are designed to comply with ASSE Standard #1017. If a thermally sensitive valve does meet the tight tolerances of ASSE #1016, it is usually designated as a thermostatic mixing valve. See Figure 31-P.

The **solenoid valve** is a common stop valve, usually spring-loaded to close, and electrically activated. Normally-open solenoid valves are available for some services, but they are more complicated in construction, and usually more expensive than the normally-closed versions. Note that *true* solenoid valves are small — usually not larger than 3/8". The reason for this is that a large solenoid will hammer itself to destruction after relatively few operations!

Thus, **diaphragm valves** were developed. The most common versions operate with a small solenoid valve as the pilot operator. They may also be operated by manual, hydraulic, or pneumatic inputs. All these valves consist of a diaphragm, which has (when the valve is closed) equal pressures above and below the diaphragm, with the incoming line pressure introduced below the diaphragm. The valve closure disc is attached to the under side of the diaphragm. The pressurized

Figure 31-P – Watts Series 70A Hot Water Extender Tempering Valve

area of the bottom of the diaphragm is less than the *area* of the top that is exposed to pipe line pressure. The lower pressurized area is less because the area below the closure disc/seat of the valve is not pressurized. Therefore, when the valve is closed at rest, the pressure above and below the diaphragm is equal, but the *force* above is greater than the *force* below and the valve is held closed by this difference in forces. <u>Note: there is a very small hole in the diaphragm which permits the incoming pressure to pass to the region above the diaphragm in a fairly short time.</u>

There is a pilot valve on the upper chamber, and when this pilot valve is opened, the upper pressure is greatly reduced, and the force pattern in the valve reverses, the diaphragm lifts, and fluid flows through the valve. When the pilot valve is closed, the pressure pattern is gradually restored (via the equalizing hole in the diaphragm), and the main flow is stopped when the diaphragm closes over the main valve seat.

VALVE BODY MARKINGS

Valves are usually marked with critical information. The following casting symbols are often seen on valve bodies:

WOG means that the valve is suitable for use with water, oil, or gas at 70° F temperature.

WSP means that the rating is for saturated steam pressure.

Class ratings are often shown. Typical values are 125, 150, 250; but many other ratings are possible, depending upon the standard to which the valve conforms.

Flow direction, where critical, is designated by an arrow.

BACKWATER VALVE

A backwater valve is a check valve arranged and trimmed for installation in a drain line. They are used to control backflow. They are also available in a combination pattern with a manual gate valve and check valve components. Figure 31-Q shows a combination type that is installed in a hub-and-spigot cast-iron soil pipe line.

Models are available for hubless installations, as well as models for plastic drain lines.

The fact that combination units are available suggests that the check-valve-only option is not a satisfactory option in many cases!

Figure 31-R shows a drain backwater check valve only.

MATERIALS OF CONSTRUCTION FOR VALVES

Valves used for plumbing work typically are made of the following materials:

BODY

Copper alloy (brass and bronze) for small valves for use at moderate temperatures and pressures. These alloys are limited to not more than 8% lead if they are installed in piping systems that convey potable water.

Cast-iron for larger sizes (over 2") and pressures up to 150 p.s.i. and 350° F.

Cast-steel for more severe services than cast-iron.

Stainless steel for more corrosive services.

Plastic for lighter duties and pressures, or where corrosion protection is critical. (Some heavy duty metallic valves are lined with plastic or epoxies for corrosion control.)

TRIM (SEAT RING, DISC OR FACING, AND STEM)

Trim materials available for valves generally used in

IN-LINE MANUAL SHUT-OFF GATE VALVE

FUNCTION: Functions as a drainage control valve; flapper type backwater valve provides normal protection against sewage backflow into building. Manually closed "spade" type gate valve provides additional complete "closure" protection during emergency storm conditions or when building is completely shut down.

REGULARLY FURNISHED:
Duco Cast Iron Body and Cover with Removable Wheel Handle. Bronze Gate and Flapper Valve. Cast Iron Extensions as Indicated by Figure Number.

VARIATIONS:
Outlet Sized for Service Weight -SW (When Specified)

OPTIONAL MATERIALS:
Galvanized Cast Iron -G

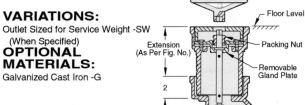

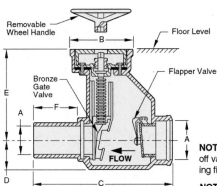

Fig. 7151C .....WITH 12 in. EXTENSION (04 in. size)
Fig. 7151Y......WITH 12 in. EXTENSION (04 in. size)
Fig. 7152C .....WITH 24 in. EXTENSION (04 in. size)
Fig. 7152Y......WITH 24 in. EXTENSION (04 in. size)
Fig. 7153C .....WITH 9 1/2 in. EXTENSION (06 in. size)
Fig. 7153Y......WITH 9 1/2 in. EXTENSION (06 in. size)
Fig. 7154C .....WITH 21 1/2 in. EXTENSION (06 in. size)
Fig. 7154Y......WITH 21 1/2 in. EXTENSION ((06 in. size)

Fig. 7150C .....CAULK INLET & NO-HUB OUTLET
Fig. 7150Y .....NO-HUB INLET & OUTLET

NOTE: During periods when manual shut-off valve is closed, use of building plumbing fixtures and drains must be avoided.

NOTE: Fig. 7150C and 7150Y are accessible from finished floor. Non-rising stem with spade-shaped gate assures positive closure.

NOTE: These valves offer protection against backwater surges. Backflow is prevented when valve is not obstructed by debris or sludge. Use for gravity flow only, not for pressurized applications.

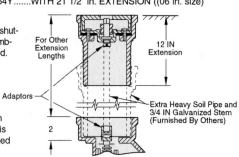

Fig. 7155C .....WITH EXTENSION ADAPTOR
Fig. 7155Y......NO-HUB 04 & 06 IN. SIZES ONLY

A SIZE	B DIA	C	D	E	F
04	8	18	3 3/4	12	5
06	10 1/2	20	4 1/2	14 1/2	5

7150 SERIES

Figure 31-Q – J. R. Smith 7150 Series In-Line Manual Shut-Off Gate Valve

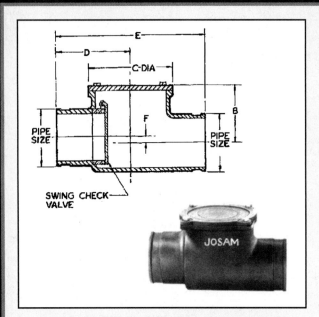

Figure 31-R – Josam Series 67400 No-Hub Swing-Check Type Backwater Valve

plumbing work include the following:

    Bronze
    Stainless steel
    Chromium steels
    Nickel-copper alloys
    Synthetic materials

The material selected for these trim options is usually based on cost, because most plumbing applications do not involve severe problems. When conditions are not typical, however, the following list of considerations must be reviewed:

Tensile properties, chemical stability, and resistance to corrosion

A temperature expansion coefficient that is similar to the coefficient of the valve body

Property differences in the disc and seat which will prevent them from binding together

Strength and hardness of the closure materials

Resistance to potential corrosiveness of the fluid in the piping

LESSON 32

BUILDING PLANS AND DRAWINGS

Any building project, from a tool shed to a multi-story high rise, would be difficult to impossible to produce if construction plans were not used. In order to acquaint you with the concepts used in construction plans, Lesson 32 is designed to introduce the following subjects:

1. Basic rules of reading plans and drawings in general and in relation to plumbing

2. Plumbing symbols used on plans and drawings

3. Tools and instruments used in the sketching and drafting of plumbing layouts

You will use the information in this lesson to develop sketching and drawing skills. By actually drawing the symbols and lines used, you will start to develop layout visualization, which is essential for on-the-job fabrication of piping and equipment installation. In order to do the basic sketching required, you will need the following items:

1. #2 pencil or fineline mechanical pencil

2. 30°-60°-90° right triangle

3. 45°-90° right triangle

4. Pad of coordinate (drafting) paper

5. Eraser

6. Scale ruler

Other drafting tools may be used, such as a T-square and drawing board. However, these tools may not be readily available on a job site.

PLANNING AND PROJECT DESIGN

It is important to know where we, as plumbers, fit into the building industry. The following lists the steps that usually occur to get a project started. There are many "players" in the process!

A. The owner or investor decides to construct a new building or modify an existing building.

B. The owner engages the services of *design professionals* — a planning firm, architect, engineer, and/or a construction management firm to develop the plans and specifications for the project.

C. The owner (investor) provides the design professionals with the necessary information so that a set of preliminary sketches and ideas can be prepared for the owner's approval.

D. Consultations between the owner and design professional continue until the basic design conforms to the owner's concept and budget.

E. Final plans — working drawings — are then completed. These are detailed enough to enable experienced mechanics of all the required trades to produce the desired end result.

Depending upon the complexity of the project, the *design professional* team may include one or more of the following specialists to assist in the completion of the working drawings and specifications:

ARCHITECT: The architect is generally in overall charge of design development and assures that the project is attractive, functional, and complete. General contractors sometimes have the necessary personnel to provide these services.

MECHANICAL ENGINEER: The mechanical engineer designs plumbing, heating, air conditioning, special process piping, and fire protection systems. These many specialties usually involve more than one individual. The mechanical contractor may provide the engineering as well as the installation.

ELECTRICAL ENGINEER: The electrical engineer designs power, lighting, communications, alarm, and similar systems. As with the mechanical part, the electrical contractor sometimes serves as both design engineer and installer.

STRUCTURAL ENGINEER: The structural engineer designs foundations, roofs, trusses, and building structural elements.

CIVIL ENGINEER: The civil engineer designs grades, site elevations and details, streets and highways, large water and sewer main extensions, railway systems, etc.

KITCHEN PLANNING SPECIALISTS: If necessary, this specialist will be called in to design and plan layout and equipment for commercial kitchen and food preparation establishments.

MATERIALS HANDLING ENGINEERS: In cases where large industrial plants need special systems designed to carry production materials, it is necessary to engage a specialist in the field.

OTHER SPECIALISTS: These specialists may be chemical engineers, process system engineers, interior decorators, or office interior designers.

The owner will initiate construction when the plans have been completed and the following details have been accomplished:

A.   Financing is arranged

B.   Approval from local zoning authorities has been obtained

C.   Assurance that the design conforms to all required codes and laws — usually accomplished by submitting plans to the applicable review board(s)

D.   Awarding contracts to the contractors

The owner may choose to act as his own general contractor or hire a general contractor to coordinate the work.

The plumbing contractor, unless part of the design team, does not become involved in the project until the General Contractor, (or the architect, in the case of **SEPARATE CONTRACTS**), solicits bids for the plumbing on the project. The contractor usually has an *estimator* take off the plans and prepare a bid. It is important that this estimator be familiar with plans and drawing methods in order to reach a price to do the job which is low enough to actually get the job and high enough to make a profit. Many estimators are plumbers who perform office work because of their ability to read and understand construction plans, and who are able to prepare a cost estimate from a set of plans.

REMEMBER: Most construction contracts are awarded to the low bidder.

When bidding a job, the plumbing contractor usually supplies a written proposal to the owner reflecting the estimated cost of installing the plumbing system according to the plans and specifications.

To develop the price for the proposal, the plumbing contractor (or the estimator) must first formulate a list for all material required to perform the job. This list is called a material "take-off." The labor required to complete the job is then estimated; subcontract quotations on all parts of the job not installed by the contractor's internal work force are obtained; and all these items then combined to arrive at the estimated job cost. After the estimated cost is known, the responsible managers of the contracting organization determine the desired margin (also called profit) that these persons believe is necessary to enter into a contract to perform the work.

If successful with the project bid, the contractor must have mechanics, equipment, and financial resources that are capable of installing the job according to the plans and specifications.

If you are going to advance to be a journeyman capable of installing and laying-out work or being an estimator, designer, or shop owner, you must be proficient in reading plans and drawings, so that your interpretations of the job requirements will be correct.

FUNDAMENTALS OF CONSTRUCTION DRAWINGS

Historical Notes on Physical Production of Plans

Construction prints now are black (or blue) line on white background but often they are still called *blueprints* because the earliest reproduction materials and methods produced white lines on a blue background.

The architect or engineer produced the original drawing on a high-quality transparent paper or plastic film. This original drawing (called a *tracing*) was as detailed as the job particulars required.

Sepias are a convenient print made (at about double the cost of a regular print) on transparent material. The original building and layout were drawn by the architect, but before the general construction notes were applied, sepias were made for the mechanical, electrical, and/or any other special designer. In this way, each designer was saved the time of executing the common material; and each specialist was assured of having precisely the same beginning layout.

More Recent Methods for Producing Plans

As early as the 1980's, progressive architects and engineers used CAD (Computer-Aided Design/Drafting) systems to produce original tracings. The computers of the times were slow, expensive, and were limited in memory and storage. However, these systems enabled changes to be made on the design in the computer, and a new paper copy made with very little effort.

Once the tracing was made (by the oldest methods or the newer ones), copies were made in whatever quantities were required. To make copies, the tracing was placed against a paper that has a light-sensitive emulsion coating.

This two-sheet combination was then passed through a machine whose bright light exposed the emulsion coating through the tracing. The tracing was removed and the copy paper sent through a developing process that removed the emulsion that was exposed to the light, and changed unexposed emulsion that was not illuminated to a black (or blue) image. Any coating that was covered with a pencil or ink mark on the original was changed and retained — the coating on all other areas was removed.

The earliest chemicals in this process retained the exposed areas and removed the unexposed, yielding the once-familiar blue-background, white-line print. The chemicals used now retain the unexposed area resulting in prints that are the more readable black-line (or blue-line) on white background.

This reproducing process can be repeated any number of times to make as many prints as desired, each of which is exactly like the original drawing. The original tracings can be stored in a safe place for long life and new prints made at any time.

Current Methods for Producing Plans

Now, just about all drawings are produced in computers (which have become much more capable and much less expensive). The cost of the drafting programs has also come down, while the capabilities have increased. The size and competence of the designer workforce is also greatly enhanced from those early days of the 1980's.

The most recent methods have evolved from the system described above to making plots on plain paper (which is essentially opaque) and making copies on large copy machines. These large machines are somewhat expensive, but all but the smallest design companies find them worth the cost. The copies are less expensive than the treated paper versions, and the flexibility of being able to copy just about anything, any time, is perceived as valuable by most designers and contractors.

Current Methods for Distributing Plans

Plans for larger projects are often placed on the Internet and contractors are expected to be able to download them and make their own copies. In this way, owners (and architects) are able to avoid the cost of producing, say, 100 sets of plans that could run to many pages. The cost of actual delivery of the many sets is also avoided.

The consequence of all this is that contractors must invest in computers, drafting programs, plotters, copy machines, *and personnel training* to "get into this game".

It is also possible for all the design professionals to exchange electronic information with the each other so that the basic drawing can be used by everyone who has a CAD system.

ALPHABET OF LINES

Craftsmen throughout the country are able to interpret plans because standards and rules have been established for architects and draftsmen. According to *Building Trades Blueprint Reading and Sketching,....*

instead of words, this "trade language" is made up of an alphabet of lines. The weight or thickness of the line and the way it is constructed, either solidly or as a combination of broken lines, are the methods used to convey specific meanings to those who are using the set of plans...

The *alphabet of lines* is used in the development of a drawing to distinguish between the objects described, characteristics of the objects, instructions and notes, and other details. The following lines are used to represent specific information. Each line will be shown in its part of the text as well as in Figure 32-K at the end of this lesson. We have included Figure 32-K so that the alphabet of lines can be seen in one illustration to help you note the differences between the lines. (Please note that the difficulties of precise reproductions in this book makes the differences between line thicknesses less pronounced on these pages than they actually are on a drawing print.)

OBJECT LINE

An object line is the visible edge perceived when an object is observed. It is a solid line (also called "continuous") with enough width to be bold on the drawing. See Figure 32-A.

Object lines appear thicker and darker than most other lines.

CENTER LINE

A center line locates the center of a circular object or part of an object. It also indicates that the object

Figure 32-A – Object Line

(or part) is circular or cylindrical.

Center lines appear as thin lines made of short and long dashes. Short dashes should be about $1/8$" in length; long dashes are drawn from $3/4$" up, depending on object sizes. Figure 32-B shows two examples.

DIMENSION AND EXTENSION LINES

Dimension lines show the extent of a length being displayed.

Dimension lines appear as thin solid lines. These lines contain a dimension number — usually midway between the ends. Arrowheads (or "tick" marks) are placed at the end of dimension lines.

Figure 32-B – Two Center Lines

HIDDEN LINES

Hidden lines show edges not seen in the view represented, such as hollow spots, internal passages, or where one object is above another object in the view being shown.

See Figure 32-D.

Extension lines are used in conjunction with dimension lines. They begin $1/16$" to $1/8$" from the edge of the object to $1/8$" to $3/16$" past the dimension line.

Figure 32-C – Two Hidden (or Dashed) Lines

Hidden lines are shown by short dashes approximately $1/8$" long with equal spaces between them.

Like dimension lines, extension lines appear as thin solid lines. Figure 32-D shows examples.

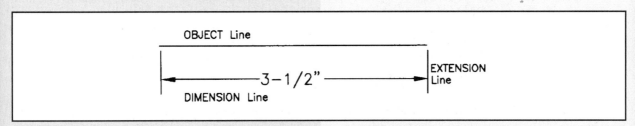

Figure 32-D – Dimension and Extension Lines

However, the dash length may vary if the drawing is larger. Hidden lines should be a little thinner than object lines. Some drawings may use two hidden lines for different purposes, as shown in Figure 32-C. Sometimes, these are called dashed lines if the dash is longer than the space between the dashes. The use of all these variations is the same: to show hidden edges in the drawing.

LEADER LINES

Leader lines connect notes to an object. They are another form of extension lines.

These lines appear as solid thin dark lines. See Figure 32-E.

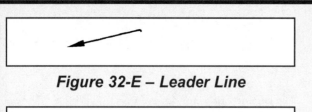

Figure 32-E – Leader Line

Figure 32-F – Cutting Plane Line

PHANTOM LINES

Phantom lines are used to show moveable position of parts.

Phantom lines appear as a long dash followed by two short dashes. These lines are thin and dark. See Figure 32-G.

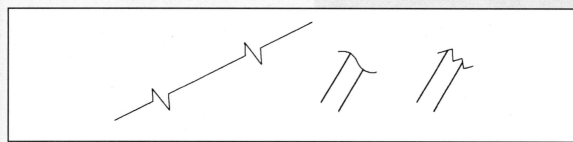

Figure 32-G – Phantom Line

CUTTING PLANE LINES

Cutting plane lines are used to show where a section is "removed" for examination.

Cutting plane lines appear as a thick line with arrowheads showing which area is to be examined. Two acceptable standards are used to show cutting

BROKEN LINES

Broken (or break) lines are used to show that the object continues beyond the portion shown in the plan.

Short broken lines appear as free hand irregular or reverse curve solid lines. Large break-away sec-

Figure 32-H – Long Break and Two Types of Short Break

plane lines: one standard states that the dashes should be of equal length; the other standard suggests alternating long segment and two short dashes. Cutting plane lines are the thickest lines on a drawing. See Figure 32-F.

tions are shown by thin line segments from ¾" to 1½" long with a free hand "Z" connecting adjacent ends.

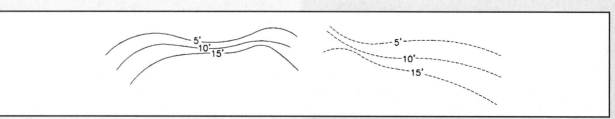

Figure 32-I – Before and After Contour Lines

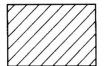 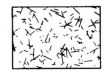

Figure 32-J – Typical Cross-Hatch Patterns

CONTOUR LINES

Contour lines are found on site plans and are used to show lines of equal elevations of the land. Existing grade contour lines appear as a free-hand continuous thin line with the elevation height noted somewhere along the line; new contours are represented by dashed lines that show the desired grade after the construction is completed. Some designers reverse this symbolism: dashed contour lines for existing grades, continuous lines for final grades.

Contour lines are usually identified in terms of the number of feet above a reference datum and are expressed in increments of feet. If several contour lines are drawn close together, the grade must slope steeply. In the same light, if the lines are widely spaced, the grade is slight.

CROSS-HATCH LINES

Cross-hatch lines are used in various patterns to show a cut surface. Cross-hatch lines are also used to mark-off any item for emphasis or clarity, e.g., supply ducts distinguished from return ducts, etc.

Cross-hatch lines are the thinnest lines on a drawing

DRAWINGS

You will see (and use) this **alphabet of lines** in two basic types of drawings throughout your career — drawings that depict products, and construction — usually called "Plans".

PRODUCT DRAWING

This drawing is usually furnished by the manufacturer of a product. It is used when parts or equipment are detailed to help understand installation or function. Object lines, center lines, hidden lines, leaders, extension lines, cutting plane lines, phantom lines, and cross-hatch lines are found in product drawings. The line technique and quality reflect information relative to the product shown.

CONSTRUCTION PLANS

Construction plans are used to detail the project being built. These drawings consist of plot plans, contour plans, landscape plans, floor plans, mechanical plans, and corresponding elevations and cross-sections, all of which must be considered by the mechanic.

Plot Plan

The plot plan is a birds-eye view of the site which outlines the position of the present and proposed structures. All visible items at the site, such as fences, sidewalks, garages, and other appurtenances will be shown on this drawing.

CONTOUR PLAN

The contour plan is a special plot plan which shows grade elevations. Sometimes contour information is included on the plot plan. All measurements in terms of the grade can be taken from some benchmark. The benchmark is critical to the building construction as the elevation of sewers, utilities, water supplies, etc., is taken from this reference point. Sometimes the benchmark is taken as sea level, sometimes at a nearby USC&GS (United States Coast & Geodetic Survey) marker, or other fixed object near the project. For remodeling projects, it may be the original building main floor elevation. Benchmarks are frequently taken as an arbitrary value of 100' elevation.

The contour plan is used by the plumber to be sure that buried lines have sufficient cover to protect them from freezing, vehicle loads, or gardening operations as the case may be.

LANDSCAPE PLAN

Landscape plans are important to the mechanic if renovation work is done. Sewers, water mains, septic systems, private wells, and other related piping should be as accessible as possible; therefore, they should not be placed under sidewalks, driveways, or major plantings.

FLOOR PLAN

Floor plans are a detailed presentation of the building structure itself, with little or no depiction of the surrounding site. Each floor is shown with shape, room dimensions, size of partitions, etc. The floor plan will show fixture placements in kitchens, baths, etc. It will also detail the placement of heating units, boilers, water heaters, and other appliances.

SPECIAL PURPOSE PLANS

The special purpose plans take the floor plans one step further. Placement of piping, ducts, plumbing, electrical, and other equipment is detailed. These plans are used any time the basic building floor plan gets too "busy" with overlapping details. Remember that no one wants more drawing sheets than are necessary to adequately describe the work that must be done, so all these special purpose drawings are used to avoid confusion, not just as an exercise in using up paper!

DRAWING LINES

When drawing lines, the following information must be remembered:

Object lines must be drawn with a slightly rounded pencil which will produce a thick, black line. When using a fine-lead mechanical pencil, draw two lines next to each other. With a little bit of care, you can develop a heavy line without gaps that would show that it was drawn twice.

Center lines are produced with a sharp pencil and light pressure for thin, black lines. Remember the length of the dashes (long: $3/4$" to $1 1/2$" — short: $1/16$" to $1/8$"), and the spaces between are about the same as the short dashes.

Hidden lines should be drawn using a slightly rounded point thereby making a medium black line (dashes $1/8$" long with equal spaces between).

Dimension lines, leaders, and extension lines should be drawn with a sharp pencil to produce a thin, black line.

Cutting plane lines are thick and dark; therefore, the pencil should have a rounded point. The line may have to be formed with adjacent pencil strokes to get the desired width. A mechanical pencil may require multiple strokes.

Phantom lines are drawn with thin, dark dashes; therefore, the pencil should be sharp.

Broken lines are drawn in a free hand method and should be a thick solid line (rounded pencil point) for small objects or a thin sharp line (sharp point) for larger objects.

Contour lines are drawn using a sharp pencil to construct a thin dark line.

It is recommended when sketching preliminary drawings that a #2 pencil on coordinate paper (graph paper) be used. Spend time sketching lines at this point in the lesson. Emphasis should be on memorizing line representation, rather than on line technique.

An alternative favored by some draftsmen is to use mechanical pencils of 0.5mm diameter for fine lines, and 0.7mm or greater for heavier lines. For extremely wide lines (cutting plane), it may be necessary to draw two or three overlapping lines. The graphites for these pencils are available in different hardness, so by varying hardness and diameter, the draftsman can obtain the line widths and intensities desired. (This advice is OK for the office where you may have all these options — out in the field you usually will have to make do with one or two pencils.)

In the earliest days of computer-aided drafting (1980's to mid '90's), these line-width conventions might be ignored in some cases because of the limitations of the equipment available to put the computer information on paper.

Computer-produced drawings are generally more readable, so some sacrifice in these line conventions was not a bad trade-off. However, as this is written (in 2005), routine plotters for generating computer drawings can make just about any line width the designer would like, so these width options are commonly seen in serious construction drawing sets.

All the lines in this series of lessons were drawn by computer and put on paper by a reasonably inexpensive drawing machine called a plotter. The example lines were then "scanned" into the computer data for this text.

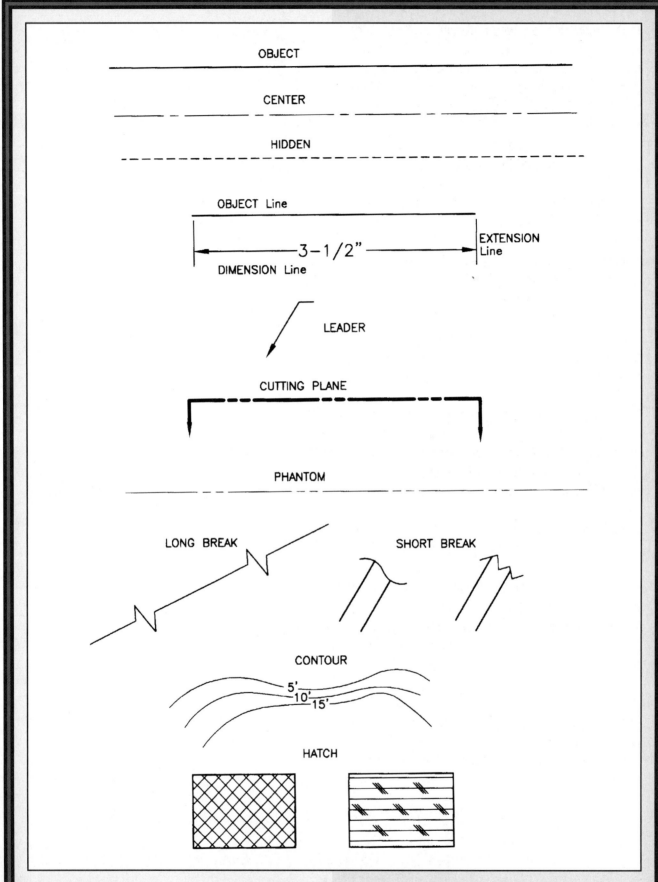

Figure 32-K – Alphabet of Lines

LESSON 33

SCALE RULERS

CONCEPT OF SCALE

The projects we build, and even the equipment we install, are physically too large to be represented by full-sized drawings. As a matter of convenience and economy, we depict our work with drawings that are reduced in size from the actual product or building.

To interpret these drawings, we measure lengths with a Scale Ruler — a ruler calibrated or marked in terms of standard scales commonly used on drawings. A very popular configuration is a 12" ruler with a triangular cross section. This shape provides six edges (the front and back of each corner of the triangle), and scales are marked from each direction, so twelve scales in all can be shown on one ruler. Figure 33-A shows several examples.

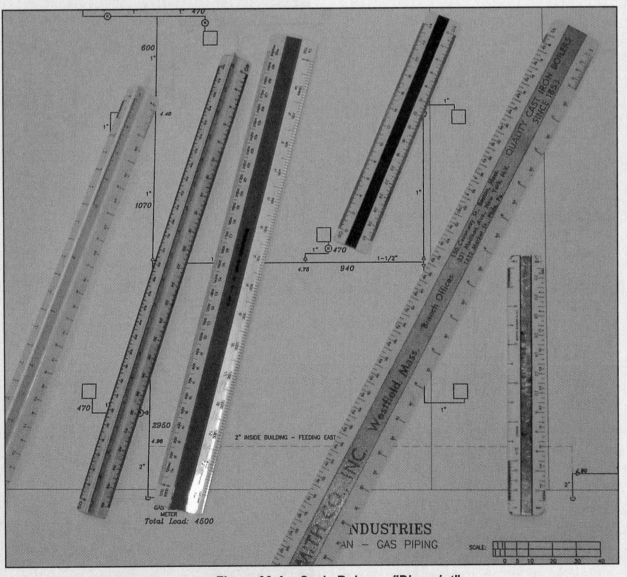

Figure 33-A – Scale Ruler on "Blueprint"

Typical scales for architectural applications are marked as follows:

One edge is full-sized in inches, 3/32 & 3/16, 1/8 & 1/4, 1/2 & 1, 3/8 & 3/4, and 1-1/2 & 3 for the six edges. Note that the scales on any particular edge are in the ratio of 1 to 2.

The use of such tools makes it easy to use the same scale as used by the person who prepared the drawings. Each sheet will indicate the scale to use on that sheet. If there are details of different scales, these will also be marked, usually just below the title of the detail.

Drawings that show an entire building floor plan use scales depending on building size. Portions of the building may be shown in

$$\frac{1"}{16} = 1'-0" \qquad \frac{1"}{8} = 1'-0" \quad \text{or} \quad \frac{1"}{4} = 1'-0"$$

$$\frac{1"}{4} = 1'-0" \qquad \frac{3"}{8} = 1'-0" \quad \text{or} \quad \frac{1"}{2} = 1'-0"$$

Details that are complicated (e.g., door or window headers) may be shown as large as

$$1\frac{1"}{2} = 1'-0" \qquad 3" = 1'-0"$$

Sometimes, these scales are expressed this way:

$$1" = 16' \quad 1" = 8' \quad 1" = 4' \ etc$$

Scales used by civil engineers for large areas may be selected from these values:

$$1" = 10' \quad 1" = 20' \quad 1" = 30' \quad 1" = 40' \quad 1" = 50' \quad 1" = 60'$$

The scale $1/8" = 1'-0"$ is read as **one-eighth inch equals one foot**. This means that every eighth of an inch drawn corresponds to one foot in actual size. If a pipeline is drawn two inches long on a plan marked $1/8" = 1'-0"$, the actual length of that pipeline is 16 feet.

Scale rulers are available for all the examples given and for many more. You can measure lengths with a folding ruler and convert by dividing by the scale, but this procedure is slow and laborious. Using a scale ruler will speed the work and reduce the chance for error.

High-quality scale rulers have a white ivory or plastic overlay on the edges with numbers and markings etched into their surfaces so that they are highly visible and long-lasting. These marks will not wear off as easily as on the models with the markings simply stamped on the edges.

Rulers should be kept clean and free of smudges and should be stored away in protective cases when not in use.

Scale rulers should be used for measurement purposes only and should not be used as a straight edge for drawing lines.

Every contractor's office uses scale rulers to develop an estimate for a job. As described in Lesson 32, the first thing required is to make a list of the quantities of materials needed to install the work. This list is priced; that is, the cost of the material is calculated and the labor estimate is also computed from this list. You can see that if the wrong scale is used, this material list can overstate or understate the amounts of pipe, insulation, hangers, excavation, etc., that the job will need.

LIMITATIONS OF SCALING DRAWINGS

Because of tolerances in drawing and reproduction methods, scale rulers cannot be used to determine *precise* locations or dimensions of building details. These locations must be shown by indicating dimensions with extension lines and dimension lines on the drawing.

Critical measurements, such as the location of plumbing-containing (wet) walls, fixture rough ins, etc., should be noted in the architect's dimensioned detail sections.

For non-critical measurements, where tolerances are plus or minus a few inches, scaling directly from architectural plans is acceptable.

See Figure 33-B for an example floor plan which utilizes the *Alphabet of Lines* and scale, and includes some insights into the above remarks.

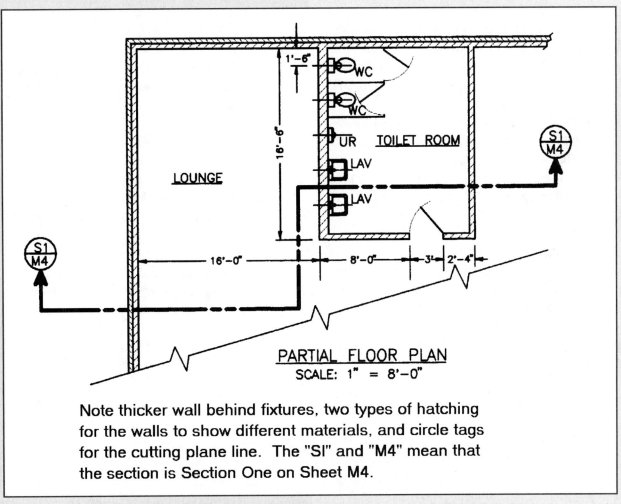

Figure 33-B – The Alphabet of Lines, Scale, and Meaning of Notes

Some drawings will be marked **NOT TO SCALE**. This means that general arrangement only is shown. You should immediately refer to the architect's notes, specifications, and/or detail sections for explanation of certain piping or equipment situations. Sometimes **NOT TO SCALE** drawings seem plausible but cannot be accomplished on the job. Be very careful with such drawings!

PROBLEMS WITH PRINTED SCALE ON DRAWING SHEET

Because of the widespread use of computer-generated drawings, most designers use a "Scale Bar" on their drawings rather than print a numerical value for the scale. The reason for this is that drawings can be easily plotted to different scales by the plotting methods that are in general use. Thus, if a drawing states $1/4$" = 1'-0", but it is plotted to $1/8$" = 1'-0", everyone who relies on the printed scale will be in error! Such reduced-size plots are commonly produced for checking by the other professionals in (or outside) the office, for progress review by the owner, for presentation to plan-review officials, and sometimes for more economical presentations to contractors for bidding purposes.

It is essential that all workers on the job use the same dimensions and starting point when laying out the job. Be sure that you discuss these layout considerations with the carpenters, masons, and electricians on the job. On large projects, the job superintendent will give directions.

During the actual construction, the mechanics on the job will also need scale rulers to measure job components or to interpret special shop drawings of prefabricated subassemblies.

FURTHER DESCRIPTION OF VARIOUS RULERS

FOLDING RULE

The folding rule is a device usually 72" in length, with $1/16$" increments dividing the ruler throughout its length. Usually made of wood, it is made to fold into 6" or 8" sections. It is calibrated on both sides for convenient use; and may be equipped with brass extension leaders for internal measurements. It is used to measure and lay out work to full size. Some folding rules are 96" long. See also Lesson 3.

SCALE RULERS

As described above, the triangular scale ruler usually has eleven scales marked on its edges.

Scale rulers are also made in flat form where the edges taper down to the paper. This arrangement produces a "top" and "bottom" where only the top is useful. Since only two edges are available, only four different scales can be shown. It is preferred by most engineers and architects, however, when the variety of scales on the triangular version is not needed because there is less "flipping around" with this version.

When using a scale ruler to make a drawing, follow these recommended procedures:

1. Select a scale that will produce a representation that will fit the tracing paper, and will also permit sufficient detail so that the job will be adequately described.

2. If both needs cannot be met with one selection, use a small scale to display the total building, and larger scales to show required details. As you encounter plans for typical jobs, you will see that this method has to be used on most projects.

3. Use the scale ruler to mark the measurements after you have drawn the lines with a straight-edge.

4. Be sure to note the scale used, preferably with a scale bar.

When reading a plan using a scale ruler, use the following steps:

1. Select the proper scale on the scale ruler.

2. To measure a particular segment, place the ruler so that one end of the segment is at or near the zero mark.

3. Adjust the ruler so that the other end of the segment is at a whole number mark on the scale.

4. Read the fractional part of the dimension past the zero mark on the scale, readjusting if necessary so that one end of the segment is within one foot past the zero mark, and the other end of the segment is at a whole number mark.

ARCHITECTS SCALE RULER

This item is used predominantly for buildings, as it is laid out in feet along its entire length. All measurements taken with the Architects scale ruler are in feet. There is one foot divided into inch calibrations just past the zero, or starting point, on each scale. The Architects scale ruler is the scale most frequently used by plumbing mechanics.

The scales used most often are:

$$1/8" = 1'-0" \quad 1/4" = 1'-0" \quad 1/2" = 1'-0"$$

EXAMPLE

Use the ¼" = 1'-0" scale.

Determine the total length of a line which must be drawn on a paper to correspond to a length of 30 feet.

Reading from right to left on a scale ruler, locate the numbers 0 and 30.

Mark the length onto a drawing.

Solving the problem mathematically, if ¼" = 1'0",

$$\frac{\frac{1"}{4}}{1'-0"} = \frac{x}{30'-0"} \qquad \frac{\frac{1"}{4} \times 30'}{1'} = x \qquad 7\frac{1"}{2} = x$$

Therefore, this length is only 7½" long when measured on the drawing sheet.

ENGINEERS SCALE RULER

Engineers scale rulers are available for either the Civil Engineers scale or the Mechanical Engineers scale. The Civil Engineers (also known as the graphic scale) is divided into decimal parts. Scales are listed as 10,

20, 30, 40, 50, and 60 which correspond to 1"=10', 1"=20', etc., or 1"=100', etc. This ruler is predominantly used as a topographers (map makers) scale, however, it can also be used to measure inches in decimal parts.

The Mechanical Engineers (people who design machines, not plumbing and heating systems in buildings) scale ruler is used to measure drawings to a one-eighth, one-quarter, one-half, or full scale. If a half scale is used, the drawing is actually one-half of the object's true size.

These rulers are marked in inches and inch increments, rather than feet.

METRIC SCALE RULER

A common scale ruler calibrated in metric units is 30 cm in length (approximately one foot). It is divided into millimeters and centimeters for metric measurements. Metric scales use scale ratios of 1:10 or 1:100 rather than English standard measurements such as ¼" = 1'-0".

The metric scale can be used for architectural drafting applications. Buildings designed in Europe will be in metric measurements. Some equipment drawings or highway projects may have metric dimensions.

EXAMPLE

If a building measures 50' x 75', all lengths could be converted (15.24m x 25.56m) to metric units, then placed on paper using the 1:50 ratio. Since the scale lines are subdivided into portions of 10, all partial sections are easier to lay out on the drawing.

Metric scale rulers come in a variety of ratios. For working drawings, use the following metric scales which are referenced to the nearest English standard sizes.

Customary Inch Ratios	Approximate Metric Scale
1:192 ($\frac{1"}{16}$ = 1' - 0")	1:200
1:96 ($\frac{1"}{8}$ = 1' - 0")	1:100
1:48 ($\frac{1"}{4}$ = 1' - 0")	1:50

PROBLEMS

Determine lengths represented for scales given:

$1/4" = 1'-0"$ $1/8" = 1'-0"$ $1/16" = 1'-0"$

_____

_____

_____

_____

Calculate the area of the rectangle for the scales given.

Area = _____ sq. ft. ($1/4"$ = 1'-0")   Area = _____ sq. ft. ($1/8"$ = 1'-0")

Area = _____ sq. ft. (1" = 100'-0")

LESSON 34

SKETCHING — FREEHAND AND WITH DRAFTING TOOLS

DRAWING TERMS

Certain terms are used when referring to plans and prints. So that we all use these terms to mean the same thing, study the following definitions and use the words accordingly:

PLANS (GENERAL) *Plans* is a term commonly used to refer to any drawing(s) or print(s).

PLAN (VIEW) A plan view is a drawing where the observer is overhead looking down toward a horizontal plane.

ELEVATION An elevation is a view of a vertical plane, usually also described in terms of the direction the view faces.

SECTION A section view shows a particular feature or features as though the object were cut-through at a *cutting plane*. Used to show detailed arrangements of parts, they can be plan or elevation views.

WORKING DRAWINGS

Nearly all mechanics will use a drawing — called a working (or shop) drawing — to install a plumbing system. For small jobs, this drawing may be a rudimentary sketch; and for larger jobs, it could be the set of architect's plans. Most of the time, it is necessary to make additional detailed drawings to be certain that the work is installed as intended. Frequently, these supplemental drawings will be sketches of a small segment of the work. These drawings may be sketched by the journeyman or foreman. They all serve the same purpose: to ensure that the work is done properly.

The formal plans and the working drawings use piping layout symbols to depict the required installation, like those shown in Table 34-A.

Figure 34-A includes a floor plan of a single-story house showing fixture locations and an elevation showing piping installed in relation to the floor, ceiling, and roof.

Two vertical systems extend from the horizontal pipe up through the roof. The vertical pipe lines are called stacks; the lowest horizontal pipe is called the building drain. When this line exits the building, it is called the building sewer. Note that the pipes are represented with a solid line where they are drains — i.e., convey wastes — and with a dashed line where they are vents — i.e., carry only gases.

As you study this sketch, note that the vents extend upward from the point where liquid wastes enter the drain pipe, and that there is a trap at each fixture, except water closets (which contain an integral trap). Note also the cleanouts at the stack bases, and the roof flashings where the vents pass through the roof (VTR).

The floor plan shows the usual symbols for bathtub, tank-type water closet, lavatory, and sink. These symbols will vary somewhat from one draftsman to another, but most of them use symbols similar to those shown.

Study Figure 34-B. This figure shows the plan views of the basement and first floor of a residence. Figure 34-C shows an elevation view of the waste and vent piping. The plan views show water piping; the elevation view shows drain and vent piping.

On the elevation view, all the fixtures are indicated, even though they are at opposite ends of the house. The advantage of this practice is that all fixtures are indicated in one view — the disadvantage is the confusion possible. For larger jobs or for more fixtures, separate elevation views would be made of each stack. Table 34-A displays many of the typical symbols used on plumbing and piping

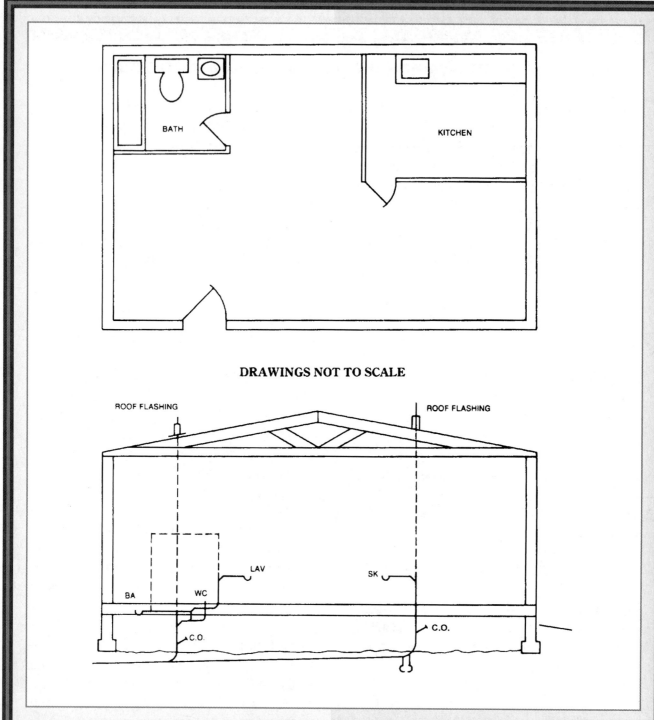

Figure 34-A – Floor Plan and Riser Diagram Rough Sketch

drawings. Because there is little standardization on these symbols, most plans will include a legend that will identify all the symbols used by the draftsman who prepared the drawings.

The elevation view in Figure 34-C shows the piping extending from the back of the house to the front as lines, whereas anything extending from side to side in the house is *lost*, because it can only be represented by points, which are covered by the lines. We will develop a solution to this problem in later lessons.

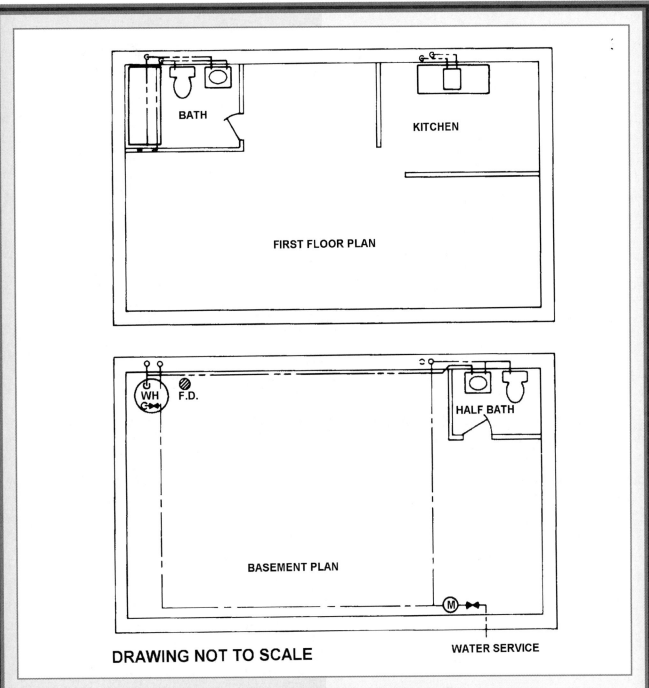

Figure 34-B – Plan Views of a Typical Dwelling

Make sketches of the piping required for a small restroom, as directed by your instructor.

SKETCHING — FREEHAND

APPEARANCE OF LINES

We are concerned with the *appearance* of a line in the building as it shows up in various plan views. The discussion that follows illustrates the problem.

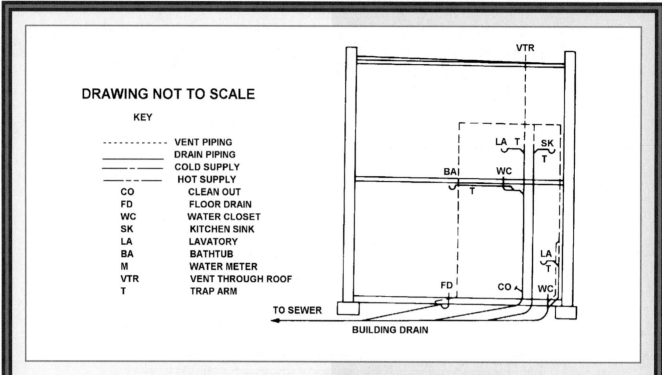

Figure 34-C – Elevation of Dwelling Shown in Figure 34-B

Vertical piping and some horizontal piping will appear as lines in elevation views, whereas vertical piping will be a point or small circle in plan views.

Horizontal piping parallel to the elevation view being drawn will appear as lines, whereas piping perpendicular to the view will be points or circles.

Symbol	Description	Line	Description
	BATHTUB—RECESSED	————————	DRAIN LINE
	BATHTUB—FREE STANDING	— — — — — —	VENT LINE
	BATHTUB—CORNER	—— — —— — ——	COLD SUPPLY
	SHOWER STALL	———— — ———— —	HOT SUPPLY
	URINAL—WALL MOUNTED	— — — — — —	HOT RECIRCULATION
	LAVATORY (VANITY BOWL)	— VAC — VAC —	VACUUM LINE
	KITCHEN SINK AND DRAINBOARD	— FG —	FUEL GAS
	WATER HEATER	— FP — FP —	FIRE PROTECTION
	WATER CLOSET	— A — A —	COMPRESSED AIR
	LAUNDRY TUBS		
	FLOOR DRAIN		
	ROOF DRAIN		
	METER		
	HOT WATER TANK		
	SILL COCK		

Many other symbols are used; study each print for a schedule of equipment symbols.

Table 34-A – Typical Plumbing Symbols

DETAILS

Figure 34-C shows these plumbing details:

BULIDING DRAIN The building drain is defined as the lowest horizontal piping of a plumbing drainage system which receives the waste from soil, waste, and/or other stacks inside the building and terminates at least three feet outside the outer face of the building wall.

TRAP ARM The trap arm is that portion of a fixture drain between the trap outlet and its vent.

HORIZONTAL DRAIN A horizontal drain is any horizontal pipe which carries waste water or water-borne wastes in a building drainage system.

Other details include the floor drain (**FD**), cleanout, water meter, and water piping.

Some presentations will employ the scheme used here — i.e., water piping in one view, waste and vent in another — to avoid confusion. Other plans will show water and waste on the same view; therefore, estimators and mechanics must study these plans very carefully.

Show fixture locations in Figure 34-E for the fixtures shown in Figure 34-D. This single-family dwelling has a bathroom, laundry, and kitchen that have to be indicated in an elevation view to show plumbing stacks. We will develop these stacks in the next few pages.

Select an elevation view that shows the bathroom on the left, laundry in the center, and kitchen on the right. You will need a #2 pencil and eraser for the initial sketches. As we continue, we will develop the complete drawing using a straight-edge and triangles.

Figures 34-D and 34-E are printed on a sheet (page 409 and 410) that can be removed to do this exercise. Figure 34-E is printed over a graph paper background to assist you in keeping the lines straight.

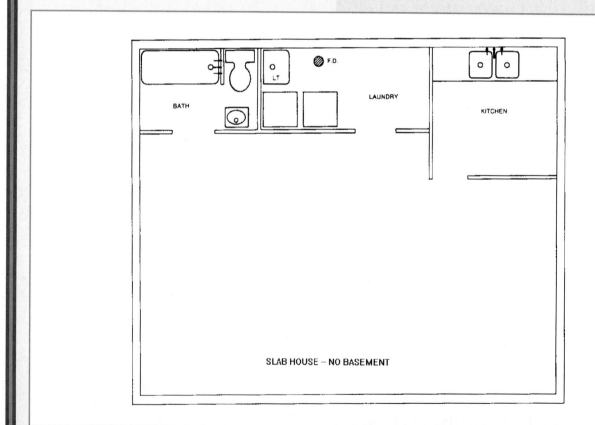

Figure 34-D – Layout for Student Problem

We will continue this sketch in the next section, after we discuss venting considerations.

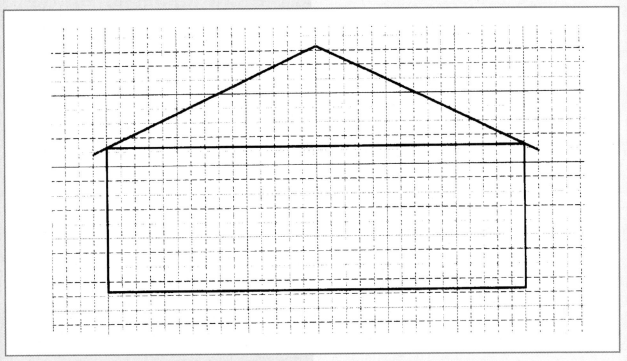

Figure 34-E – Freehand Sketch to be Developed by Student of Elevation View of Drainage on Graph Paper

VENTING

The vent is a pipe or system of pipes of a plumbing system which supply or remove air to relieve pressures below or above atmospheric. The vent pipe is a **breather** for the plumbing system. Vent pipe is usually shown by a dashed line on a drawing.

The vent piping system is a system of pipes originating from the fixture drains. Venting system rules vary in different locations, therefore, local codes should be studied for details which deal with venting. Despite the variations in fixture and system venting requirements, all venting systems are designed to provide air circulation so that the maximum pressure variation within the total drainage piping system is 1" water column from atmospheric pressure. If this criterion is observed, waste products will drain properly, and sewer gases will be contained within the piping system and be properly vented outdoors.

Different venting terms will be encountered during your training such as *yoke, relief, wet, dry, back, loop, circuit, stack,* etc. These terms describe certain venting configurations, all of which will be examined more extensively in later lessons. It is important to realize that each venting configuration is performing the task of maintaining neutral pressures in the system and provides an escape system for sewer gases.

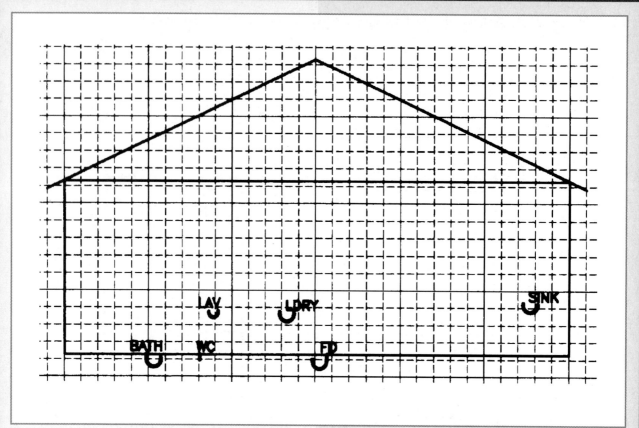

Figure 34-F – Beginning Drain Sketch for Figure 34-E

Continue to place fixtures on Figure 34-E, based on Figure 34-D. Your fixture location sketch should look like Figure 34-F.

SKETCHING — (CONTINUED)

Single line sketching is a necessity for understanding piping and plumbing. You should be able to **see** a job in your mind before doing any part of it. Plans assist you to understand the installation.

After you have items located similar to Figure 34-F, continue your work to show the piping, which should result in a sketch something like Figure 34-G. Follow these concepts:

1. Drainage pipes are drawn with continuous lines.

2. Vent pipes are represented by dashed lines.

3. Vent pipes, depending upon the length of the run or upon the local code, may extend through the roof, or may tie into the stack or other vent.

4. Vent pipes passing through the roof should be shown as extending according to local code and should be covered with an approved flashing material.

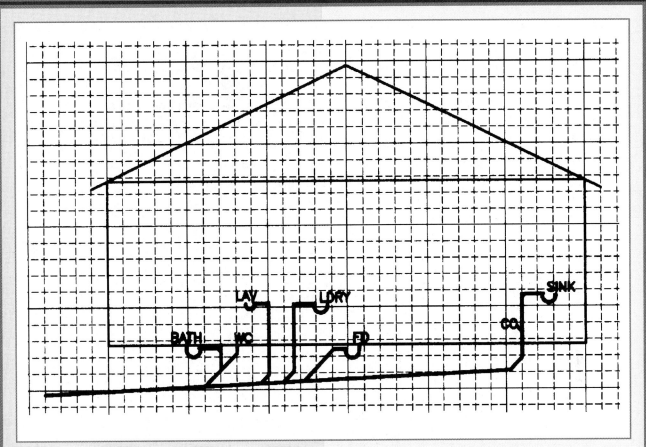

Figure 34-G – Drain Sketch – Continue With Freehand Sketch of Vents

Be sure that you slope all horizontal pipes down slightly in the direction of flow. Locate cleanouts on the stacks in accessible locations.

Make the drain pipes solid and dark.

Draw the drainage piping (building drain flowing to the left of the drawing) and connect all items. Be sure to show a radiused entry into tees and 45° entry into wyes for this drainage piping

At this point, you have prepared a <u>freehand</u> drainage diagram of a single story house which includes the following fixtures:

1. Kitchen sink
2. Laundry tub
3. Washing machine standpipe
4. Water closet
5. Lavatory
6. Bathtub

The sketch should appear similar to Figure 34-G. Continue on this figure to show the venting.

ADDING VENTS

The most important idea to realize at this point is how venting is shown on the plans, and how to draw vents.

Beginning at the tub trap arm, extend a vent up through the roof. The lavatory and laundry trap arms are extended up above the rim (flood level rim) of each fixture and then tied into the vent that rises from the tub drain. A dry vent may not be horizontal until it is at least 6" above the flood level rim of the fixture being vented. Some jurisdictions require that the floor drain also must tie to that vent group — other jurisdictions do not require this connection. Ask your instructor for the code requirement in your area.

The final sketch should look like Figure 34 - H

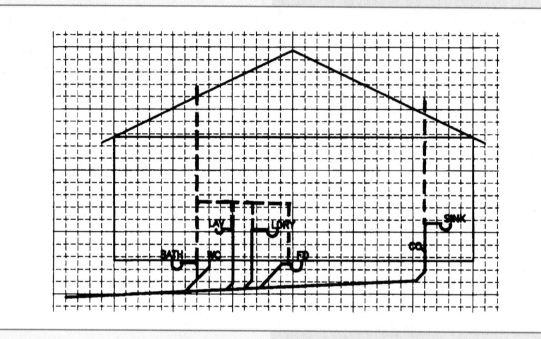

Figure 34-H – Completed Sketch

SKETCHING WITH DRAFTING AIDS

Using drafting tools, the student should make the same drawing developed from freehand sketches in Figure 34-E to G into a more finished form.

Sketches are drawn to highlight a particular job detail, to help the workers visualize a portion of the project, or to show inspectors or architect's representatives how you plan to proceed with the job. To make an improved drawing of your initial freehand sketch, you should use standard methods and symbols. Draw them in a neat, professional manner.

It is helpful to use a drafting table and tee square also, but if only occasional sketches are required, the following list of tools will be sufficient.

The drafting tools necessary to make quality sketches include the following:

1. 0.5 mm pencil
2. 0.7 mm pencil
3. #2 pencil
4. Eraser
5. 30°-60°-90° triangle
6. 45°-45°-90° triangle
7. Graph paper (also called coordinate paper)

If a 0.5mm or 0.7mm mechanical pencil is not available, a standard #2 pencil will produce a dark line for best reproduction. Sharpen the pencil and use it as suggested by the following table for various lines:

LINE DESIRED	#2 PENCIL POINT	LINE CHARACTER
Object Line	Rounded	Dark and Thick
Hidden Line	Slightly Rounded	Dark and Medium
Contour Line	Sharp	Dark and Thin
Dimension Line	Sharp	Dark and Thin

Draw a uniform line by moving the pencil along the straight-edge at a constant angle. Rotate the pencil as you move along the straight-edge.

Use the light-blue gridwork of lines on graph paper as a reference frame; the grid will not reproduce in the usual copying processes. Such preprinted paper is available in many line spacings and types for various applications.

Note that the sheets prepared for this lesson have highly visible graph gridwork for training purposes.

Scale rulers are used to draw the sketch to scale. However, the graph paper grid can often be used to eliminate the need for a scale ruler.

Pink Pearl or similar erasers are used to correct errors and to generally keep the sketch clean and professional-appearing. These erasers do a good cleaning job and yet do not destroy the paper surface. Clean the eraser by rubbing it on a blank sheet before using it to clean your sketch.

Figure 34-I – Paper Mate Pink Pearl Eraser

Two plastic triangles complete the basic equipment list. They can be used to lay out standard angles, as a straight edge for line drawing, and together to draw parallel lines. For best results, pick up the triangle before you move it on the sketch — sliding it over the sketch will produce unsightly smudges. Keep the triangle surfaces clean to eliminate smudging the drawing.

Drafting aids such as circle, ellipse, plumbing fixture, and detail templates speed the sketching process, as do conventional drafting tools — compass, dividers, bow compass, bow dividers, and beam compass. Such professional devices are not necessary to produce good looking sketches, however.

Refer again to Figure 34-H. A building section elevation can be drawn using light lines. Lightly sketch the piping in the elevation view, sloping the horizontal lines and drawing the vertical lines as vertical. After checking these lines, highlight them so they are dark and thick. Use the normal line convention for drains (continuous) and vents (dashed). Place only enough notes on the sketch to make the details clear.

Thus, the free-hand sketch completed as Figure 34-H can be improved with simple drafting aids to the sketch shown in Figure 34-J. Work at these methods to develop facility with these simple drafting tools.

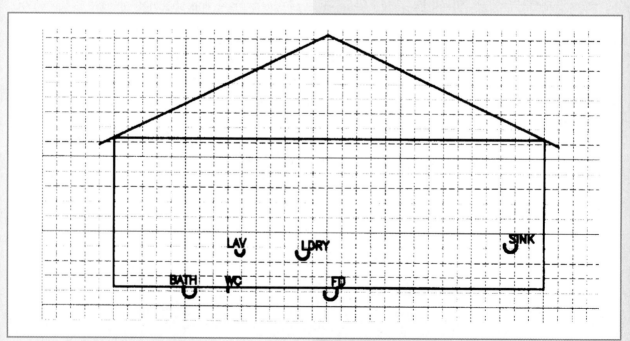

Figure 34-J – Start for Student Piping Layout Sketch Using Drafting Tools

LESSON 35

SYMBOLS AND DETAIL SKETCHING

In our study of plan and drawing reading to this point, we have been concerned with the lines used to represent objects; buildings; and details, such as piping; with distinctions based only on the function of the line (waste, vent, cold water, etc.). When more detail is needed, we must use symbols to indicate the method of construction, exact fittings, or the material of the piping. Such detail is needed in sketches for large pump installations, softener interconnecting piping, other complex commercial or industrial piping jobs, or appliance installations.

Such details are also needed for prefabricated assemblies, such as in apartment or hotel bathrooms. They are likewise needed for any other application where space is critical or where the designer must determine the exact arrangement of fittings.

Figure 35-A shows symbols frequently used to detail fittings, devices, and joints. A few of these symbols signify a particular material (soil pipe = cast-iron), but most of the symbols can be applied to several materials, in various applications.

Usually, these detail symbols are used when the person making the sketch wants to control the precise selection of fittings and their arrangement.

Draw sketches showing these combinations:

Exercise 35-A	A horizontal tank, valved inlet and drain on the bottom, outlet on top.

Exercise 35-B	A centrifugal pump with flanged valve and strainer on inlet, flanged valve and check valve on outlet.

Exercise 35-C	A closet bend below a floor connecting to an 1/8th bend and wye in the stack.

Exercise 35-D	A hot water and cold water riser connecting through valves to mains on the basement ceiling.

Keep in mind that the symbols shown in Figure 35-A are not the only ones possible for the products shown. All drawings should show a schedule of symbols used for the project to avoid misunderstandings.

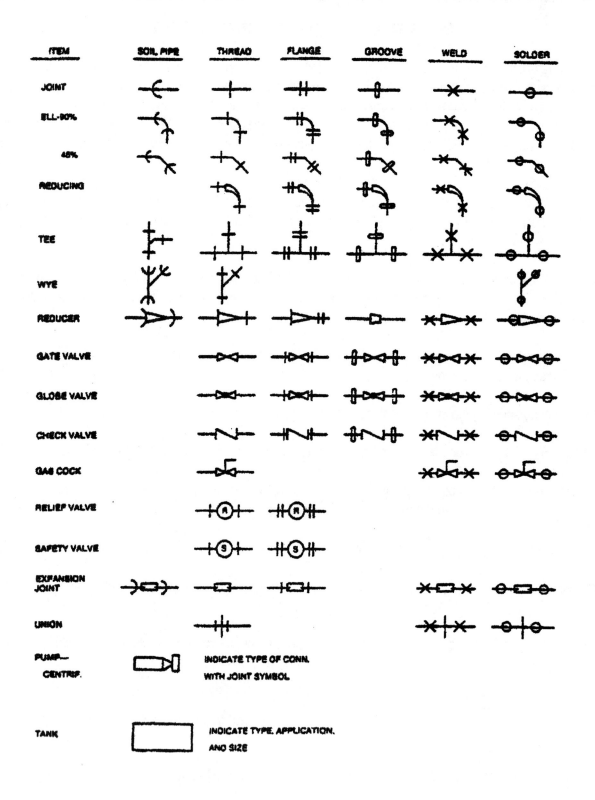

Figure 35-A – Piping Symbols and Definitions for Draftsmen (1)

Symbol	Description	Abbrev.
——— — ———	STORM, DRAIN, RAINWATER DRAIN	SD
———————	SOIL, WASTE OR SANITARY SEWER	S
— — — —	SUB-SOIL DRAIN, FOOTING DRAIN	SSD, FD
– – – – –	VENT	V
——— AW ———	ACID WASTE	AW
– – AV – – –	ACID VENT	AV
——— D ———	INDIRECT DRAIN	D
——— — ———	COLD WATER	CW
——— SW – – –	SOFT COLD WATER	SW
– – DWS – –	CHILLED DRINKING WATER SUPPLY	DWS
——— DWR ———	CHILLED DRINKING WATER RETURN	DWR
——— — — ———	HOT WATER	HW
——— — — ———	HOT WATER RETURN	HWR
——— TEMP.° – – –	HOT WATER	TEMP. S
——— TEMP.° – – –	HOT WATER RETURN	TEMP. R
♀	AQUASTAT	AQ
——— CL ———	CHLORINATED WATER	CL
——— DI ———	DISTILLED WATER	DI
——— DE ———	DEIONIZED WATER	DE
– – ATV – – –	ATMOSPHERIC VENT (STEAM OR HOT VAPOR)	ATV
——— PD ———	PUMP DISCHARGE LINE	PD
——— F ———	FIRE PROTECTION WATER SUPPLY	F
——— SP ———	AUTOMATIC FIRE SPRINKLER	SP
——— G ———	GAS—LOW PRESSURE	G
——— MG ———	GAS—MEDIUM PRESSURE	MG
——— HG ———	GAS—HIGH PRESSURE	HG
——— FOS ———	FUEL OIL SUPPLY	FOS
——— FOR ———	FUEL OIL RETURN	FOR
– – – FOV – – –	FUEL OIL VENT	FOV

Figure 35-A – Piping Symbols and Definitions for Draftsmen (2)

Symbol	Description	Abbr.
——— RG ———	REGULAR GASOLINE	RG
——— NLG ———	NON-LEADED GASOLINE	NLG
——— PG ———	PREMIUM GASOLINE	PG
— — GV — —	GASOLINE VENT	GV
——— LO ———	LUBRICATING OIL	LO
— — LOV — —	LUBRICATING OIL VENT	LOV
——— WO ———	WASTE OIL	WO
— — WOV — —	WASTE OIL VENT	WOV
——— A ———	COMPRESSED AIR	A
——— X#A ———	COMPRESSED AIR—X#	X#A
——— MA ———	MEDICAL COMPRESSED AIR	MA
——— LA ———	LABORATORY COMPRESSED AIR	LA
——— V ———	VACUUM	V
——— MV ———	MEDICAL VACUUM	MV
——— LV ———	LABORATORY VACUUM	LV
——— VC ———	VACUUM CLEANING	VC
——— O ———	OXYGEN	O
——— LOX ———	LIQUID OXYGEN	LOX
——— N ———	NITROGEN	N
——— LN ———	LIQUID NITROGEN	LN
——— NO ———	NITROUS OXIDE	NO
——— H ———	HYDROGEN	H
——— HE ———	HELIUM	HE
——— AC ———	ACETYLENE	AC
——— AR ———	ARGON	AR
——— LPG ———	LIQUEFIED PETROLEUM GAS	LPG
——— INW ———	INDUSTRIAL WASTE	INW
——/——/——	LOW PRESSURE STEAM	LPS

Figure 35-A – Piping Symbols and Definitions for Draftsmen (3)

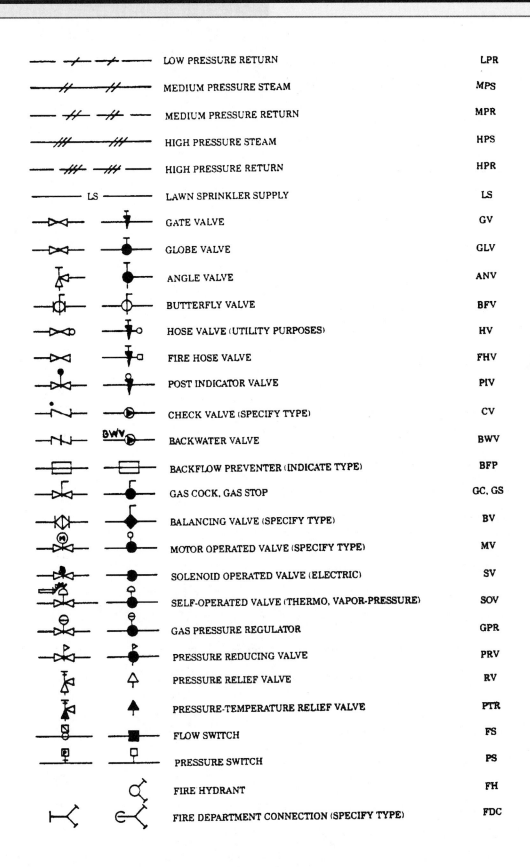

Figure 35-A – Piping Symbols and Definitions for Draftsmen (4)

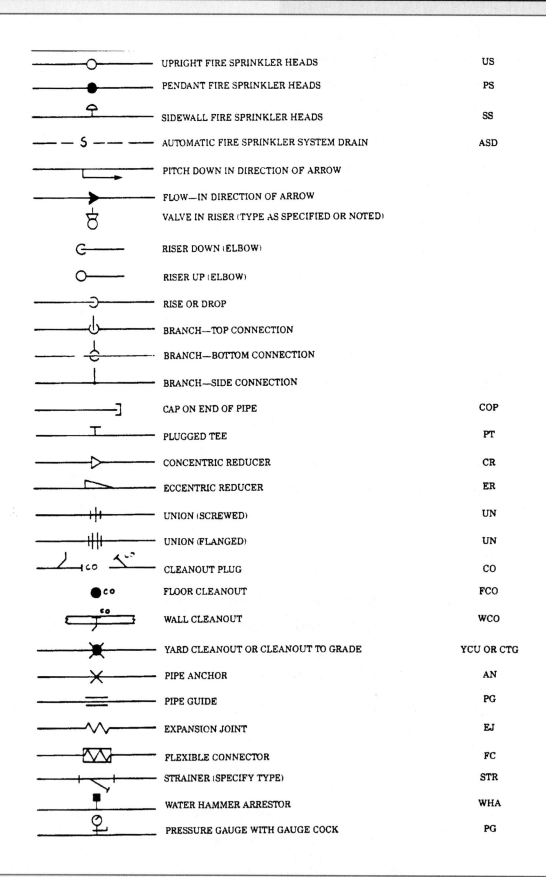

Figure 35-A – Piping Symbols and Definitions for Draftsmen (5)

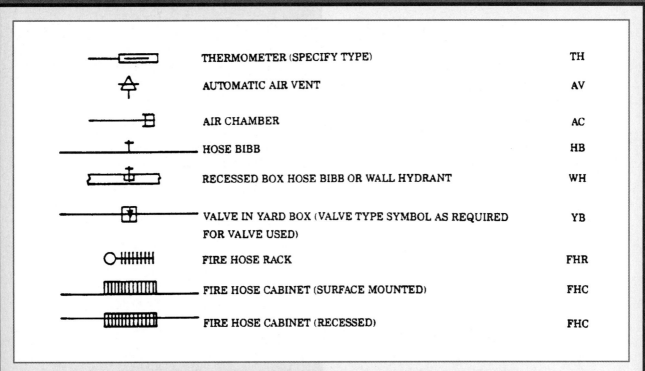

Figure 35-A – Piping Symbols and Definitions for Draftsmen (6)

Using the symbols shown in Figure 35-A, make a detailed sketch of the drainage and vent piping required for the fixtures shown in Figure 35-B.

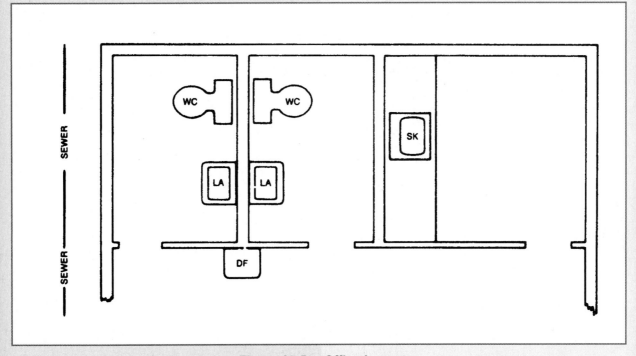

Figure 35-B – Office Layout

Use the drafting tools and aids described in Lesson 34 to produce a finished sketch.

Use notes to indicate that the piping below the floor is to be hub-and-spigot soil pipe, that the above-floor waste piping is to be PVC plastic, and that the vent piping is to be copper.

Show plan and elevation views, fittings and joint types, and connection to the sewer.

The instructor will have drawings that show the development intended for Figures 35-C and 35-D.

Figure 35-C – Office Layout DWV Piping Rough Sketch Plan View (To Be Developed by Student)

Figure 35-D – Office Layout DWV Piping Rough Sketch Elevation View (To Be Developed by Student)

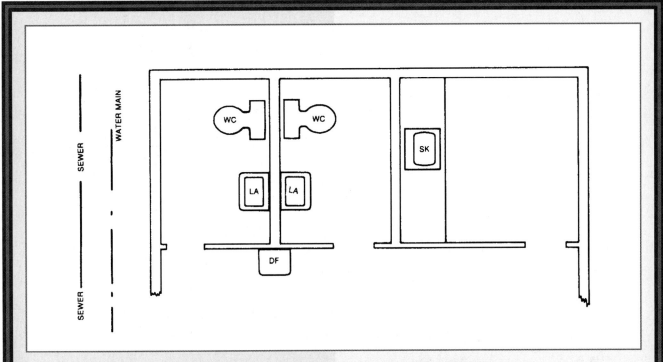

Figure 35-E – Office Layout (Full Basement Below)

In the first part of this lesson, we developed the detailed sketch of the drainage system for the floor plan of Figure 35-B. Next, we will make a sketch of the water piping for this plan.

Keep in mind that for all but the simplest layouts, you should make the water piping sketch separate from the DWV sketch. With separate sketches, it is possible to miss seeing a conflict of piping; but, with some experience, you will know where conflicts can occur. Clarity resulting from separate sketches is the overriding consideration in these detail drawings.

When making this sketch, use notes to indicate that the cold water piping is galvanized steel, that the hot water piping is copper, and that the recirculating hot water line is CPVC.

Include plan and elevation views, indicate the fittings and show joints, and show water meter and heater located in a basement below the restrooms.

Your instructor will have drawings that show what Figures 35-F and 35-G should look like after you have developed the piping details.

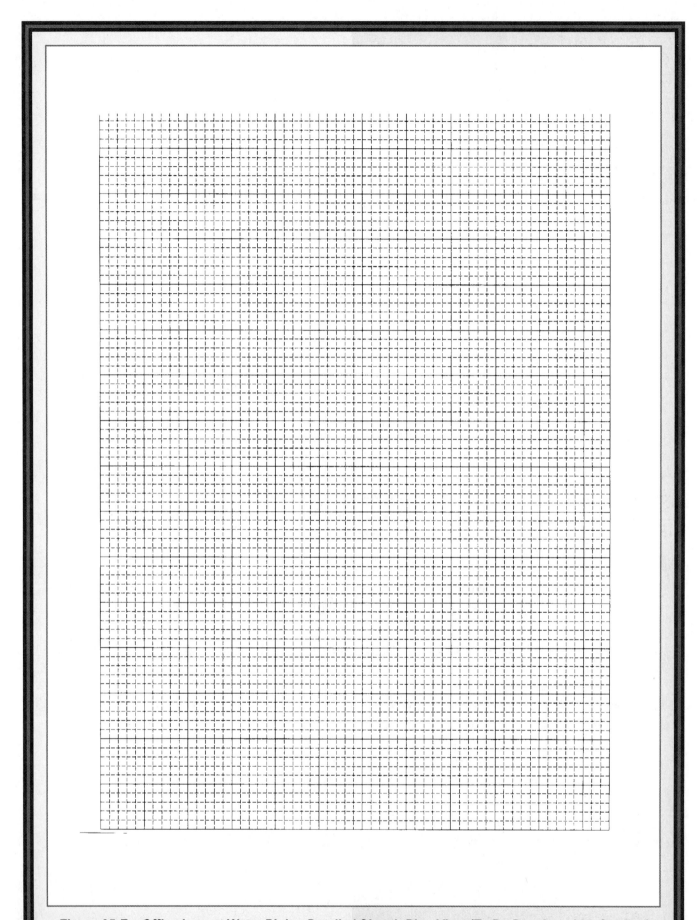

Figure 35-F – Office Layout Water Piping Detailed Sketch Plan View (To Be Developed by Student)

*Figure 35-G – Office Layout Water Piping Detailed Sketch Elevation View
(To Be Developed by Student)*

LESSON 36

PROJECTIONS USED FOR DRAWINGS — ADVANTAGES AND DISADVANTAGES

ORTHOGRAPHIC PROJECTION

Orthographic projection is the name given to the system of views of an object that result if we imagine the object placed inside a transparent rectangular box, the faces of the object projected to the box, and the sides folded out to form a flat set of views. Six regular views are thus developed:

1. Front and back

2. Top and bottom

3. Right and left sides

In most cases, front, top, and a side are sufficient to display the object. In some cases, oblique views can aid in identifying the object. See Figure 36-A. Note that any line that is on the side of the object being shown will appear true size in the drawing of that side.

With this system, a basketball appears as a circle in each view, so you read the drawing as indicating a sphere. An unsharpened pencil would appear as a rectangle in the top and front views, and as a circle with a center dot in the side view.

Place a ¾" sweat coupling on a table and sketch the three views. The top and front views will be rectangles, and the side view will be a circle.

DETAILS SHOWN IN VIEWS

When making orthographic sketches, draw only the views needed to describe the object. Draw all object lines, center lines, and hidden lines if it helps describe the object. Use hidden lines to indicate details that are behind, or hidden by, the face shown in the view. Remember that object lines are heavy and dark. Center lines are thin lines with alternating long and short dashes. Hidden lines are medium lines of short dashes.

The arrangement of views is as shown below:

		Top View	
Rear View	Left Side	Front View	Right Side
		Bottom View	

The block with a cut-off corner shown in Figure 36-A is depicted with four views to illustrate the method. Note that the only view that shows the cut surface in true size is the view that is perpendicular to that surface.

Except for very complicated shapes, three views are usually adequate to display the object. The conventional method is to present the largest view, with the maximum amount of information, as the front view — thus, the top and side views are supplementary.

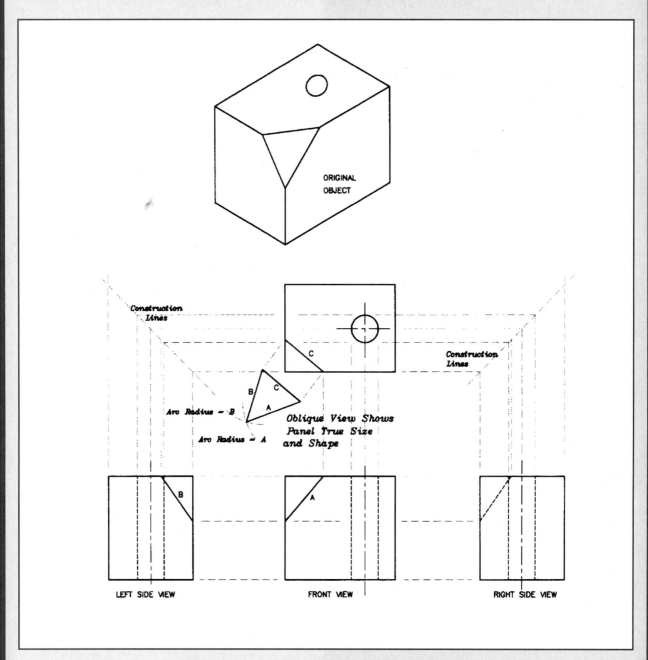

Figure 36-A – Orthographic View Arrangement

Although orthographic sketches are commonly used to describe objects, they can also be used for showing complicated piping arrangements.

EXAMPLE

Figure 36-B shows a copper fitting. The side view could represent a tee or an ell, but the front view shows that the fitting is a tee. Note that center lines shown in the front and side view suggest a cylindrical form for the side branch. We could say that we don't need the top view — that we know the object is a tee — so that the top view adds nothing to our information, but most people would prefer to have the top view, *just to be sure*.

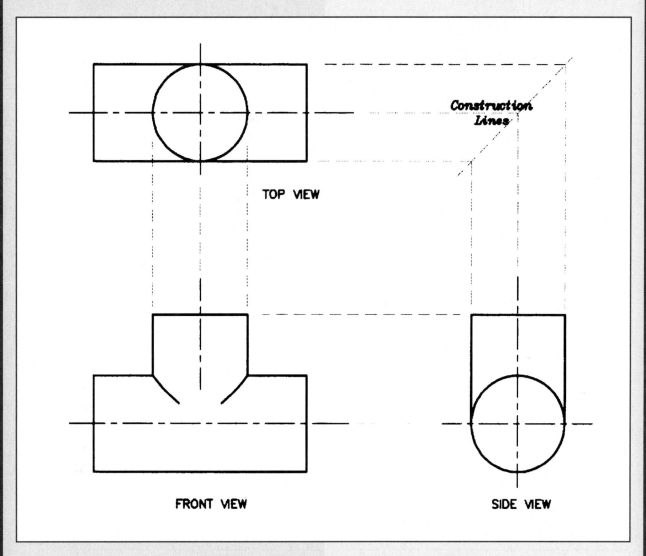

Figure 36-B – Orthographic Projection of a Tee

Figures 36-A and 36-B show the method of projecting the top view to the side. First, a line in the top view is projected to the 45° construction line. This intersection is projected to the side view. In this way, depths can be projected from the top view and heights from the front view, so that no measuring is required to construct the side view.

This system is cumbersome and time-consuming to use in piping installations, unless the project is so complicated that the effort is justified by the increased understanding that is obtained.

PRINCIPAL ADVANTAGE OF ORTHOGRAPHIC METHOD

The principal advantage of orthographic projection is that views are shown true size and shape. **Each view**, however is only two dimensional. Thus, the reader of the drawings must have considerable experience and time to deduce the shape and arrangement of the sides and thus of the piece. This characteristic is a problem especially with something like a piping system, where it is difficult to determine the elevation of piping in plan views, or where a particular item is in depth direction in elevation views.

We will now describe systems of drawing that ease these problems.

ISOMETRIC DRAWING

After orthographic views, the most often-used view type is called *isometric*. Another type, discussed below, is the *oblique* view. The type of view that is most natural is *perspective*, but it is very hard to draw (for non-artists) and cannot be scaled. It is possible to draw isometric and oblique views to scale in some parts of the views, but this is very seldom done.

The isometric system lays out the "X", "Y", and "Z" axes 120° apart as shown in Figure 36-C. As with all pictorial drawing, there are distortions with this system. The advantages of the isometric system are that it produces a representation of the object that looks good, and it is relatively easy to draw.

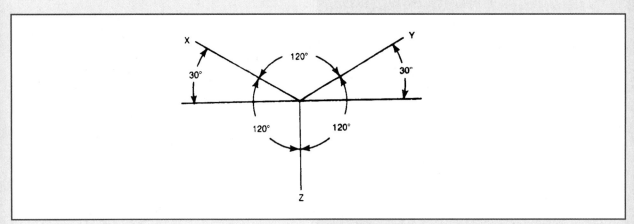

Figure 36-C – Isometric Axes

Consider the chisel shown in Figure 36-D.

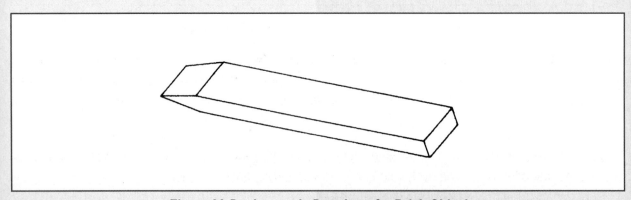

Figure 36-D – Isometric Drawing of a Brick Chisel

Draw all left-right lines parallel to the "X" axis, all front-back lines parallel to the "Y" axis, and all vertical lines parallel to the "Z" axis.

First, draw a line (on the "X" or "Y" axis) to form one long line of the object. Continue with lines in the three axes to form the object. After it is lightly sketched, erase unneeded lines and make the principal object lines very heavy.

Note that the end of the chisel, which is actually rectangular, is a parallelogram in the isometric sketch. Also, circles appear as ellipses. Even with these distortions, isometric sketches *look right*.

Study the bathroom sketched in Figure 36-E and the accompanying isometric drawing of the drain and vent piping. The relationship of the lines is clear.

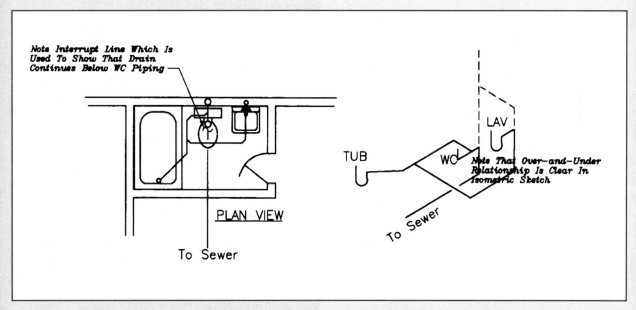

Figure 36-E – Isometric Sketch of Waste and Vent Piping of a Typical Bathroom

To draw the isometric view, first draw the building drain along the "Y" axis, and then a vertical line for the stack. Continue with the details to produce the isometric shown in the figure.

Look at a similar bathroom sketched in Figure 36-F, with the piping indicated in each orthographic view. You will find it very difficult to visualize the relationships from these views. Draw the isometric views of the waste and vent piping for this bathroom. The building drain goes to the left, to change things somewhat from Figure 36-E. There is another change to notice, namely, pipe "P" is not parallel to the "X" or "Y" axis. As you draw this isometric, you must locate the two ends of "P" and connect them with a straight line. In this case, you will find that the line appears very short — almost a dot.

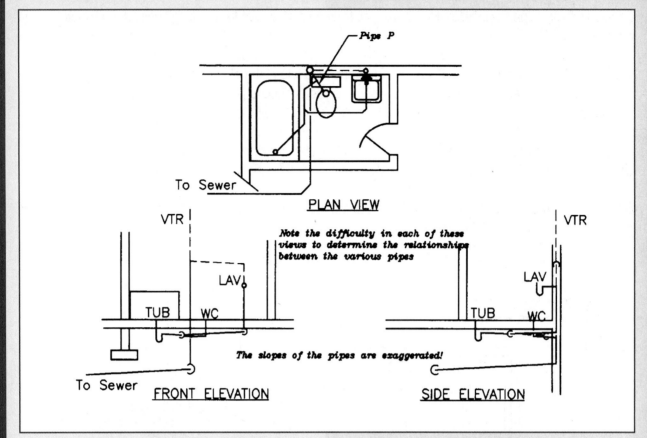

Figure 36-F – Orthographic Views of Typical Bath DWV Piping

Remember that isometric views help to visualize the three-dimensional relationships of a piping system. Drawing an isometric sketch will help you visualize what you have to do should you be presented with a plan showing equipment location and you are required to design the piping.

ISOMETRIC SKETCH PADS

Office and drafting supply stores have pads of isometric graph paper. A typical example is shown in Figure 36-H. This figure had to be especially drawn for presentation here, because the grid on normal graph paper is printed in a color that does not show up in most copying processes. Note that the lines are arranged in three directions: vertical, 120° right, and 120° left.

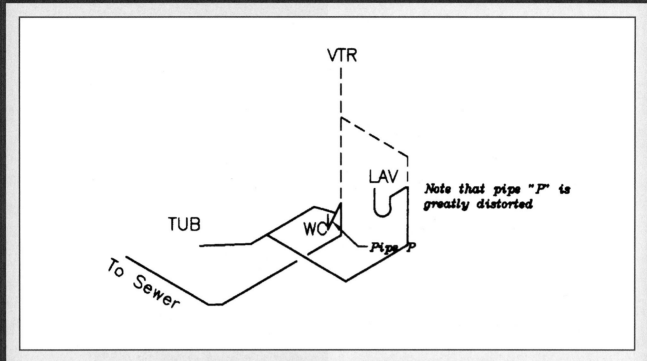

Figure 36-G – Isometric Sketch of Layout in Figure 36-F

Figure 36-G illustrates the way the grid is used to draw a bathtub placed in a corner, with trim shown. First, draw a rectangle using the grid patterns as guides. Next, place vertical lines on the three corners where they would be visible to the observer, and connect the lower ends of these lines together. You now have the outline of the tub.

Third, extend the corners of the back walls (and ceiling). Notice how strongly this suggests that you are looking into a room. Last, sketch the opening into the tub sump, and locate the faucet spout, handles, and shower head.

Similar steps can be used to represent anything that you need to show three-dimensional relationships.

Figure 36-H – Example of Isometric Graph Paper

Add the further details to Figure 36-I to arrive at an isometric drawing that shows the tub well, the faucet, spout, and the shower head.

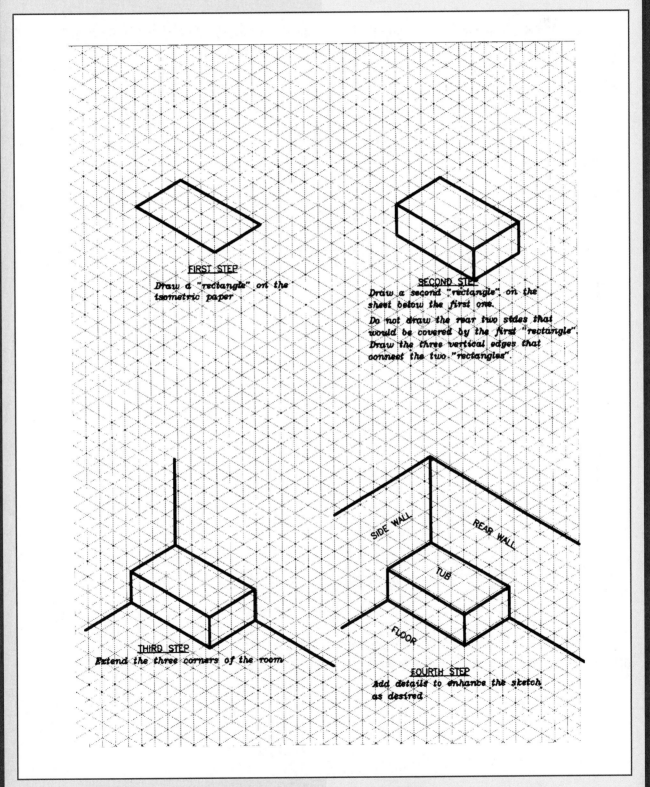

Figure 36-I – Isometric Sketch of Bathtub
Note the Drawing is Not Complete

OBLIQUE SKETCHES

Pictorial drawings help you visualize piping in three-dimensions. The isometric method was described above.

The oblique method overcomes a **problem** with the isometric. Technically, any system of projection that is not orthographic is an oblique system. However, we will reserve the term oblique for a system where the "X" and "Z" axes are 90° apart, and the "Y" axis is 135° from each of the other two. This system has the advantage that the front view is true for size and shape, and depths ("Y" axis) are distorted. To make a pictorial view of an object with this method is satisfactory for shallow depth items, otherwise the distortion is apparent and unpleasant.

This criticism of this method is less significant for piping arrangements — again we are interested in a one-view presentation of three-dimensional relationships.

Figure 36-J shows the three axes.

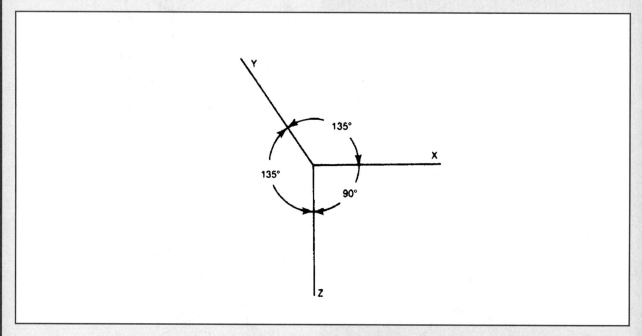

Figure 36-J – Oblique Axes

Figures 36-K, 36-L, and 36-M show the development of the **oblique** view of the chisel sketched (with the **isometric** method) in Figure 36-B. Notice in 36-K, the end of the chisel is drawn true size and shape.

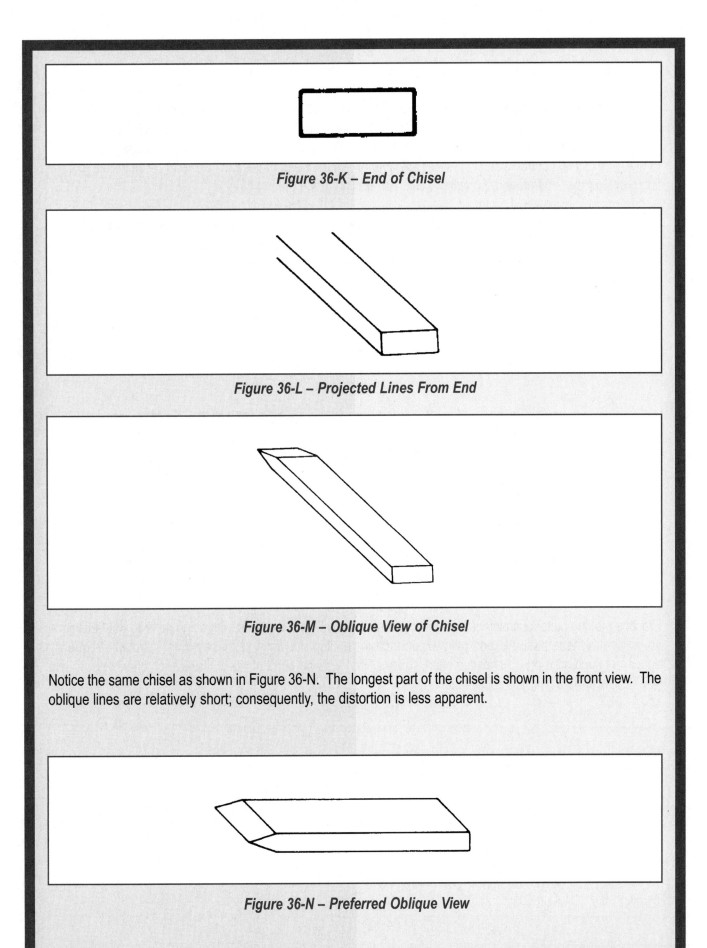

Figure 36-K – End of Chisel

Figure 36-L – Projected Lines From End

Figure 36-M – Oblique View of Chisel

Notice the same chisel as shown in Figure 36-N. The longest part of the chisel is shown in the front view. The oblique lines are relatively short; consequently, the distortion is less apparent.

Figure 36-N – Preferred Oblique View

EXAMPLES

Consider the bathroom layout shown in Figure 36-F. The oblique drawing for this bathroom can be drawn by making all left-right lines parallel to the "X" axis, all depth lines parallel to the "Y" axis, and all vertical lines parallel to the "Z" axis. Notice that although all lines in the "X" and "Z" directions are true size, the lines parallel to the "Y" axis are not true size. Lines that are not parallel to one of these axes will also be distorted, e.g. the first piece off the tub trap and the water closet connection to the stack.

Make an oblique drawing based on Figure 36-F. Figure 36-O below is one version of such a drawing.

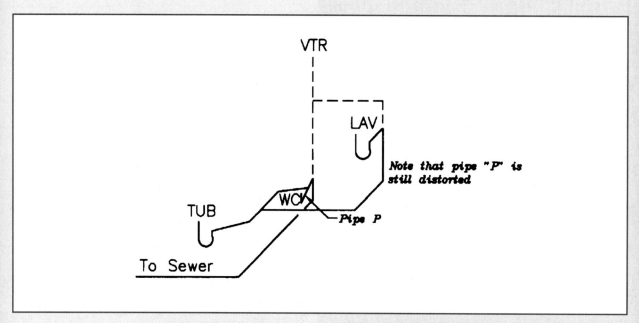

Figure 36-O – Oblique View of Layout in Figure 36-F

Practice making pictorial drawings using the isometric and oblique methods. By doing so, you will familiarize yourself with these methods and gain experience in reading drawings of these types. You will frequently encounter such drawings. Almost as often, you will find jobs where the ability to make such drawings will help you visualize the piping you must install.

APPENDIX A

OVERVIEW OF HAND TOOLS

INTRODUCTION

Lesson 3 describes tools that are more or less special to plumbing work. This Appendix presents a review of common tools used at home and in industry that most people are at least somewhat familiar with. Tools are described as English or metric, as the case may be.

The competent mechanic knows that tools are necessary to perform most tasks, and he/she therefore treats tools properly and respectfully. In this way, the plumber obtains maximum life and utility from the tools.

Manual hand tools may be divided into six categories:

1. Measuring tools (linear)

2. Testing tools (horizontal and vertical)

3. Sawing and cutting tools

4. Boring tools

5. Assembly tools

6. Cutting tools (metal)

MEASURING (LINEAR)

A measuring tool is used to establish length. It is usually marked in feet and inches, subdivided to $1/16$ inches, and available in lengths up to 100'. Examples include:

FOLDING RULE

Made of wood or metal, these tools are marked down to $1/32$". They are available in six-foot or eight-foot lengths, and may be called zig-zag rules. Joints must be lubricated when they become stiff and difficult to work. Failure to lubricate may lead to tool breakage.

One option available with folding rulers is "inside" marking or "outside" marking. If, when the ruler is completely closed, the visible first blade is marked from zero to 6", it is an *outside* ruler. If, when the ruler is completely closed, the visible first blade is marked from 66" to 72", it is an *inside* ruler.

The point to these two options is that when laying work on sheet metal (or flat gasket material) an inside ruler works best (the ruler lays flat on the material and the markings proceed from zero up.) On the other hand, when measuring a length along a pipe or along a wall, the outside marking is more convenient.

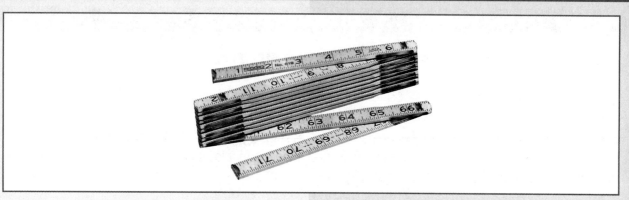

Figure A-A – RIDGID Wood Folding Rule

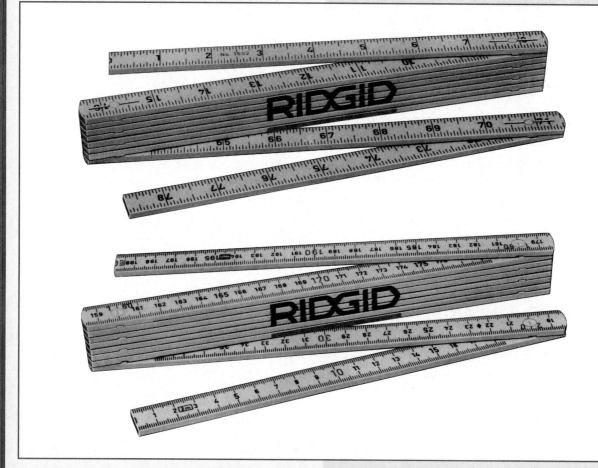

Figure A-B – RIDGID Fiberglass Folding Rule

CLOTH OR STEEL TAPE

They have markings down to $1/16$ inch or 1/100ths of a foot. They are available in 25', 50', and 100' lengths. They have a toothed hook on the end for gripping.

These tapes are exposed to having dirt cling to the tape, so they should be cleaned after each use. Just a small amount of dirt over much of the tape will make it impossible to get the tape back within the holder!

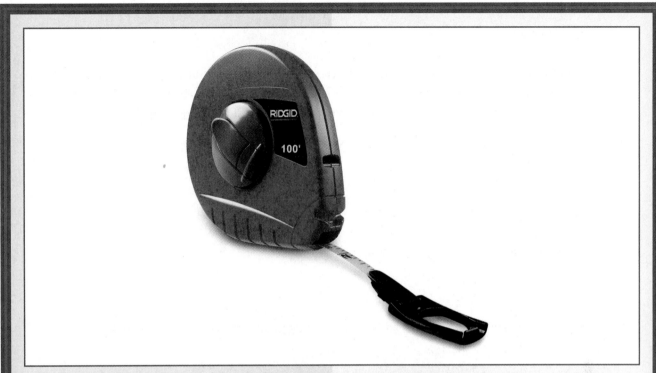

Figure A-C – RIDGID Long Steel Tape

SPRING-LOCK TAPE

They are marked down to $1/16$ inch, and available in lengths up to 25'. They are effective for a wide variety of measurements.

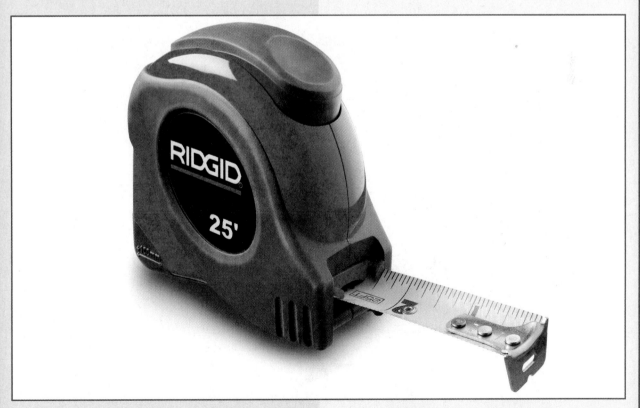

Figure A-D – RIDGID Locking Steel Tape

METRIC NOTE

The basic configurations of measuring devices are available in metric calibration. Most measurements in building applications are in millimeters (1/1000 meter). You should realize that metric calculations, where everything is a decimal, are generally much easier than English units of subdivided fractions: $1/2$, $1/4$, , $1/16$, etc., =1'.

TESTING (Horizontal and Vertical)

Testing tools are used to check measurements and workmanship. Examples include:

(Other tools of this type are described in Lesson 3.)

PLUMB BOB

Conical steel or brass weight, hung from a string. They are used to define a vertical reference line. They are handy for layout of stacks and risers.

Figure A-E – Stanley Brass Plumb Bob

SAWING AND CUTTING

SAWS

Saws especially suited to plumbing work are described in Lesson 3.

Sawing and cutting tools are used to cut wood, plastic, or metal. Saws are available with different number of teeth per inch and different hardness of the saw material itself. Check the manufacturer's recommendations for best performance with each material.

Manual cutting is performed with gentle, smooth down strokes using nearly the full length of the blade for the cut. For best results and personal safety, stand firmly when cutting, keep blade sharp and lubricated, watch for metal embedded in wood, clamp or hold all work firmly to prevent injury, and clean up all debris when finished. Examples of general-purpose saws include the following:

CROSS-CUT

Best all-purpose saw for lumber products. The cross-cut saw is used for straight cuts across the grain.

RIPPING

Similar in arrangement to cross-cut, it has different type teeth that favor straight cuts with the grain in lumber.

Figure A-F – IRWIN Wooden Handle Carpenter Saw

BACK SAW AND MITER BOX

Used mainly for accurate angular cuts so that trim pieces will fit properly. Material must be supported so that cuts will be true.

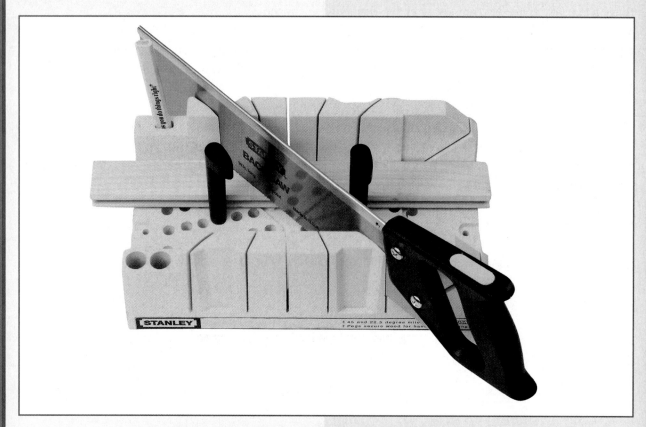

Figure A-G – Stanley Clamping Mitre Box with Saw

HACKSAW

General purpose saw for cutting metal. The number of teeth per inch to be used is determined by the thickness of material to be cut and the metal being cut. Hold with lead hand on the handle and other hand on the frame end.

CHISELS

Chisels are usually used to remove material from the piece being worked. A chisel should be struck with a ball peen hammer. Chisel heads will become enlarged (called a mushroom) with use and small pieces can break off. The **mushroom** must be ground off to prevent flying fragments. It may be wise to use gloves when using a chisel. Eye protection is a must! Examples of chisels include:

WOOD CHISEL

Thin, very sharp tools, used to cut thin slices parallel to the grain in wood. They are also used to clean gaskets from iron or steel flange faces.

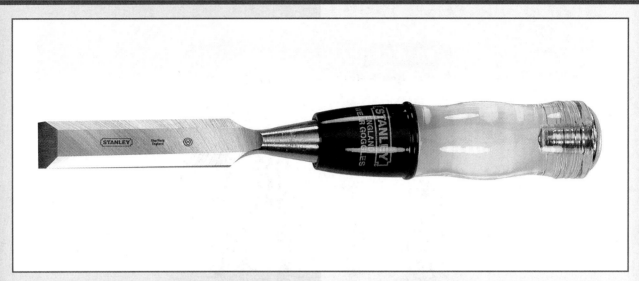

Figure A-H – Stanley Wood Chisel – Short Blade

COLD CHISEL

Used to cut iron, steel, or concrete, they may also be used to force flanges apart after unbolting.

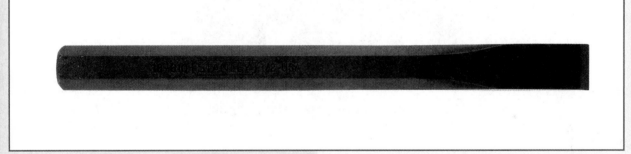

Figure A-I – Stanley 6" Cold Chisel

FILES

Files are used to remove small amounts of material for careful fitting and shaping. The pointed end (tang) of the file should be fitted with a wooden or plastic handle to avoid injury. Some files are provided with cutting surfaces appropriate for wood, others are appropriate for metals.

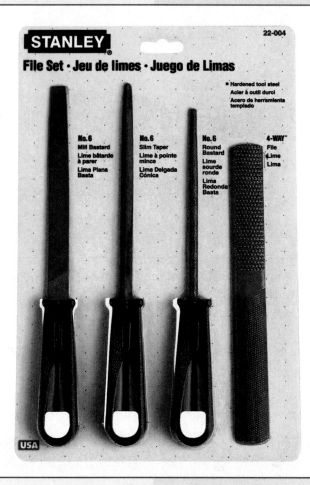

Figure A-J – Stanley 4-Piece File Set

METRIC NOTE

There is nothing about the cutting tools described here that is peculiar to English or ISO measurements. It is likely, however, that if metric units become more commonly used, these tools will be slightly different. One exception: chisels are made to specific widths, so a 1" wood chisel will become 25 mm, etc.

BORING

Boring tools are used to form holes. Examples include:

BRACE AND BIT

Manual tool for wood boring, they are useful when power is not available. Also useful when only a few holes are needed. Bits must be sharp; any encounter with metal will dull bits.

HAND DRILL

Used for a few holes (up to ½" diameter). They are useful where power is not available. Avoid excessive pressure or you will break drill bits.

Other powered tools include the following:

Figure A-K – RIDGID Model 122 Copper Cutting and Prep Machine

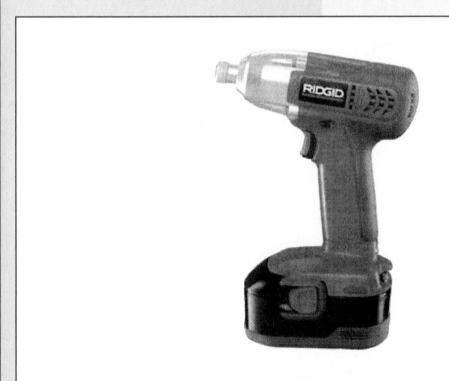

Figure A-L – RIDGID 14.4V Impact Driver

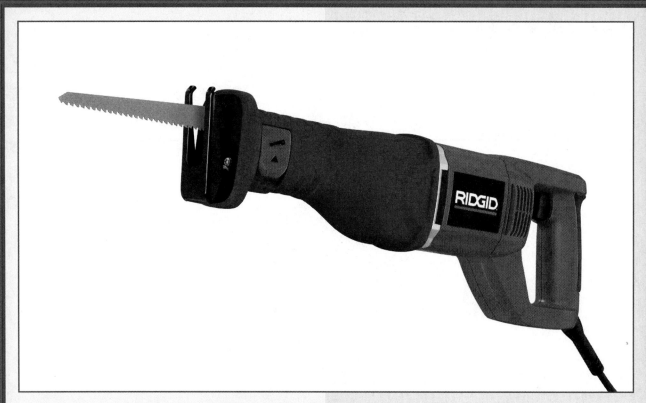

Figure A-M – RIDGID Reciprocating Saw

APPENDIX B

OVERVIEW OF HAND TOOLS (continued)

ASSEMBLY

Assembly tools include screwdrivers, hammers, and wrenches in many forms. The standard, general-purpose versions are described in the following:

SCREWDRIVERS

A screwdriver is a hand tool for driving screws; it has a tip that fits into the head of a screw. It is important that the correct, proper-fitting screwdriver be used and that the screwdriver be properly positioned on the head of the screw. Depending on the material and screw-type, a pilot hole for the screw may have to be drilled or punched first. Examples of screwdrivers include:

FLAT BLADE

These tools are used for "regular" screws, i.e., those with a single straight slot in the head. The screwdrivers are available in a very wide range of sizes as the illustration suggests.

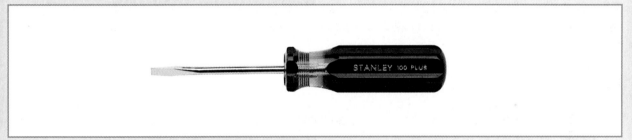

Figure B-A – Stanley 3/16" 100 Plus® Square Blade Standard Tip Screwdriver

PHILLIPS

Phillips screwdrivers are one of the oldest variations from the straight blade type. This type, plus matching screw heads, greatly reduces the risk of marring adjacent surfaces because it is held centered on the screw due to the fact that the turning surfaces are at right angles to each other.

Figure B-B – Stanley 2-Point 100 PLUS® Stubby Phillips Tip Screwdriver

OTHER TYPES

Many special purpose screw or bolt heads are available, including square shank, Allen types, Torx, with matching drive tools. Many of you will have encountered these in motor vehicles, etc.

Figure B-C – Stanley ProDriver T20x4" TORX

HAMMERS

A hammers is a hand tool with a handle and an attached head of metal or other rigid material; it is used for striking or pounding. Hammer weights vary from 6 ounces to 16 pounds. Handles must be tight to the head to transfer the swinging motion of the arm to the impact blow on the work. Various handle arrangements are available; the least expensive are wood, set with wedges; the heaviest duty are fiberglass set in epoxy. Handles must be checked periodically to prevent injury from a flying hammer head. Examples of hammers include:

BALL PEEN (PEIN)

A ball peen (pein) hammer has a striking head and a ball end for peening (swaging, spreading or riveting); it is used for striking impact tools, working metal, or breaking cast-iron fittings. Safety glasses are a must when using ball peen hammers.

SLEDGE

Large hammer with two striking surfaces, they are used for breaking out concrete or breaking down walls or cast-iron parts.

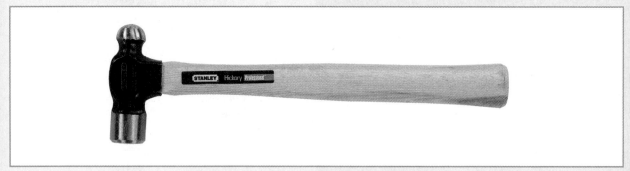

Figure B-D – Stanley 4 oz Ball Pein Hammer

Figure B-E – Stanley 10 lb. Hickory Handle Sledge Hammer

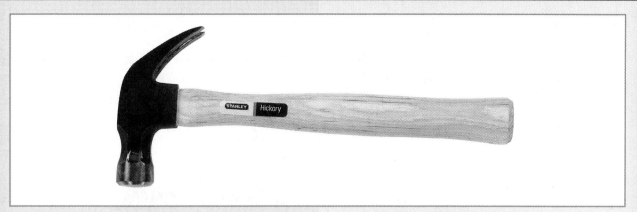

Figure B-F – Stanley 7 oz Curved Claw Wood Handle Nail Hammer

NO BOUNCE

Hollowed out head filled with sand or ball bearing material, they are used to deliver more impact to the work. They do less damage to the work (as in driving a shaft from a bearing).

CLAW

Striking face and 2-prong claw for removing nails or prying. Use block of wood under claw for leverage. They are used for nailing and general wood working

Figure B-G – Lead Hammer (Cook Hammer Company, Inc.)

LEAD MAUL

Used to strike where a force is needed but where a steel hammer may break the material. Used to assist in the assembly of cast-iron hubbed gasket joints. Most lead mauls are "home-made." Great care should be taken using this tool as lead is a heavy metal; OSHA rules and regulations should be consulted.

RAWHIDE MALLET

Used to prevent damage to surface being hammered. Rawhide has been replaced in many uses by rubber or impact-absorbent materials.

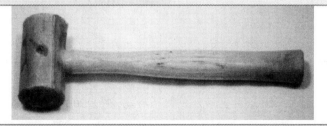

Figure B-H – Rawhide Mallet (Vaughan & Bushnell Mfg Company)

WRENCHES

Wrenches come in many sizes and shapes. The proper wrench should be used for each job. Size is indicated by measuring the distance from end of the handle to the jaw.

ADJUSTABLE

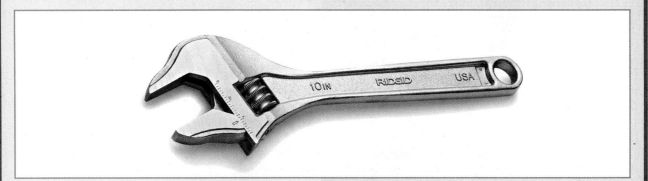

Figure B-I – RIDGID Adjustable Wrench

OPEN END/BOX END
VISE-GRIP

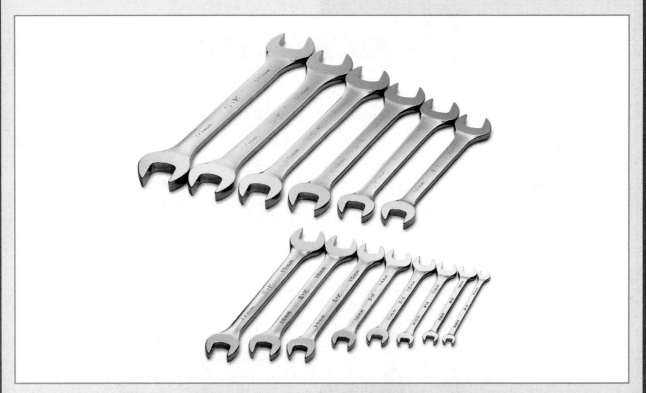

Figure B-J – SK 14 Piece SuperKrome Metric Open End Wrench Set

The remaining figures show miscellaneous equipment. You will probably add even more tools to your pouch or tool box as time goes on.

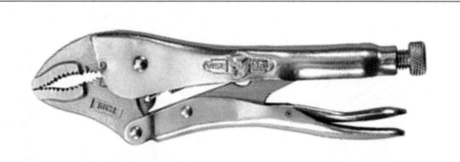

Figure B-K – Irwin Curved Jaw Locking Pliers

Figure B-L – Stanley 4-7/8" Center Punch

Figure B-M – Telephone

Figure B-N – Fire Extinguisher

Figure B-O – Two-Way Radio

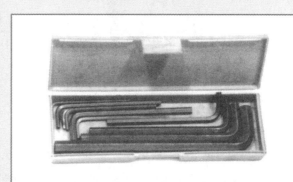

Figure B-P – Allen Wrench Set

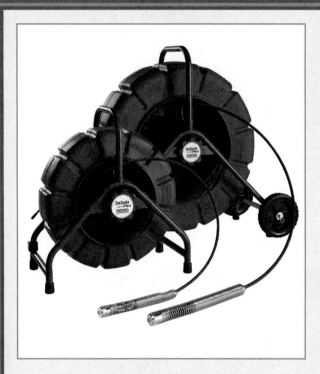

Figure B-Q – RIDGID SeeSnake Plus

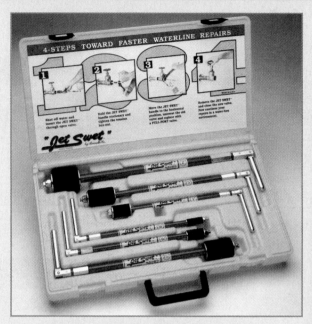

Figure B-S – "Jet Swet™" by Brenelle Co., LLC

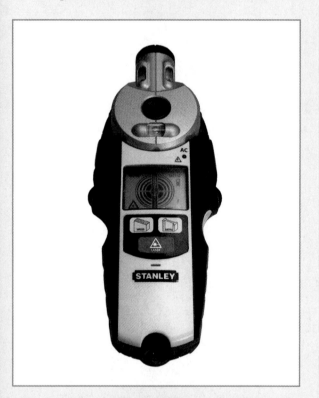

Figure B-R – Stanley IntelliLaser™ Pro Laser Line Level/Stud Finder

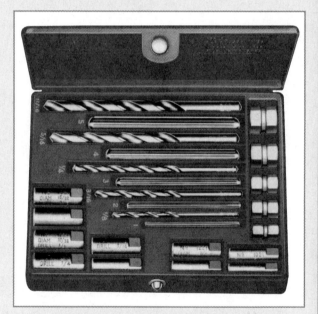

Figure B-T – RIDGID Extractors/Twist Drills

"For people who take pride in their work...tools to be proud of"

STRIKING TOOL SAFETY

Vaughan is committed to the education of consumers on the safe use of striking tools. A major part of this effort is advising consumers on how to select and use the correct tool for the job at hand. Through our advertising and collateral materials we urge retailers to caution their customers to always wear safety goggles when using striking tools. A warning label carrying the same message is affixed to each tool handle, and we publish a safety poster showing correct tools to use for various tasks, and listing important Do's and Dont's concerning the use of striking tools.

ALWAYS USE THE CORRECT TOOL!

Hammers are arguably the most abused, most misused of all hand tools. Injuries can be caused by trying to strike too heavy a blow with a lightweight hammer, by using a damaged hammer and by using the wrong style of hammer for the task. There is no such thing as a "universal" hammer, and users are well advised to use only tools that have been designed for the job.

Ball Pein Hammer-For riveting, center punching and bending or shaping soft metal. Choose a size to match the task.

Hand Drilling Hammer-The only hammer to use with star drills, masonry nails, steel chisels and nail pullers. Easy to handle; packs plenty of punch.

Rawhide Mallet-For use in furniture assembly, shaping soft sheet metals or any task that requires non-marring blows.

Brick Hammer-Designed for cutting and setting bricks or blocks, and for chipping mortar.

Curved Claw or Nail Hammer-For use with non-hardened, common or finishing nails only. Choose 16 or 20 oz. weights for general carpentry; lighter weights for model work or fine cabinetry.

Straight Claw or Rip Hammer-For use with non-hardened, common or finishing nails only. Choose weights from 20 to 32 oz. for framing and ripping.

BASIC RULES

Here are some basic rules for proper use of striking tools:

- ALWAYS WEAR SAFETY GOGGLES

- ALWAYS strike the surface squarely--avoid glancing blows

- DO NOT use claw hammers or hatchets on concrete, stone or hardened metal objects (such as nail pullers, pry bars, cold chisels, punches, star drills, drift pins, masonry nails, clevis or hitch pins, etc.)

Figure B-U – Striking Toll Safety-Vaughn

Shingling Hatchet-The Pro's choice for installing shakes and shingles. Handy gauge pin sets exposed length of shingle. Milled and crowned face sets nails cleanly.

- ALWAYS use a hammer of the right size and weight for the job

- NEVER use a brick hammer to strike metal

Wallboard Tool-Multi-purpose tool scores wallboard, makes cutouts and sets nails with a perfect dimple.

- Do not use a claw hammer to strike a cold chisel, punch or star drill.

- Never strike any hammer or hatchet with or against another hammer or hatchet.

NEVER USE DAMAGED TOOLS

Discard hammer or hatchet with chipped or battered face.

Replace loose or cracked handle with replacement handle of same size and quality.

Discard hammer or hatchet with mushroomed striking face.

Discard hammer with cracked claw or eye section.

A well maintained, quality tool is a safer tool!

PROTECTIVE EYEWEAR

- Adjustable temples and lens angle for comfortable fit
- Polycarbonate lenses are anti-fog, anti-static, anti-scratch & chemical resistant
- Excellent UV absorption
- Side shields and brow guards provide extra protection
- Available in conventional and wrap-around lens styles with clear, tinted and mirrored lenses
- Supplied in reusable, clear plastic protective case

Safety Goggles
Extra large, to fit comfortably over eyeglasses or sunglasses. Flexible frame conforms to facial contours.

APPENDIX C

LEAD SAFETY

GENERAL REMARKS

Lead is one of the "heavy" metals. One characteristic of heavy metals on living organisms (human beings, fish, other animals, plants) is that it tends to accumulate in the tissues and is not eliminated, or it is eliminated very slowly.

The result of this is that lead is very difficult for biologists to establish "safe limits" for lead exposure in human beings. First of all, lead is much more damaging to the very young (young children), and to persons that are medically fragile – which can be any of us at one time or another!

As construction workers, there are many possible modes of exposure to lead, and plumbers are just one of many construction workers who can be exposed.

The materials that follow are a reprint of *Lead in Construction* by the Occupational Safety and Health Administration, usually known as OSHA.

It is recommended that you read through this material to become acquainted with the hazards of lead exposure. While the material is printed with a view toward the employer, it is up to the individual to let the employer know when lead may be a hazard on a job. Without information the employer cannot act.

Important points covered include:

Symptoms of Chronic Overexposure
Reproductive Risks
Chelating Agents
Exposure Limits
Applicability to Construction
Worker Protections

More information may be found on the OSHA web site: www.osha.gov.

This informational booklet provides a general overview of a particular topic related to OSHA standards. It does not alter or determine compliance responsibilities in OSHA standards or the *Occupational Safety and Health Act of 1970*. Because interpretations and enforcement policy may change over time, you should consult current OSHA administrative interpretations and decisions by the Occupational Safety and Health Review Commission and the Courts for additional guidance on OSHA compliance requirements.

This publication is in the public domain and may be reproduced, fully or partially, without permission. Source credit is requested but not required.

This information is available to sensory impaired individuals upon request.
Voice phone: (202) 693-1999; teletypewriter (TTY) number: (877) 889-5627.

Lead in Construction

U.S. Department of Labor

Occupational Safety and Health Administration

OSHA 3142-09R
2003

Contents

Health Hazards of Lead Exposure...3
Worker Exposure...5
Construction Workers and Lead Exposure...5
OSHA's Lead Standard...6
Employer Responsibilities...8
    Hazard Assessment...9
    Medical Surveillance...12
    Medical Removal Provisions...14
    Recordkeeping...16
Exposure Reduction and Employee Protection...18
    Engineering Controls...18
    Housekeeping and Personal Hygiene...21
    Protective Clothing and Equipment...24
    Respiratory Protection...26
    Employee Information and Training...29
OSHA Assistance, Services, and Products...30
OSHA Regional Office Directory...36

Health Hazards of Lead Exposure

Pure lead (Pb) is a heavy metal at room temperature and pressure. A basic chemical element, it can combine with various other substances to form numerous lead compounds.

Lead has been poisoning workers for thousands of years. Lead can damage the central nervous system, cardiovascular system, reproductive system, hematological system, and kidneys. When absorbed into the body in high enough doses, lead can be toxic.

In addition, workers' lead exposure can harm their children's development.

Short-term (acute) overexposure—as short as days--can cause acute encephalopathy, a condition affecting the brain that develops quickly into seizures, coma, and death from cardiorespiratory arrest. Short-term occupational exposures of this type are highly unusual but not impossible.

Extended, long-term (chronic) overexposure can result in severe damage to the central nervous system, particularly the brain. It can also damage the blood-forming, urinary, and reproductive systems. There is no sharp dividing line between rapidly developing acute effects of lead and chronic effects that take longer to develop.

SYMPTOMS OF CHRONIC OVEREXPOSURE

Some of the common symptoms include:

- Loss of appetite;
- Constipation;
- Nausea;
- Excessive tiredness;
- Headache;
- Fine tremors;
- Colic with severe abdominal pain;
- Metallic taste in the mouth;
- Weakness;
- Nervous irritability;
- Hyperactivity;

- Muscle and joint pain or soreness;
- Anxiety;
- Pallor;
- Insomnia;
- Numbness; and
- Dizziness.

REPRODUCTIVE RISKS

Lead is toxic to both male and female reproductive systems. Lead can alter the structure of sperm cells and there is evidence of miscarriage and stillbirth in women exposed to lead or whose partners have been exposed. Children born to parents who were exposed to excess lead levels are more likely to have birth defects, mental retardation, or behavioral disorders or to die during the first year of childhood.

Workers who desire medical advice about reproductive issues related to lead should contact qualified medical personnel to arrange for a job evaluation and medical followup--particularly if they are pregnant or actively seeking to have a child. Employers whose employees may be exposed to lead and who have been contacted by employees with concerns about reproductive issues must make medical examinations and consultations available.

CHELATING AGENTS

Under certain limited circumstances, a physician may prescribe special drugs called chelating agents to reduce the amount of lead absorbed in body tissues. Using chelation as a preventive measure--that is, to lower blood level but continue to expose a worker--is prohibited and therapeutic or diagnostic chelations of lead that are required must be done under the supervision of a licensed physician in a clinical setting, with thorough and appropriate medical monitoring. The employee must be notified in writing before treatment of potential consequences and allowed to obtain a second opinion.

Worker Exposure

Lead is most commonly absorbed into the body by inhalation. When workers breathe in lead as a dust, fume, or mist, their lungs and upper respiratory tract absorb it into the body. They can also absorb lead through the digestive system if it enters the mouth and is ingested.

A significant portion of the lead inhaled or ingested gets into the bloodstream. Once in the bloodstream, lead circulates through the body and is stored in various organs and body tissues. Some of this lead is filtered out of the body quickly and excreted, but some remains in the blood and tissues. As exposure continues, the amount stored will increase if the body absorbs more lead than it excretes. The lead stored in the tissue can slowly cause irreversible damage, first to individual cells, then to organs and whole body systems.

Construction Workers and Lead Exposure

HOW LEAD IS USED

In construction, lead is used frequently for roofs, cornices, tank linings, and electrical conduits. In plumbing, soft solder, used chiefly for soldering tinplate and copper pipe joints, is an alloy of lead and tin. Soft solder has been banned for many uses in the United States. In addition, the Consumer Product Safety Commission bans the use of lead-based paint in residences. Because lead-based paint inhibits the rusting and corrosion of iron and steel, however, lead continues to be used on bridges, railways, ships, lighthouses, and other steel structures, although substitute coatings are available.

Construction projects vary in their scope and potential for exposing workers to lead and other hazards. Projects such as removing paint from a few interior residential doors may involve limited exposure. Others projects, however, may involve removing or stripping substantial quantities of lead-based paints on large bridges and other structures.

MOST VULNERABLE WORKERS

Workers potentially at risk for lead exposure include those involved in iron work; demolition work; painting; lead-based paint abatement; plumbing; heating and air conditioning maintenance and repair; electrical work; and carpentry, renovation, and remodeling work. Plumbers, welders, and painters are among those workers most exposed to lead. Significant lead exposures also can arise from removing paint from surfaces previously coated with lead-based paint such as bridges, residences being renovated, and structures being demolished or salvaged. With the increase in highway work, bridge repair, residential lead abatement, and residential remodeling, the potential for exposure to lead-based paint has become more common.

Workers at the highest risk of lead exposure are those involved in:

- Abrasive blasting and
- Welding, cutting, and burning on steel structures.

Other operations with the potential to expose workers to lead include:

- Lead burning;
- Using lead-containing mortar;
- Power tool cleaning without dust collection systems;
- Rivet busting;
- Cleanup activities where dry expendable abrasives are used;
- Movement and removal of abrasive blasting enclosures;
- Manual dry scraping and sanding;
- Manual demolition of structures;
- Heat-gun applications;
- Power tool cleaning with dust collection systems; and
- Spray painting with lead-based paint.

OSHA's Lead Standard

OSHA's Lead Standard for the Construction Industry, Title 29 Code of Federal Regulations 1926.62, covers lead in a variety of forms, including metallic lead, all inorganic lead compounds, and organic lead soaps.

EXPOSURE LIMITS

The standard establishes maximum limits of exposure to lead for all workers covered, including a permissible exposure limit (PEL) and action level (AL).

The PEL sets the maximum worker exposure to lead: 50 micrograms of lead per cubic meter of air (50µg/m3) averaged over an eight-hour period. If employees are exposed to lead for more than eight hours in a workday, their allowable exposure as a TWA for that day must be reduced according to this formula:

Employee exposure (in µg/m3) = 400 divided by the hours worked in the day.

The AL, regardless of respirator use, is an airborne concentration of 30µg/m3, averaged over an eight-hour period. The AL is the level at which an employer must begin specific compliance activities outlined in the standard.

APPLICABILITY TO CONSTRUCTION

OSHA's lead in construction standard applies to all construction work where an employee may be exposed to lead. All work related to construction, alteration, or repair, including painting and decorating, is included. Under this standard, construction includes, but is not limited to:

- Demolition or salvage of structures where lead or materials containing lead are present;
- Removal or encapsulation of materials containing lead;
- New construction, alteration, repair, or renovation of structures, substrates, or portions or materials containing lead;
- Installation of products containing lead;
- Lead contamination from emergency cleanup;
- Transportation, disposal, storage, or containment of lead or materials containing lead where construction activities are performed; and
- Maintenance operations associated with these construction activities.

Employer Responsibilities

WORKER PROTECTIONS

Employers of construction workers are responsible for developing and implementing a worker protection program. At a minimum, the employer's worker protection program for employees exposed to lead above the PEL should include:

- Hazard determination, including exposure assessment;
- Medical surveillance and provisions for medical removal;
- Job-specific compliance programs;
- Engineering and work practice controls;
- Respiratory protection;
- Protective clothing and equipment;
- Housekeeping;
- Hygiene facilities and practices;
- Signs;
- Employee information and training; and
- Recordkeeping.

Because lead is a cumulative and persistent toxic substance and health effects may result from exposure over prolonged periods, employers must use these precautions where feasible to minimize employee exposure to lead.

The employer should, as needed, consult a qualified safety and health professional to develop and implement an effective, site-specific worker protection program. These professionals may work independently or may be associated with an insurance carrier, trade organization, or onsite consultation program.

ELEMENTS OF A COMPLIANCE PROGRAM

For each job where employee exposure exceeds the PEL, the employer must establish and implement a written compliance program to reduce employee exposure to the PEL or below. The compliance program must provide for frequent and regular inspections of job sites, materials, and equipment by a competent person. Written programs, which must be reviewed and updated at least every six months, must include:

- A description of each activity in which lead is emitted (such as equipment used, material involved, controls in place, crew size, employee job responsibilities, operating procedures, and maintenance practices);
- The means to be used to achieve compliance and engineering plans and studies used to determine the engineering controls selected where they are required;
- Information on the technology considered to meet the PEL;
- Air monitoring data that document the source of lead emissions;
- A detailed schedule for implementing the program, including copies of documentation (such as purchase orders for equipment, construction contracts);
- A work practice program;
- An administrative control schedule, if applicable; and
- Arrangements made among contractors on multi-contractor sites to inform employees of potential lead exposure.

Hazard Assessment

An employer is required to conduct an initial employee exposure assessment of whether employees are exposed to lead at or above the AL based on:

- Any information, observation, or calculation that indicates employee exposure to lead;
- Any previous measurements of airborne lead; and
- Any employee complaints of symptoms attributable to lead exposure.

Objective data and historical measurements of lead may be used to satisfy the standard's initial monitoring requirements.

INITIAL EMPLOYEE EXPOSURE ASSESSMENT

Initial monitoring may be limited to a representative sample of those employees exposed to the greatest concentrations of airborne lead. Representative exposure sampling is permitted when there are a number of employees performing the same job, with lead exposure of similar duration and level, under essentially the same conditions. For employees engaged in similar work, the standard requires that the members of the group reasonably expected to have the highest exposure levels be monitored. This result is then attributed to the other employees of the group.

The employer must establish and maintain an accurate record documenting the nature and relevancy of previous exposure data. Instead of performing initial monitoring, the employer may in some cases rely on objective data that demonstrate that a particular lead-containing material or product cannot result in employee exposure at or above the action level when it is processed, used, or handled.

BIOLOGICAL MONITORING TESTS

Analysis of blood lead samples must be conducted by an OSHA-approved lab and be accurate (to a confidence level of 95 percent) within plus or minus 15 percent, or 6 µg/dl, whichever is greater. If an employee's airborne lead level is at or above the AL for more than 30 days in any consecutive 12 months, the employer must make biological monitoring available on the following schedule:

- At least every two months for the first six months and every six months thereafter for employees exposed at or above the action level for more than 30 days annually;
- At least every two months for employees whose last blood sampling and analysis indicated a blood lead level at or above 40 µg/dl; and
- At least monthly while an employee is removed from exposure due an elevated blood lead level.

PENDING EMPLOYEE EXPOSURE ASSESSMENT

Until the employer performs an exposure assessment and documents that employees are not exposed above the PEL, OSHA requires some degree of interim protection for employees. This means providing respiratory protection, protective work clothing and equipment, hygiene facilities, biological monitoring, and training—as specified by the standards—for certain tasks prone to produce high exposure. These include:

10

- Manual demolition of structures such as dry wall, manual scraping, manual sanding, and use of a heat gun where lead-containing coatings or paints are present;
- Power tool cleaning with or without local exhaust ventilation;
- Spray painting with lead-containing paint;
- Lead burning;
- Use of lead-containing mortar;
- Abrasive blasting, rivet busting, welding, cutting, or torch-burning on any structure where lead-containing coatings or paint are present;
- Abrasive blasting enclosure movement and removal;
- Cleanup of activities where dry expendable abrasives are used; and
- Any other task the employer believes may cause exposures in excess of the PEL.

TEST RESULTS SHOWING NO OVEREXPOSURES

If the initial assessment indicates that no employee is exposed above the AL, the employer may discontinue monitoring. Further exposure testing is not required unless there is a change in processes or controls that may result in additional employees being exposed to lead at or above the AL, or may result in employees already exposed at or above the AL being exposed above the PEL. The employer must keep a written record of the determination, including the date, location within the work site, and the name and social security number of each monitored employee.

EMPLOYEE NOTIFICATION OF MONITORING RESULTS

The employer must notify each employee in writing of employee exposure assessment results within five working days of receiving them. Whenever the results indicate that the representative employee exposure, without the use of respirators, is above the PEL, the employer must include a written notice stating that the employee's exposure exceeded the PEL and describing corrective action taken or to be taken to reduce exposure to or below the PEL.

Medical Surveillance

When an employee's airborne exposure is at or above the AL for more than 30 days in any consecutive 12 months, an immediate medical consultation is required when the employee notifies the employer that he or she:

- Has developed signs or symptoms commonly associated with lead-related disease;
- Has demonstrated difficulty in breathing during respirator use or a fit test;
- Desires medical advice concerning the effects of past or current lead exposure on the employee's ability to have a healthy child; and
- Is under medical removal and has a medically appropriate need.

MEDICAL EXAMS

The best indicator of personal lead exposure is through a blood test to indicate elevated blood lead levels. A medical exam must also include:

- Detailed work and medical histories, with particular attention to past lead exposure (occupational and nonoccupational), personal habits (smoking and hygiene), and past gastro-intestinal, hematologic, renal, cardiovascular, reproductive, and neurological problems;
- A thorough physical exam, with particular attention to gums, teeth, hematologic, gastrointestinal, renal, cardiovascular, and neurological systems; evaluation of lung function if respirators are used;
- A blood pressure measurement;
- A blood sample and analysis to determine blood lead level;
    • Hemoglobin and hematocrit determinations, red cell indices, and an exam of peripheral smear morphology; and
    • Zinc protopor-phyrin; blood urea nitrogen; and serum creatinine;
- A routine urinalysis with microscopic exam; and
- Any lab or other test the examining physician deems necessary.

INFORMATION FOR THE EXAMINING PHYSICIAN

The employer must provide all examining physicians with a copy of the lead in construction standard, including all appendices, a description of the affected employee's duties as they relate to the employee's exposure, the employee's lead exposure level or anticipated exposure level, a description of personal protective equipment used or to be used, prior blood lead determinations, and all prior written medical opinions for the employee.

WHEN MONITORING SHOWS NO EMPLOYEE EXPOSURES ABOVE THE AL

Employers must make available, at no cost to the employee, initial medical surveillance for employees exposed to lead on the job at or above the action level on any one day per year. This initial medical surveillance consists of biological monitoring in the form of blood sampling and analysis for lead and zinc protoporyrin (ZPP) levels. In addition, a medical surveillance program with biological monitoring must be made available to any employee exposed at or above the action level for more than 30 days in any consecutive 12 months.

AFTER THE MEDICAL EXAMINATION

Employers must obtain and provide the employee a copy of a written opinion from each examining or consulting physician that contains only information related to occupational exposure to lead and must include:

- Whether the employee has any detected medical condition that would increase the health risk from lead exposure;
- Any special protective measures or limitations on the worker's exposure to lead,
- Any limitation on respirator use; and
- Results of the blood lead determinations.

In addition, the written statement may include a statement that the physician has informed the employee of the results of the consultation or medical examination and any medical condition that may require further examination or treatment.

The employer must instruct the physician that findings, including lab results or diagnoses unrelated to the worker's lead exposure, must not be revealed to the employer or included in the written opinion to the employer. The employer must also instruct the physician to advise employees of any medical condition, occupational or non-occupational, that necessitates further evaluation or treatment. In addition, some states also require laboratories and health care providers to report cases of elevated blood lead concentrations to their state health departments.

Medical Removal Provisions

Temporary medical removal can result from an elevated blood level or a written medical opinion. More specifically, the employer is required to remove from work an employee with a lead exposure at or above the AL each time periodic and follow-up (within two weeks of the periodic test) blood sampling tests indicate that the employee's blood level is at or above 50 µg/dl. The employer also must remove from work an employee with lead exposure at or above the AL each time a final medical determination indicates that the employee needs reduced lead exposure for medical reasons. If the physician who is implementing the employer's medical program makes a final written opinion recommending the employee's removal or other special protective measures, the employer must implement the physician's recommendation.

For an employee removed from exposure to lead at or above the AL due to a blood lead level at or above 50 µg/dl, the employer may return that employee to former job status when two consecutive blood sampling tests indicate that the employee's blood lead level is below 40 µg/dl. For an employee removed from exposure to lead due to a final medical determination, the employee must be returned when a subsequent final medical determination results in a medical finding, determination, or opinion that the employee no longer has a detected medical condition that places the employee at increased risk of lead exposure.

The employer must remove any limitations placed on employees or end any special protective measures when a subsequent final medical determination indicates they are no longer necessary. If the former position no longer exists, the employee is returned consistent with whatever job assignment discretion the employer would have had if no removal occurred.

WORKER PROTECTIONS AND BENEFITS

The employer must provide up to 18 months of medical removal protection (MRP) benefits each time an employee is removed from lead exposure or medically limited. As long as the position/job exists, the employer must maintain the earnings, seniority, and other employment rights and benefits as though the employee had not been removed from the job or otherwise medically limited. The employer may condition medical removal protection benefits on the employee's participation in followup medical surveillance.

If a removed employee files a worker's compensation claim or other compensation for lost wages due to a lead-related disability, the employer must continue medical removal protection benefits until the claim is resolved. However, the employer's MRP benefits obligation will be reduced by the amount that the employee receives from these sources. Also, the employer's MRP benefits obligation will be reduced by any income the employee receives from employment with another employer made possible by virtue of the employee's removal.

RECORDS REQUIREMENTS INVOLVING MEDICAL REMOVAL

In the case of medical removal, the employer's records must include:

- The worker's name and social security number,
- The date of each occasion that the worker was removed from current exposure to lead,
- The date when the worker was returned to the former job status,
- A brief explanation of how each removal was or is being accomplished, and
- A statement indicating whether the reason for the removal was an elevated blood lead level.

Recordkeeping

EMPLOYER REQUIREMENTS

The employer must maintain any employee exposure and medical records to document ongoing employee exposure, medical monitoring, and medical removal of workers. This data provides a baseline to evaluate the employee's health properly. Employees or former employees, their designated representatives, and OSHA must have access to exposure and medical records in accordance with 29 CFR 1910.1020. Rules of agency practice and procedure governing OSHA access to employee medical records are found in 29 CFR 1913.10.

EXPOSURE ASSESSMENT RECORDS

The employer must establish and maintain an accurate record of all monitoring and other data used to conduct employee exposure assessments as required by this standard and in accordance with 29 CFR 1910.1020. The exposure assessment records must include:

- The dates, number, duration, location, and results of each sample taken, including a description of the sampling proce-dure used to determine representative employee exposure;
- A description of the sampling and analytical methods used and evidence of their accuracy;
- The type of respiratory protection worn, if any;
- The name, social security number, and job classification of the monitored employee and all others whose exposure the mea-surement represents; and
- Environmental variables that could affect the measurement of employee exposure.

MEDICAL SURVEILLANCE RECORDS

The employer must maintain an accurate record for each employee subject to medical surveillance, including:

- The name, social security number, and description of the employee's duties;
- A copy of the physician's written opinions;

- The results of any airborne exposure monitoring done for the employee and provided to the physician; and
- Any employee medical complaints related to lead exposure.

In addition, the employer must keep or ensure that the examining physician keeps the following medical records:

- A copy of the medical examination results including medical and work history;
- A description of the laboratory procedures and a copy of any guidelines used to interpret the test results; and
- A copy of the results of biological monitoring.

The employer or physician or both must maintain medical records in accordance with 29 CFR 1910.1020.

DOCUMENTS FOR EMPLOYEES SUBJECT TO MEDICAL REMOVAL

The employer must maintain--for at least the duration of employ-ment--an accurate record for each employee subject to medical removal, including:

- The name and social security number of the employee;
- The date on each occasion that the employee was removed from current exposure to lead and the corresponding date which the employee was returned to former job status;
- A brief explanation of how each removal was or is being accomplished; and
- A statement about each removal indicating whether the reason for removal was an elevated blood level.

EMPLOYER REQUIREMENTS RELATED TO OBJECTIVE DATA

The employer must establish and maintain an accurate record documenting the nature and relevancy of objective data relied on to assess initial employee exposure in lieu of exposure monitoring. The employer must maintain the record of objective data relied on for at least 30 years.

DOCUMENTS FOR OSHA AND NIOSH REVIEW

The employer must make all records--including exposure monitor-ing, objective data, medical removal, and medical records--

available upon request to affected employees, former employees, and their designated representatives and to the OSHA Assistant Secretary and the Director of the National Institute for Occupational Safety and Health (NIOSH) for examination and copying in accordance with 29 CFR 1910.1020.

WHEN CLOSING A BUSINESS

When an employer ceases to do business, the successor employer must receive and retain all required records. If no successor is available, these records must be sent to the Director of NIOSH.

Exposure Reduction and Employee Protection

The most effective way to protect workers is to minimize their exposure through engineering controls, good work practices and training, and use of personal protective clothing and equipment, including respirators, where required. The employer needs to designate a competent person capable of identifying existing and predictable lead hazards and who is authorized to take prompt corrective measures to eliminate such problems. The employer should, as needed, consult a qualified safety and health professional to develop and implement an effective worker protection program. These professionals may work independently or may be associated with an insurance carrier, trade organization, or onsite consultation program.

Engineering Controls

Engineering measures include local and general exhaust ventilation, process and equipment modification, material substitution, component replacement, and isolation or automation. Examples of recommended engineering controls that can help reduce worker exposure to lead are described as follows.

EXHAUST VENTILATION

Equip power tools used to remove lead-based paint with dust collection shrouds or other attachments so that paint is exhausted through a high-efficiency particulate air (HEPA) vacuum system. For operations such as welding, cutting/burning, or heating, use local exhaust ventilation. Use HEPA vacuums during cleanup operations.

For abrasive blasting operations, build a containment structure that is designed to optimize the flow of clean ventilation air past the workers' breathing zones. This will help reduce the exposure to airborne lead and increase visibility. Maintain the affected area under negative pressure to reduce the chances that lead dust will contaminate areas outside the enclosure. Equip the containment structure with an adequately sized dust collector to control emissions of particulate matter into the environment.

ENCLOSURE OR ENCAPSULATION

One way to reduce the lead inhalation or ingestion hazard posed by lead-based paint is to encapsulate it with a material that bonds to the surface, such as acrylic or epoxy coating or flexible wall coverings. Another option is to enclose it using systems such as gypsum wallboard, plywood paneling, and aluminum, vinyl, or wood exterior siding. Floors coated with lead-based paint can be covered using vinyl tile or linoleum.

The building owner or other responsible person should oversee the custodial and maintenance staffs and contractors during all activities involving enclosed or encapsulated lead-based paint. This will minimize the potential for an inadvertent lead release during maintenance, renovation, or demolition.

SUBSTITUTION

Choose materials and chemicals that do not contain lead for construction projects. Among the options are:

- Use zinc-containing primers covered by an epoxy intermediate coat and polyurethane topcoat instead of lead-containing coatings.
- Substitute mobile hydraulic shears for torch cutting under certain circumstances.
- Consider surface preparation equipment such as needle guns with multiple reciprocating needles completely enclosed within an adjustable shroud, instead of abrasive blasting under certain

19

conditions. The shroud captures dust and debris at the cutting edge and can be equipped with a HEPA vacuum filtration with a self-drumming feature. One such commercial unit can remove lead-based paint from flat steel and concrete surfaces, outside edges, inside corners, and pipes.

- Choose chemical strippers in lieu of hand scraping with a heat gun for work on building exteriors, surfaces involving carvings or molding, or intricate iron work. Chemical removal generates less airborne lead dust. (Be aware, however, that these strippers themselves can be hazardous and that the employer must review the material safety data sheets (MSDSs) for these stripping agents to obtain information on their hazards.)

COMPONENT REPLACEMENT

Replace lead-based painted building components such as windows, doors, and trim with new components free of lead-containing paint. Another option is to remove the paint offsite and then repaint the components with zinc-based paint before replacing them.

PROCESS OR EQUIPMENT MODIFICATION

When applying lead paints or other lead-containing coatings, use a brush or roller rather than a sprayer. This application method introduces little or no paint mist into the air to present a lead inhalation hazard. (Note that there is a ban on the use of lead-based paint in residential housing.)

Use non-silica-containing abrasives such as steel or iron shot/grit sand instead of sand in abrasive blasting operations when practical. The free silica portion of the dust presents a respiratory health hazard.

When appropriate for the conditions, choose blasting techniques that are less dusty than open-air abrasive blasting. These include hydro- or wet-blasting using high-pressure water with or without an abrasive or surrounding the blast nozzle with a ring of water, and vacuum blasting where a vacuum hood for material removal is positioned around the exterior of the blasting nozzle.

When using a heat gun to remove lead-based paints in residential housing units, be sure it is of the flameless electrical softener type. Heat guns should have electronically controlled temperature settings to allow usage below 700 degrees F. Equip heat guns with various nozzles to cover all common applications and to limit the size of the heated work area.

When using abrasive blasting with a vacuum hood on exterior building surfaces, ensure that the configuration of the heads on the blasting nozzle match the configuration of the substrate so that the vacuum is effective in containing debris.

Ensure that HEPA vacuum cleaners have the appropriate attachments for use on unusual surfaces. Proper use of brushes of various sizes, crevice and angular tools, when needed, will enhance the quality of the HEPA-vacuuming process and help reduce the amount of lead dust released into the air.

ISOLATION

Although it is not feasible to enclose and ventilate some abrasive blasting operations completely, it is possible to isolate many operations to help reduce the potential for lead exposure. Isolation consists of keeping employees not involved in the blasting operations as far away from the work area as possible, reducing the risk of exposure.

Housekeeping and Personal Hygiene

Lead is a cumulative and persistent toxic substance that poses a serious health risk. A rigorous housekeeping program and the observance of basic personal hygiene practices will minimize employee exposure to lead. In addition, these two elements of the worker protection program help prevent workers from taking lead-contaminated dust out of the worksite and into their homes where it can extend the workers' exposures and potentially affect their families' health.

HOUSEKEEPING PRACTICES

An effective housekeeping program involves a regular schedule to remove accumulations of lead dust and lead-containing debris. The schedule should be adapted to exposure conditions at a particular worksite. OSHA's Lead Standard for Construction requires employers to maintain all surfaces as free of lead contamination as practicable. Vacuuming lead dust with HEPA-filtered equipment or wetting the dust with water before sweeping are effective control measures. Compressed air may not be used to remove lead from contaminated surfaces unless a ventilation system is in place to capture the dust generated by the compressed air.

In addition, put all lead-containing debris and contaminated items accumulated for disposal into sealed, impermeable bags or other closed impermeable containers. Label bags and containers as lead-containing waste. These measures provide additional help in controlling exposure.

PERSONAL HYGIENE PRACTICES

Emphasize workers' personal hygiene such as washing their hands and face after work and before eating to minimize their exposure to lead. Provide and ensure that workers use washing facilities. Provide clean change areas and readily accessible eating areas. If possible, provide a parking area where cars will not be contaminated with lead. These measures:

- Reduce workers' exposure to lead and the likelihood that they will ingest lead,
- Ensure that the exposure does not extend beyond the worksite,
- Reduce the movement of lead from the worksite, and
- Provide added protection to employees and their families.

CHANGE AREAS

The employer must provide a clean change area for employees whose airborne exposure to lead is above the PEL. The area must be equipped with storage facilities for street clothes and a separate area with facilities for the removal and storage of lead-contaminated protective work clothing and equipment. This separation prevents cross contamination of the employee's street and work clothing.

Employees must use a clean change area for taking off street clothes, suiting up in clean protective work clothing, donning respirators before beginning work, and dressing in street clothes after work. No lead-contaminated items should enter this area.

Work clothing must not be worn away from the jobsite. Under no circumstances should lead-contaminated work clothes be laundered at home or taken from the worksite, except to be laundered professionally or for disposal following applicable federal, state, and local regulations.

SHOWERS AND WASHING FACILITIES

When feasible, showers must be provided for use by employees whose airborne exposure to lead is above the permissible exposure limit so they can shower before leaving the worksite. Where showers are provided, employees must change out of their work clothes and shower before changing into their street clothes and leaving the worksite. If employees do not change into clean clothing before leaving the worksite, they may contaminate their homes and automobiles with lead dust, extending their exposure and exposing other members of their household to lead.

In addition, employers must provide adequate washing facilities for their workers. These facilities must be close to the worksite and furnished with water, soap, and clean towels so employees can remove lead contamination from their skin.

Contaminated water from washing facilities and showers must be disposed of in accordance with applicable local, state, or federal regulations.

PERSONAL PRACTICES

The employer must ensure that employees do not enter lunchroom facilities or eating areas with protective work clothing or equipment unless surface lead dust has been removed. HEPA vacuuming and use of a downdraft booth are examples of cleaning methods that limit the dispersion of lead dust from contaminated work clothing.

In all areas where employees are exposed to lead above the PEL, employees must observe the prohibition on the presence and consumption or use of food, beverages, tobacco products, and cosmetics. Employees whose airborne exposure to lead is above the PEL must wash their hands and face before eating, drinking, smoking, or applying cosmetics.

END-OF-DAY PROCEDURES

Employers must ensure that workers who are exposed to lead above the permissible exposure limit follow these procedures at the end of their workday:

- Place contaminated clothes, including work shoes and personal protective equipment to be cleaned, laundered, or disposed of, in a properly labeled closed container.
- Take a shower and wash their hair. Where showers are not provided, employees must wash their hands and face at the end of the workshift.
- Change into street clothes in clean change areas.

Protective Clothing and Equipment

EMPLOYER REQUIREMENTS

Employers must provide workers who are exposed to lead above the PEL or for whom the possibility of skin or eye irritation exists with clean, dry protective work clothing and equipment that are appropriate for the hazard. Employers must provide these items at no cost to employees. Appropriate protective work clothing and equipment used on construction sites includes:

- Coveralls or other full-body work clothing;
- Gloves, hats, and shoes or disposable shoe coverlets;
- Vented goggles or face shields with protective spectacles or goggles;
- Welding or abrasive blasting helmets; and
- Respirators.

Clean work clothing must be issued daily for employees whose exposure levels to lead are above 200 µg/m3, weekly if exposures are above the PEL but at or below 200 µg/m3 or where the possibility of skin or eye irritation exists.

HANDLING CONTAMINATED PROTECTIVE CLOTHING

Workers must not be allowed to leave the worksite wearing lead-contaminated protective clothing or equipment. This is an essential step in reducing the movement of lead contamination from the workplace into the worker's home and provides added protection for employees and their families.

Disposable coveralls and separate shoe covers may be used, if appropriate, to avoid the need for laundering. Workers must remove protective clothing in change rooms provided for that purpose.

Employers must ensure that employees leave the respirator use area to wash their faces and respirator facepieces as necessary. In addition, employers may require their employees to use HEPA vacuuming, damp wiping, or another suitable cleaning method before removing a respirator to clear loose particle contamination on the respirator and at the face-mask seal.

Place contaminated clothing that is to be cleaned, laundered, or disposed of by the employer in closed containers. Label containers with the warning: "Caution: Clothing contaminated with lead. Do not remove dust by blowing or shaking. Dispose of lead-contaminated wash water in accordance with applicable local, state, or federal regulations."

Workers responsible for handling contaminated clothing, including those in laundry services or subcontractors, must be informed in writing of the potential health hazard of lead exposure. At no time shall lead be removed from protective clothing or equipment by brushing, shaking, or blowing. These actions disperse the lead into the work area.

PREVENTING HEAT STRESS

Workers wearing protective clothing, particularly in hot environments or within containment structures, can face a risk from heat stress if proper control measures are not used.

Heat stress is caused by several interacting factors, including environmental conditions, type of protective clothing worn, the work activity required and anticipated work rate, and individual employee characteristics such as age, weight, and fitness level. When heat stress is a concern, the employer should choose lighter, less insulating protective clothing over heavier clothing, as long as it provides adequate protection. Other measures the employer can take include: discussing the possibility of heat stress and its signs and symptoms with all workers; using appropriate work/rest regimens; and providing heat stress monitoring that includes measuring employees' heart rates, body temperatures, and weight loss. Employers must provide a source of water or electrolyte drink in a non-contaminated eating and drinking area close to the work area so workers can drink often throughout the day. Workers must wash their hands and face before drinking any fluid if their airborne exposure is above the PEL.

Respiratory Protection

Although engineering and work practice controls are the primary means of protecting workers from exposure to lead, source control at construction sites sometimes is insufficient to control exposure. In these cases, airborne lead concentrations may be high or may vary widely. Respirators often must be used to supplement engineering controls and work practices to reduce worker lead exposures below the PEL. When respirators are required, employers must provide them at no cost to workers.

The standard requires that respirators be used during periods when an employee's exposure to lead exceeds the PEL, including

- Periods necessary to install or implement engineering or work practice controls, and
- Work operations for which engineering and work practice controls are insufficient to reduce employee exposures to or below the PEL.

Respirators also must be provided upon employee request. A requested respirator is included as a requirement to provide increased protection for those employees who wish to reduce their lead burden below what is required by the standard, particularly if they intend to have children in the near future. In addition, respirators must be used when performing previously indicated high exposure or "trigger" tasks, before completion of the initial assessment.

PROVIDING ADEQUATE RESPIRATORY PROTECTION

Before any employee first starts wearing a respirator in the work environment, the employer must perform a fit test. For all employees wearing negative or positive pressure tight-fitting facepiece respirators, the employer must perform either qualitative or quantitative fit tests using an OSHA-accepted fit testing protocol. In addition, employees must be fit tested whenever a different respirator facepiece is used, and at least annually thereafter.

Where daily airborne exposure to lead exceeds 50 µg/m3, affected workers must don respirators before entering the work area and should not remove them until they leave the high-exposure area or have completed a decontamination procedure. Employers must assure that the respirator issued to the employee is selected and fitted properly to ensure minimum leakage through the facepiece-to-face seal.

RESPIRATORY PROTECTION PROGRAMS

When respirators are required at a worksite, the employer must establish a respiratory protection program in accordance with the OSHA standard on respiratory protection, 29 CFR 1910.134. At a minimum, an acceptable respirator program for lead must include:

- Procedures for selecting respirators appropriate to the hazard;
- Fit testing procedures;
- Procedures for proper use of respirators in routine and reasonably foreseeable emergency situations, including cartridge change schedules;
- Procedures and schedules for cleaning, disinfecting, storing, inspecting, repairing, discarding, and otherwise maintaining respirators;
- Training of employees in the respiratory hazard to which they are potentially exposed during routine and emergency situations;
- Training of employees in the proper use of respirators, including putting on and removing them, any limitations of their use, and their maintenance;

- Procedures for regularly evaluating the effectiveness of the program;
- Procedures to ensure air quality when supplied air is used;
- A written program and designation of a program administrator; and
- Recordkeeping procedures.

In addition, the construction industry lead standard stipulates medical evaluations of employees required to use respirators.

If an employee has difficulty in breathing during a fit test or while using a respirator, the employer must make a medical examination available to that employee to determine whether he or she can wear a respirator safely.

SELECTING A RESPIRATOR

The employer must select the appropriate respirator from Table 1 of the lead standard, 29 CFR 1926.62(f)(3)(i). The employer must provide a powered air-purifying respirator when an employee chooses to use this respirator and it will provide the employee adequate protection. A NIOSH-certified respirator must be selected and used in compliance with the conditions of its certification. In addition, if exposure monitoring or experience indicates airborne exposures to contaminants other than lead such as silica, solvents, or polyurethane coatings, these exposures must be considered when selecting respiratory protection.

Select type CE respirators approved by NIOSH for abrasive blasting operations. Currently, there are two kinds of CE respirators with the following assigned protection factors (APFs): a continuous-flow respirator with a loose-fitting hood, APF 25; and a full facepiece supplied-air respirator operated in a positive-pressure mode, APF 2,000. (Note: OSHA recognizes Bullard Helmets, Models 77 and 88 (1995); Clemco Appollo, Models 20 and 60 (1997); and 3M Model 8100 (1998) as having APFs of 1,000.)

For any airline respirator, it is important to follow the manufacturer's instructions regarding air quality, air pressure, and inside diameter and length of hoses. Be aware that using longer hoses or smaller inside diameter hoses than the manufacturer specifies or hoses with bends or kinks may reduce or restrict the airflow to a respirator.

Employee Information and Training

The employer must inform employees about lead hazards according to the requirement of OSHA's Hazard Communication standard for the construction industry, 29 CFR 1926.59, including-- but not limited to--the requirements for warning signs and labels, material safety data sheets (MSDSs), and employee information and training. (Refer to 29 CFR 1910.1200.)

PROGRAM REQUIREMENTS

Employers must institute an information and training program and ensure that all employees subject to exposure to lead or lead compounds at or above the action level on any day participate. Also covered under information and training are employees who may suffer skin or eye irritation from lead compounds. Initial training must be provided before the initial job assignment. Training must be repeated at least annually and, in brief summary, must include:

- The content of the OSHA lead standard and its appendices;
- The specific nature of operations that could lead to lead expo-sure above the action level;
- The purpose, proper selection, fit, use, and limitations of respirators;
- The purpose and a description of the medical surveillance program, and the medical removal protection program;
- Information concerning the adverse health effects associated with excessive lead exposure;
- The engineering and work practice controls associated with employees' job assignments;
- The contents of any lead-related compliance plan in effect;
- Instructions to employees that chelating agents must not be used routinely to remove lead from their bodies and when necessary only under medical supervision and at the direction of a licensed physician; and

- The right to access records under "Access to Employee Exposure and Medical Records," 29 CFR 1910.1020.

All materials relating to the training program and a copy of the standard and its appendices must be made readily available to all affected employees.

WARNING SIGNS

Employers are required to post these warning signs in each work area where employee exposure to lead is above the PEL:

- WARNING
- LEAD WORK AREA
- POISON
- NO SMOKING OR EATING

All signs must be well lit and kept clean so that they are easily visible. Statements that contradict or detract from the signs' meaning are prohibited. Signs required by other statutes, regulations, or ordinances, however, may be posted in addition to, or in combination with, this sign.

OSHA Assistance, Services, and Products

OSHA can provide extensive help through a variety of programs, including assistance about safety and health programs, state plans, workplace consultations, voluntary protection programs, strategic partnerships, alliances, and training and education. An overall commitment to workplace safety and health can add value to your business, to your workplace, and to your life.

How does safety and health management system assistance help employers and employees?

Working in a safe and healthful environment can stimulate innovation and creativity and result in increased performance and higher productivity. The key to a safe and healthful work environment is a comprehensive safety and health management system.

OSHA has electronic compliance assistance tools, or eTools, on its website that walks users through the steps required to develop a comprehensive safety and health program. The eTools are posted at www.osha.gov, and are based on guidelines that identify four general elements critical to a successful safety and health management system:

- Management leadership and employee involvement,
- Worksite analysis,
- Hazard prevention and control, and
- Safety and health training.

STATE PROGRAMS

The Occupational Safety and Health Act of 1970 (OSH Act) encourages states to develop and operate their own job safety and health plans. OSHA approves and monitors these plans and funds up to 50 percent of each program's operating costs. State plans must provide standards and enforcement programs, as well as voluntary compliance activities, that are at least as effective as federal OSHA's.

Currently, 26 states and territories have their own plans. Twenty-three cover both private and public (state and local government) employees and three states, Connecticut, New Jersey, and New York, cover only the public sector. For more information on state plans, see the list at the end of this publication, or visit OSHA's website at www.osha.gov.

CONSULTATION ASSISTANCE

Consultation assistance is available on request to employers who want help establishing and maintaining a safe and healthful workplace. Funded largely by OSHA, the service is provided at no cost to small employers and is delivered by state authorities through professional safety and health consultants.

SAFETY AND HEALTH ACHIEVEMENT RECOGNITION PROGRAM

Under the consultation program, certain exemplary employers may request participation in OSHA's Safety and Health Achievement Recognition Program (SHARP). Eligibility for participation includes, but is not limited to, receiving a full-service, comprehensive consultation visit, correcting all identified hazards, and developing an effective safety and health management system.

Employers accepted into SHARP may receive an exemption from programmed inspections (not complaint or accident investigation inspections) for 1 year initially, or 2 years upon renewal. For more information about consultation assistance, see the list of consultation projects at the end of this publication.

VOLUNTARY PROTECTION PROGRAMS

Voluntary Protection Programs (VPP) are designed to recognize outstanding achievements by companies that have developed and implemented effective safety and health management programs. There are three VPP programs: Star, Merit, and Demonstration. All are designed to

- Recognize who that have successfully developed and implemented effective and comprehensive safety and health management programs;
- Encourage these employers to continuously improve their safety and health management programs;
- Motivate other employers to achieve excellent safety and health results in the same outstanding way; and
- Establish a cooperative relationship between employers, employees, and OSHA.

VPP participation can bring many benefits to employers and employees, including fewer worker fatalities, injuries, and illnesses; lost-workday case rates generally 50 percent below industry averages; and lower workers' compensation and other injury- and illness-related costs. In addition, many VPP sites report improved employee motivation to work safely, leading to a better quality of life at work; positive community recognition and interaction; further improvement and revitalization of already-good safety and health programs; and a positive relationship with OSHA.

After a site applies for the program, OSHA reviews an employer's VPP application and conducts a VPP onsite evaluation to verify that the site's safety and health management programs are operating effectively. OSHA conducts onsite evaluations on a regular basis.

Sites participating in VPP are not scheduled for regular, programmed inspections. OSHA does, however, handle any employee complaints, serious accidents, or significant chemical releases that may occur at VPP sites according to routine enforcement procedures.

Additional information on VPP is available from OSHA regional offices listed at the end of this booklet. Also, see "Cooperative Programs" on OSHA's website.

COOPERATIVE PARTNERSHIPS

OSHA has learned firsthand that voluntary, cooperative partnerships with employers, employees, and unions can be a useful alternative to traditional enforcement and an effective way to reduce worker deaths, injuries, and illnesses. This is especially true when a partnership leads to the development and implementation of a comprehensive workplace safety and health management system.

ALLIANCE PROGRAM

Alliances enable organizations committed to workplace safety and health to collaborate with OSHA to prevent injuries and illnesses in the workplace. OSHA and its allies work together to reach out to, educate, and lead the nation's employers and their employees in improving and advancing workplace safety and health.

Alliances are open to all, including trade or professional organizations, businesses, labor organizations, educational institutions, and government agencies. In some cases, organizations may be building on existing relationships with OSHA through other cooperative programs.

There are few formal program requirements for alliances, which are less structured than other cooperative agreements, and the agreements do not include an enforcement component. However, OSHA and the participating organizations must define, implement, and meet a set of short- and long-term goals that fall into three categories: training and education; outreach and communication; and promotion of the national dialogue on workplace safety and health.

STRATEGIC PARTNERSHIP PROGRAM

OSHA Strategic Partnerships are agreements among labor, management, and government to improve workplace safety and health. These partnerships encourage, assist, and recognize the efforts of the partners to eliminate serious workplace hazards and achieve a high level of worker safety and health. Whereas OSHA's Consultation Program and VPP entail one-on-one relationships between OSHA and individual worksites, most strategic partnerships build cooperative relationships with groups of employers and employees.

For more information about this program, contact your nearest OSHA office or visit our website.

OCCUPATIONAL SAFETY AND HEALTH TRAINING

The OSHA Training Institute in Arlington Heights, Ill., provides basic and advanced training and education in safety and health for federal and state compliance officers, state consultants, other federal agency personnel, and private-sector employers, employees, and their representatives.

TRAINING GRANTS

OSHA awards grants to nonprofit organizations to provide safety and health training and education to employers and workers in the workplace. Grants often focus on high-risk activities or hazards or may help nonprofit organizations in training, education, and outreach.

OSHA expects each grantee to develop a program that addresses a safety and health topic named by OSHA, recruit workers and employers for the training, and conduct the training. Grantees are also expected to follow up with students to find out how they applied the training in their workplaces.

For more information contact OSHA Office of Training and Education, 2020 Arlington Heights Rd., Arlington Heights, IL 60005; or call (847) 297-4810.

34

OTHER ASSISTANCE MATERIALS

OSHA has a variety of materials and tools on its website at www.osha.gov. These include eTools such as Expert Advisors and Electronic Compliance Assistance Tools, information on specific health and safety topics, regulations, directives, publications, videos, and other information for employers and employees.

OSHA also has an extensive publications program. For a list of items, visit OSHA's website at www.osha.gov or contact the OSHA Publications Office, U.S. Department of Labor, 200 Constitution Avenue, NW, N-3101, Washington, DC 20210. Telephone (202) 693-1888 or fax to (202) 693-2498.

In addition, OSHA's CD-ROM includes standards, interpretations, directives, and more. It is available for sale from the U.S. Government Printing Office. To order, write to the Superintendent of Documents, U.S. Government Printing Office, Washington, DC 20402, or phone (202) 512-1800.

IN CASE OF AN EMERGENCY OR TO FILE A COMPLAINT

To report an emergency, file a complaint, or seek OSHA advice, assistance, or products, call (800) 321-OSHA or contact your nearest OSHA regional office listed at the end of this publication. The teletypewriter (TTY) number is (877) 889-5627.

Employees can also file a complaint online and get more information on OSHA federal and state programs by visiting OSHA's website at www.osha.gov.

GUIDE TO ILLUSTRATIONS BY MANUFACTURER

American Standard Customer Care
P. O. Box 6820
1 Centennial Plaza
Piscataway, NJ 08855-6820
800-442-1902
http://www.americanstandard-us.com

AMETEK U. S. Gauge
820 Pennsylvania Blvd.
Feasterville, PA 19053
863-534-1504
http://www.ametekusg.com

Bennette Design Group, Inc.
P. O. Box 326
Millis, MA 02054
800-746-1090
http://www.plumberspad

Brenelle Company, LLC
P. O. Box 80064
Las Vegas, NV 89180-0064
800-727-1018
http://www.brenelle.com

Charlotte Pipe and Foundry Company
P. O. Box 35430
Charlotte, North Carolina 28235
800-438-6091
http://www.charlottepipe.com

Copper Development Association, Inc.
260 Madison Avenue
New York, New York 10016
212-251-7200
http://www.copper.org

COB Industries, Inc.
P. O. Box 361175
Melbourne, FL 32936-1175
800-431-1311
http://www.cob-industries.com

Delta Faucet Company
Corporate Headquarters
55 E. 111th Street
P. O. Box 40980
Indianapolis, IN 46280
800-345-3358
http://www.deltafaucet.com

FALCON Waterfree Technologies, LLC
1593 Galbraith Avenue S.E.
Grand Rapids, MI 49546
866-275-3781
http://www.falconwaterfree.com

Halsey Taylor
2222 Camden Court
Oak Brook, IL 60523
630-574-3500
http://www.halseytaylor.com

Hammond Valve
2375 South Burrell Street
Milwaukee, WI 53207-1519
414-486-3210
http://www.hammondvalve.com

IRWIN Industrial Tool Company
92 Grant Street
Wilmington, OH 45177-0829
800-464-7946
http://www.irwin.com

Josam Company
525 W. U. S. Highway 20
Michigan City, IN 46360
800-365-6726
http://www.josam.com

KOHLER Plumbing
444 Highway Drive
Kohler, WI 53044
800-456-4537
http://www.kohler.com

Mephisto Tool Co.
313 Union Turnpike (Route 66)
P. O. Box 16
Hudson, New York   12534
518-828-1563
http://www.mmephistotool.com

Miller Electric Manufacturing Co.
1635 W. Spencer Street
P. O. Box 1079
Appleton, WI   54912-1079
920-734-9821
http://www.millerwelds.com

Milwaukee Electric Tool Corporation
13135 West Lisbon Road
Brookfield, WI   53005-2550
262-781-3600
http://www.milwaukeetool.com

National Standard Plumbing Code Illustrated 2003
Published by the Plumbing-Heating-Cooling Contractors – National Association
180 S. Washington Street
P. O. Box 6080
Falls Church, VA   22046-1148
1-800-813-7061
http://www.phccweb.org

NIBCO, Inc.
World Headquarters
1516 Middlebury Street
Elkhart, IN   46516-4740
574-295-3000
http://www.nibco.com

Paper Mate
Sanford Corporation
Attn: Consumer Affairs
2711 Washington Boulevard
Bellwood, IL   60104
800-323-0749
http://www.papermate.com

Parker Hannifin Corporation
Parker Instrumentation
Climate Systems Division
10801 Rose Avenue
New Haven, Indiana   46774
800-C-PARKER
http://www.parker.com

PLUMBEREX Specialty Products, Inc.
P. O. Box 1684
Palm Springs, CA 92263
760-343-7363
www.plumberex.com

Plumbing Dictionary – Fifth Edition
Published by American Society of Sanitary Engineering
Suite A
901 Canterbury Road
Westlake, Ohio   44145-1480
publication.sales@asse-plumbing.org

Ridge Tool Company
RIDGID Tools
400 Clark Street
Elyria, Ohio   44035
1-888-743-4333
http://www.ridgid.com

St. Joseph County - South Bend Building Department
Suite 100
205 W. Jefferson Boulevard
South Bend, Indiana   46601
574-235-9554

S•K Hand Tool Corporation
9500 W. 55th Street, Suite B
McCook, IL   60525-3605
800 U CALL SK
http://www.skhandtool.com

Sloan Valve Co.
World Headquarters
10500 Seymour Avenue
Franklin Park, IL 60131
800-982-5839
http://www.sloanvalve.com

Jay R. Smith Mfg. Co.
2781 Gunter Park Dr. East
Montgomery, AL 36109-1405
(334) 277-8520
http://www.jrsmith.com

Stanley Tools Product Group
480 Myrtle Street
New Britain, CT 06053
http://www.stanleytools.com

Stockham
9200 New Trails Drive, Suite 200
The Woodlands, TX 77381
800-STOCKHAM
http://www.stockham.com

Symmons Industries, Inc.
31 Brooks Drive
Braintree, MA 02184-3804
800-SYMMONS
http://www.symmons.com

Victaulic Company of America
P. O. Box 31
4901 Kesslersville Road
Easton, PA 18044-0031
610-559-3457
http://www.victaulic.com

Watts Regulator Company
815 Chestnut Street
North Andover, MA 01845-6098
978-688-1811
http://www.wattsreg.com

WDI International, Inc.
3 Musick
Irvine, CA 92618
949-250-4576
http://wdi-ecoflush.com

Wolverine Brass, Inc.
2951 Hwy. 501 E
Conway, SC 29526
800-944-9292
http://www.wolverinebrass.com

Woodford Manufacturing
2121 Waynoka Road
Colorado Springs, CO 80915
800-765-4115
http://www.woodfordmfg.com